ansformations Through Space and Time

NATO ASI Series

Advanced Science Institutes Series

A Series presenting the results of activities sponsored by the NATO Science Committee, which aims at the dissemination of advanced scientific and technological knowledge, with a view to strengthening links between scientific communities.

The Series is published by an international board of publishers in conjunction with the NATO Scientific Affairs Division

A	Life Sciences	Plenum Publishing Corporation
B	Physics	London and New York
C	Mathematical and Physical Sciences	D. Reidel Publishing Company Dordrecht and Boston
D	Behavioural and Social Sciences	Martinus Nijhoff Publishers Dordrecht/Boston/Lancaster
E	Applied Sciences	
F	Computer and Systems Sciences	Springer-Verlag Berlin/Heidelberg/New York
G	Ecological Sciences	

Series D: Behavioural and Social Sciences – No. 29

Transformations Through Space and Time

An Analysis of Nonlinear Structures, Bifurcation Points and Autoregressive Dependencies

edited by

Daniel A. Griffith
Department of Geography
State University of New York at Buffalo
Buffalo, New York
USA

Robert P. Haining
Department of Geography
University of Sheffield
Sheffield
England

1986 **Martinus Nijhoff Publishers**
Dordrecht / Boston / Lancaster
Published in cooperation with NATO Scientific Affairs Division

Proceedings of the NATO Advanced Study Institute on "Transformations Through Space and Time", Hanstholm, Denmark, August 3-14, 1985

Library of Congress Cataloging in Publication Data

ISBN 90-247-3362-6 (this volume)
ISBN 90-247-2688-3 (series)

Distributors for the United States and Canada: Kluwer Academic Publishers, 190 Old Derby Street, Hingham, MA 02043, USA

Distributors for the UK and Ireland: Kluwer Academic Publishers, MTP Press Ltd, Falcon House, Queen Square, Lancaster LA1 1RN, UK

Distributors for all other countries: Kluwer Academic Publishers Group, Distribution Center, P.O. Box 322, 3300 AH Dordrecht, The Netherlands

Printed in The Netherlands

PREFACE

In recent years there has been a growing concern for the development of both efficient and effective ways to handle space-time problems. Such developments should be theoretically as well as empirically oriented. Regardless of which of these two arenas one enters, the impression is quickly gained that contemporary work on dynamic and evolutionary models has not proved to be as illuminating and rewarding as first anticipated. Historically speaking, the single, most important lesson this avenue of research has provided, is that linear models are woefully inadequate when dominant non-linear trends and relationships prevail, and that independent activities and actions are all but non-existent in the real-world. Meanwhile, one prominent implication stemming from this literature is that the easiest modelling tasks are those of specifying good dynamic space-time models. Somewhat more problematic are the statistical questions of model specification, parameter estimation, and model validation, whereas even more problematic is the operationalization of evolutionary conceptual models. A timely next step in spatial analysis would seem to be a return to basics, with a pronounced focus both on specific problems (and data) and on the mechanisms that transform phenomena through space and/or time. It appears that these transformation mechanisms must embrace both non-linear and autoregressive formalisms. Given, also, the variety of geographic forms, they must allow for bifurcation points to emerge, too.

A better understanding of the transformation mechanisms will help dynamic and evolutionary spatial modeling. This last class of models is especially in need of enlargement, and accompanying its expansion should be clearer insights into the complexity of geographical organization. Because coming to grips with this issue requires a firm grasp of dynamic and evolutionary spatial modelling problems, it seemed appropriate to provide a forum for intensive interaction among scholars of spatial analysis. A NATO Advanced Studies Institute was held in July of 1980 at the Chateau de Bonas, in Bonas, France, to address critical issues associated with dynamic spatial modelling research. The successful and fruitful cooperation of European and North American scientists at this first Institute provided an impetus for organizing another one that focused on evolving geographical structures. This second NATO Institute was held in July of 1982 at <i Cappuccini>, in San Miniato, Italy, to address fundamental questions in the formulation and calibration of evolutionary geographical models. Again, and in turn, the successful and fruitful cooperation of European and North American scholars at this second Institute furnished the motivation for organizing a third one that would focus on transformations through space and time. One important consensus reached in these two previous Institutes maintains that space and time domains are inseparable, and that for many important classes of events the simultaneous treatment of space and time is an important criterion for an acceptable model. Another conclusion has been that although numerous dynamic spatial models have been formulated, far more attention needs to be paid to the development of evolutionary spatial models, which are substantially fewer in number. This third Institute, the basis for this volume, held at the Hotel Hanstholm, Denmark, drew upon the scholarship and findings of these preceding two Institutes in order to take yet another step in the examination of space-time processes, patterns and structures.

This third Institute had as its general objective the achievement of a more informed and deeper understanding of space-time processes and patterns for socio-economic phenomena, and how these patterns may be more accurately described and predicted. Hence it was far less ambitious in terms of the goals that were set for it. Its four specific aims were:

(1) to acquire a more comprehensive understanding of space-time phenomena;

(2) to identify and describe important real-world space-time processes;

(3) to exchange ideas and perspectives, in a multidisciplinary forum, that are held by quantitative geographers, spatial economists, regional scientists, planners and statisticians from different NATO countries, concerning the mechanisms that transform phenomena over space and through time; and,

(4) to advance the understanding of space-time transformations, especially as they pertain to spatial structure, spatial interaction and urban dynamics.

Accordingly, the Institute focused on theories and empirically based mathematical models of space-time transformations. International exchanges of ideas are both a valuable and a productive means of reviewing the state-of-the-art as well as stimulating fresh ideas that lead to a sound and lasting contribution to knowledge. In pursuit of this goal the lectures were organized around the following four principal themes: (a) transformations of geographical structures, (b) transformations of urban systems, (c) transformations involving interaction over space, and (d) transformations involving autoregressive dependencies.

The mathematical formalisms behind all four themes are non-linear equations, bifurcation points, and autoregressive dependencies. Each of the expositions included in this volume touches upon all three of these principal notions, regardless of which of the four generic categories a paper falls into. Non-linear systems of equations are conspicuous in the models used to explore migration, predator-prey relations, quadratic programming problems, and other features affiliated with the transforming of geographical structures, in the retail models of transformations involving interaction over space, in the models of government fiscal policy, urban systems evolution, and entropy maximization that are concerned with transformations of urban systems, and in the autoregressive models, especially the time-series ARIMA ones. Bifurcation points are alluded to quite often, and are explicitly discussed within the context of urban systems. Autoregressive dependencies particularly are inspected in papers treating spatial autocorrelation mechanisms. All in all, these three principal formalisms are frequently developed and applied in this volume.

We are indebted to the Scientific Affairs Division of the North Atlantic Treaty Organization, the National Science Foundation, and the State University of New York for providing funding of the Institute, to Drs. Robert Bennett, Bruno Dejon, Ross MacKinnon and Giovanni Rabino for serving as advisors to the Institute, and to Drs. Johannes Broecker, Ross MacKinnon, Gordon Mulligan, Keith Ord, Daniel Wartenberg, Anthony Williams and Michael Woldenberg for acting as referees during the editing of this volume. Special appreciation goes to Mr. Kjeld Olsen and his staff, of the Hotel Hanstholm, for their hospitality during the Institute, and for providing an inviting and relaxing physical environment, to the Geography Department of the State University of New York at Buffalo for providing word-processing facilities for the typesetting of this volume, and to Diane Griffith for the typing of much of this volume and her efforts in helping to put together both the Institute and this volume.

Daniel A. Griffith

and

Robert P. Haining

Buffalo, New York
February 19, 1986

TABLE OF CONTENTS

INTRODUCTION

Many spatial process models are concerned with dynamics, evolution, and hence transformations through space and time. Distinctions between static and dynamic geographical models have been outlined by Griffith and MacKinnon (1981), whereas distinctions have been drawn between dynamic and evolutionary spatial models by Griffith and Lea (1983) in the introductions to two companion volumes to this one. The objective of this introductory section is four-fold. First, the notion of a space-time transformation will be clarified. Second, a distinction will be made between spatio-temporal transformation mechanisms, on the one hand, and dynamic and evolutionary spatial models, on the other hand. Third, salient concepts associated with the topic of this book--transformations through space and time--will be examined briefly. Finally, each of the papers of this volume will be related to these key concepts.

The notion of a transformation refers to the establishment of, mathematically speaking, a functional relation between objects. This term occurs in a variety of mathematical situations, and often means simply that a change is being described by an equation or algebraic expression in order to characterize some process. When the objects in question are geometrical in nature, it is customary to label this foregoing equational or functional relation a transformation. These geometric objects may be of any sort. Clearly the functional relation that describes changes in a geographical map pattern over time qualify as transformations, according to this definition. But such transformations can take on one of two forms. First, the functional relation can transform one map into another in a parallel fashion, implying that for any set of n areal units, each being denoted by i, and an n-by-1 vector of some geographically distributed phenomenon $\underline{Y}$,

$$\underline{Y}_t = \underline{AY}_{t-1} \quad , \tag{1}$$

where $\underline{A}$ is an n-by-n matrix of numerical coefficients, and t is the time subscript.

Here the transformation is an affine one if, among other more abstract properties (Gans, 1969),

(1) matrix $\underline{A}$ is a one-to-one mapping, and is such that $\underline{Y}_{t-1} = \underline{A}^{-1}\underline{Y}_t$, and hence $\det(\underline{A}) \neq 0$;

(2) $\underline{Y}_t = \underline{AY}_{t-1}$ and $\underline{Y}_{t-1} = \underline{AY}_{t-2} \Rightarrow \underline{Y}_t = \underline{A}^2\underline{Y}_{t-2}$; and,

(3) the matrix $\underline{A}$ that maps $\underline{Y}_{t-1}$ into $\underline{Y}_t$ is unique.

In other words, this is a linear transformation. If matrix $\underline{A}$ is diagonal, then the accompanying transformation mechanism is purely temporal. If matrix $\underline{A}$ is stochastic, and has $a_{ii} = 0$ for all i, then the associated mechanism is geographically autoregressive. And, if $\underline{A}$ is a relatively dense matrix, then the mechanism is spatio-temporal.

Second, the functional relation can be such that it encompasses a more general class of transformations, known as the projective transformations. Accordingly, this transformation is a generalization of the preceding one in that matrix $\underline{A}$ is augmented, such that using matrix-partitioning notation it could be expressed as

$$\underline{A}^* = \begin{pmatrix} \underline{A}_{11} & \vdots & \underline{A}_{12} \\ \cdots & \cdots & \cdots \\ \underline{A}_{21} & \vdots & b \end{pmatrix} \quad \text{and accordingly} \quad \underline{Y}^* = \begin{pmatrix} \underline{Y} \\ \cdots \\ \xi_t \end{pmatrix} ,$$

where $\underline{A}_{11} = \underline{A}$,
$\underline{A}_{12}$ is an n-by-1 vector of coefficients,
$\underline{A}_{21}$ is a 1-by-n vector of coefficients,
b is a scalar, and
ξ_t could be a random error scalar or a translation parameter.

If vector $\underline{A}_{21} = \underline{0}$ and $\det(\underline{A}_{11}) \neq 0$, then the projective transformation characterized by equation (2) is an affine one. Now,

$$y_{i,t} = \left(\sum_{j=1}^{j=n} a^*_{ij}\, y_{j,t-1} + a^*_{i,n+1}\xi_t\right) \Big/ \left(\sum_{j=1}^{j=n} a^*_{n+1,i}\, y_{i,t-1} + b\xi_t\right) , \quad (2)$$

which is a non-linear equation when vector $\underline{A}_{21} \neq \underline{0}$ [also, when $\det(\underline{A}) \neq 0$ and $b \neq 0$]. Equation (2) permits vanishing points to exist, which is a construct critical to the existence of bifurcation points. Further, since matrix $\underline{A}_{11}$ is not diagonal, spatial autocorrelation mechanisms are present, whereas, as was mentioned earlier, since $a_{ii} \neq 0$, serial correlation mechanisms are present.

Geography has a rich background in the adaption and employment of transformations like equations (1) and (2), especially in its sub-field of cartography (see Cole and King, 1968; Harvey, 1969; Snyder, 1982). However, this volume is concerned with transformations whose purpose is to accurately describe and emulate spatio-temporal processes. The assumption of ergodicity is a convenient one in the formulation of such transformations (Harvey, 1969, pp. 128-9). This assumption exemplifies the need to invoke the scientific law of parsimony when constructing a transformation. Gould (1970, p. 44) warns us that an elaborate transformation function can be concocted, here in order to map one spatial distribution into another over time, but diminishing marginal returns to effort rapidly set in, and when all is said and done, for resulting complex functions '... we have not the faintest idea what [the transformation in question] means.'

Utilizing the class of ARIMA models, to characterize transformations through time, is presently in vogue. Traditionally, models used for characterizing transformations over space have been of many kinds. Tobler (1961) was concerned with defining transformations over space that best capture map pattern in terms of different metric spaces (those other than an Euclidean one). Rushton (1971) has addressed the problem of manipulating a set of points on a punctiform planar surface to best represent a set of local interpoint distances, a problem that seems very similar to that studied by Tobler. Taylor (1971) has further discussed the distance transformation problem solely within the context of spatial interaction phenomena, and to some degree follows the famous format set out by Box and Cox. Angel and Hyman (1972) demonstrated that many human geography theories require combinations of assumptions for which appropriate transformations over space do not exist, a caveat emptor warning to spatial practitioners who shop around for 'ready-made' models. Fourth, and of more relevance to the contents of this book, Gatrell (1979), among others, has summarized the role spatial autocorrelation models play in portraying transformations of geographic phenomena over space. Finally, Wilson (1981, p. 72-3) illustrates a fourth-degree transformation equation necessary to reconstruct the potential function for the cusp catastrophe on an urban space.

Some insights into space-time transformations can be found in Cliff and Ord (1981), and in Griffith (1981). The first two authors have proposed that one primary goal of a space-time transformation is to unravel complex patterns of autocorrelation in both space and time in order to gain insights into functional dependencies amongst areal units that are implied by the presence of non-zero autocorrelation. They then review measures of autocorrelation suitable for spatio-temporal analysis, together with ways of modelling corresponding processes. Griffith, meanwhile, emphasizes the role of assumptions regarding the underlying transformation mechanisms, especially the aforementioned ergodic one, in space-time model specification. Results reported in these two works concur to a very close degree. In part these authors imply that transformations play a very important role in the formulation of dynamic and evolutionary spatial models. Transformations are concerned with the functional forms of relationships that become embedded within a theoretical or conceptual model, whereas dynamic models build upon transformations in such a way that motion is captured, while evolutionary models build upon transformations in such a way that gradual, non-reversible development is captured. Moreover, dynamic spatial models often are written in terms of differential/difference equations, with the time variable permitting a description of change in the geographic distribution of one or more variables to be represented by a transformation between time periods t and t + 1. And, feedback effects are introduced with various types and orders of lag structures. In contrast, evolutionary models focus on movement along a trajectory toward equilibrium, movement which may be described by a dynamic model, with considerable attention being devoted to the disappearance of anomalies from and increasing disorder within the system in question. Hence evolutionary models attempt to take into account the indelible memory of a system, implying time irreversibility. Therefore, transformations help form the kernel of both dynamic and evolutionary spatial models. (Griffith and Lea, 1983)

Given the preceding discussion, the idea of a transformation through space as well as through time will be clarified. Consider a two-dimensional surface over which some phenomenon is distributed. Items can be located on this surface by noting their Cartesian coordinates (u,v). Different time slices of this surface can be denoted by adding the third coordinate of time, yielding a three-dimensional Cartesian coordinate system containing points (u,v,t). The geographical component here is the location specific context of information. The absolute arrangement of areal units that conforms to this three-dimensional space is depicted by equations (1) and (2), which constitute a transformation specification step in model building. At this point, one should recall that the linear model, equation (1), is nothing more than a special case of the non-linear one, namely equation (2). Since the general class of projective transformations represented by equation (2) is very extensive, transformation identification involves screening numerical values in order to determine appropriate entries into matrix $\underline{A}^*$. Consequently, mechanisms for transforming geographic distributions through time emphasize and embody the three ingredients of non-linearities, bifurcation points, and autoregressive dependencies, and hence are best represented by equation (2), which embodies all three of these ingredients. Equation (1) universally displays only this last trait.

This book ultimately is concerned with achieving a better understanding of spatio-temporal structures. We believe gaining this sort of understanding is a prerequisite for the establishment of comprehensive and general evolutionary spatial models. Progress in this latter area has waned, particularly due to impasses encountered by social science researchers who are attempting to pursue this line of inquiry and analysis. Presumably these impasses can be circumvented if a sound foundation were constructed for dynamic and evolutionary modelling. Establishing suitable forms of mechanisms that transform phenomena over space and through time certainly is a step in the right direction. Thus the papers of this volume seek to improve the level of knowledge scholars currently hold about mechanisms governing spatio-temporal change. Obviously the ultimate goal of this sort of undertaking is the formulation, estimation, diagnostic evaluation, and empirical testing of fully evolutionary spatial models. Judging from the papers in this volume, modelling transformations through space and time will involve the following:

(1) the nature and form of subsystem interactions, especially of a geographic origin, specified in a model,

(2) the geographical structure that governs transfers over space,

(3) autoregressive mechanisms, relative positioning of entities and

the spatial metric in which entities are located,

(4) the non-linear nature of laws of motion describing flows through space,

(5) bifurcation points, and

(6) statistical methods for exploiting the latent spatial nature of data during model calibration and parameter estimation.

These are only the more conspicuous components uncovered in this volume that need to be considered when a space-time transformation is being specified.

The task of creating clusters of papers for sections of this book has been a somewhat trying and, at times, difficult one. All of the papers of the Institute dealt with quantitative, geographical problems, and hence common problems were selected as the organizational basis here. We believe that the prominent communality running across all of these papers is the volume's global theme of transformation through space and time. Further, we feel that each of the papers included in these proceedings holds to this theme, rather than the theme appearing to have been constructed around some set of quantitative geography/regional science conference papers, once these papers were selected and collected. Perhaps some readers will disagree with our decisions and viewpoint, and go away disappointed--we hope not. After all, any collection of papers, such as this one, almost by necessity will embrace several somewhat isolated as well as a number of underlying themes. But we feel that, nevertheless, productive and illuminating subsequent research will grow from the seeds planted in this fertile book. Because topics addressed here are very much on a research frontier, the volume complements its two predecessors quite nicely, completes a useful three-volume reference set for spatial analysts, and should have its merits and success judged on the basis of quality of research it propagates.

1. REFERENCES

Angle, S. and G. Hyman, 1972, Transformations and Geographic Theory, Geographical Analysis, 4: 350-367.

Cliff, A., and J. Ord, 1981, Spatial and Temporal Analysis: Autocorrelation in Space and Time, in Quantitative Geography: A British View, edited by N. Wrigley and R. Bennett, London: Routledge and Kegan Paul, pp. 104-110.

Cole, J., and C. King, 1968, Quantitative Geography: Techniques and Theories in Geography, New York: Wiley.

Gans, D., 1969, Transformations and Geometries, New York: Appleton-Century-Crofts.

Gatrell, A., 1979, Autocorrelated Spaces, Environment and Planning A, 11: 507-516.

Gould, P., 1970, Is Statistix Inferens the Geographical Name for a Wild Goose?, Economic Geography, 46 (Supplement): 439-448.

Griffith, D., 1981, Interdependence in Space and Time: Numerical and Interpretative Considerations, in Dynamic Spatial Models, edited by D. Griffith and R. MacKinnon, Alphen aan den Rijn: Sijhoff and Noordhoff, pp. 258-287.

_____, and A. Lea (eds.), 1983, Evolving Geographical Structures, The Hague: Martinus Nijhoff.

_____, and R. MacKinnon (eds.), 1981, Dynamic Spatial Models, Alphen aan den Rijn: Sijhoff and Noordhoff.

Harvey, D., 1969, Explanation in Geography, New York: St. Martin's.

Rushton, G., 1971, Map Transformations of Point Patterns: Central Place Patterns in Areas of Variable Population Density, Papers of the Regional Science Association, 28: 111-129.

Snyder, J., 1982, Map Projections Used by the U. S. Geological Survey, 2nd ed., Washington, D. C.: United States Government Printing Office.

Taylor, P., 1971, Distance Transformations and Distance Decay Functions, Geographical Analysis, 3: 221-238.

Tobler, W., 1961, Map Transformations of Geographic Space, unpublished doctoral dissertation, Department of Geography, University of Washington.

Wilson, A., 1981, Catastrophe Theory and Bifurcation: Applications to Urban and Regional Systems, London: Croom Helm.

SECTION 1

TRANSFORMATIONS OF GEOGRAPHICAL STRUCTURES

By model structure we mean the nature and form of subsystem interactions specified in the model. By geographical structure we refer to the spatial organization of those systems. Geographical models address both system interactions and the spatial organization of that interaction. In the set of papers in this section 'geographical structure' takes on a variety of meanings. The organization of social and economic systems in space is one of the principal foci of geographical research, which includes a wide variety of spatial forms of varying temporal permanence. That variety is captured in the set of papers here, which include transport networks, price distributions, population and income distributions. One of the themes that runs through the set of papers in this section is that the transformations that are described frequently relate to transfers--of people, goods and income, for example. These transfers both shape the developing spatial forms and are shaped by them--a mutual dependence between structure and flow. It is a duality that underlies the development of much spatial/geographical theory in which we are concerned in understanding how spatial structures both influence and are influenced by processes operating in that space. These processes are specified by the transformation rules by which structure at one time period becomes structure at the next; transformation rules at one time period that may be, as suggested here, a function of the existing structural forms. In the case of Weidlich and Haag's migration models, the process is specified as a set of non-linear stochastic differential equations that relate to movements of people between regions. The spatial structure is the population distribution across the regions, and with migration rates dependent on existing configurations of population here, there is a mutual dependency between geographical structure and flow, structure and process.

It also is evident that the authors are dealing with systems for which an equilibrium form may or may not exist, for which the time paths of adjustment (between structure and flow) may be rapid or slow, and where the constituent subsystems (and spatial forms) are changing at different rates. The authors of these papers use a range of mathematical formalisms or transformation rules that can handle some of these problems.

Macmillan's paper considers the inter-relationships between theoretical computational and practical issues in spatial and spatial-temporal economic equilibrium analysis. His interest is in establishing what the spatial forms and structures look like that are associated with the equilibria of spatial and spatial-temporal theory. This is frequently a computational issue because of the complexity of the systems, not least of which is the spatial complexity of forms, a point he exemplifies from classical location theory. Although computational models show us the forms, they must be consistent with theory. His paper is a critical discussion of the spatial equilibrium price models of Takayama and Judge and the use of mathematical programming models in this context.

Curry's paper addresses the problem of developing a truly geographical theory of trade. Such a theory must recognize the continuum of exchange (that includes on the one hand the substitution that takes place between goods within a region, and on the other commodity flows between regions). The objective is to explain regional and commodity price variation, the effects of price distortions on trade, and the structure of trading links in which multilateral exchange is the basic trading relationship. The starting point for his analysis is the pure theory of spatial interaction between buyers and sellers, and he develops the links between price structures, potential gradients and commodity flows. Potential theory allows examination of inter-regional flows and the trading relations between regions. It is a powerful formalism that enables important connections to be examined between properties of flows and structural attributes. The paper develops and broadens issues in the analysis of spatial pricing presented at the San Miniato conference, taking into account the problems of the balance of payments and exchange rates. Pricing and trade in a central place system are examined.

Haining's paper also is concerned with the examination of spatial structure in response to transfer mechanisms. He considers regional income variation and the process of income transfer arising from wage expenditure (hence trade and trading relationships are an implicit element of this paper as well, although not discussed in those terms). The introduction emphasizes the importance of space in the analysis of economic events, and discusses mathematical formalisms that connect spatial structure, spatial pattern and spatial flow, emphasizing the different time periods over which adjustments occur. He reviews several wage expenditure models that have strong formal similarities with some of Curry's models, and then 'opens up' their structure in order to establish relationships between income variation and certain parameters, such as the propensity to save and spend locally. The second half of the paper deals with problems in the statistical fitting of these models to aggregate spatial income data, and concludes with an empirical study. Haining alludes to the need for micro-level surveys to supplement aggregate modeling, an issue that also is present in the paper by Weidlich and Haag in the context of migration modeling.

Sonis's paper deals with the nature of transportation flows between settlements and the implications for network structure. He investigates the set of all possible types of structurally stable optimal transportation flows associated with transport networks in a central place system. He shows that the topological structure of only a very small number of Christallerian systems correspond to optimal minimal cost flows. He includes a discussion of the Beckmann-McPherson generalization, and shows how actual, more complex, hierarchical systems can be expressed as the combinations of basic building blocks.

Weidlich and Haag's paper develops a model of migration that links macro-scale properties of migration levels to micro-level statements of individual motivations and decisions to migrate. The model is a system of stochastic non-linear differential equations. Where the probability distributions associated with such models are known to be unimodal, important analytical insights can be obtained from the mean value equations when more detailed analysis is impossible except by simulation. The importance of these models lies in their explicit connection between micro-level or behavioural attributes of the system and macro-level properties. The need to construct macro-level models that have theoretically sound micro-level foundations is an important focus of research in a large area of quantitative social science, and through the development of these sorts of models the opportunity of real progress presents itself. Of particular interest is the behaviour of some of these systems--the existence of phase transitions and system bifurcations. This theme will arise again in the section on 'Transformations Involving Interaction Over Space,' where it will appear in the context of other kinds of geographical systems. In the Weidlich and Haag model applied to dramatic nineteenth century urban growth, it is the behaviour of an agglomeration parameter that, as it shifts in value, generates a set of different spatial configurations from, at one end of the spectrum, spatial uniformity to, at the other end, spatial concentration. Of course we should not lose sight of the need to interpret those parameters that play such a key role. It is an empirical question of some importance to devise experiments that will enable us to interpret and then measure the 'agglomeration parameter' so that it can be related to changes in observed aggregate system behaviour. Even so, these models offer fertile ground for examining the behaviour of complex geographical structures and their transformations.

COMPUTABLE SPACE-TIME EQUILIBRIUM MODELS

W. D. Macmillan
University of Oxford
England

1. INTRODUCTION

Spatial and spatio-temporal economic equilibrium analysis have a long and chequered history. From the work of von Thunen, in 1826, to the present day it has had three inter-related concerns--theoretical, computational and practical. Von Thunen set himself a theoretical problem that dealt with agricultural land use in an isolated state. The problem was posed in such a way that it required computation to produce a solution, and the method of analysis employed was regarded as thoroughly practical. 'This method of analysis,' wrote von Thunen, a practising farmer and sometimes politician, 'has illuminated--and solved--so many problems in my life, and appears to me to be capable of such widespread application, that I regard it as the most important matter contained in all my work.'

Weber's seminal contribution to industrial location theory (1929), which also was rooted in practice, had similar theoretical and computational concerns. He said that his first purpose, having supposedly solved the theoretical problem of showing that an equilibrium exists, was to solve a computational problem to 'show how it looks.'

Early work in residential location theory, often conducted in a planning context, also addressed both theoretical and computational issues. However, unlike agricultural and industrial location theory, two rather separate strands emerged in the literature. The essentially theoretical approach of Alonso (1964) may be contrasted with the computational approach of Herbert and Stevens (1960). Similarly, more recent work on the existence of an equilibrium, by Schweitzer, Varaiya, and Hartwick (1976), may be contrasted with Weaton's modified version of the Herbert-Stevens computational procedure (Weaton, 1974).

In central place theory, the retreat from the highly idealised assumptions of the early authors, in order to improve both the foundations of the theory and its applicability, has been accompanied by increasing interest in the problem of computing central place patterns (see, for example, Puryear, 1975).

Throughout location and land use theory, then, theoretical and computational concerns have been coupled. It is not hard to see why. In so far as it is interesting at all to look at equilibria, it is clearly insufficient for spatial analysts to restrict themselves to proving that an equilibrium exists in specified circumstances. The whole point of adopting an explicitly spatial approach is to produce theorems about the nature, or form and structure, of spatial organisation. In Weber's terms the emphasis has to be not on the existence of equilibria but on showing what equilibrium patterns of activity look like in space. This becomes a computational problem as soon as (and sometimes before) the assumption of spatial homogeneity or quasi-homogeneity is abandoned. Of course it is perfectly possible, as contributors to the 'New Urban Economics' have ably demonstrated, to adopt a purely analytical approach. To sustain such an approach, however, a high degree of abstraction is required, which for many practical purposes is unacceptable.

This paper is concerned with the first of two problems that are of central importance when practical considerations dictate the use of computational procedures. The first of these problems has to do with the relationship between theory and computation. The second has to do with computation and practical application. It will be argued that 'models,' especially mathematical programming models, often are used in a way that confounds their theoretical and computational roles. It will be claimed that a model is supposed to both operationalise a theory and provide the means to compute operational theorems--to both embody the ideas of the theory and to produce solutions that represent predictions of the theory. This combination of tasks will be shown to be liable to lead to theoretical problems. As part of this argument, the question of the proper interpretation (and nature) of the objective function in mathematical programming models of spatial economies will be tackled.

The second problem derives from the fact that there have been few attempts to apply any but the simplest of the many spatial and spatio-temporal equilibrium models that have been developed in the literature. A set of inter-related difficulties over the estimation of production possibility set parameters appears to have been a major inhibiting factor. A new technique called 'Data Envelopment Analysis' (DEA) provides the basis for a satisfactory method of estimation. Moreover, the use of DEA can supply important theoretical insights, and has two notable contributions to make to planning. First, it allows the relative efficiencies of production units to be measured and efficiency adjustments to inputs and outputs to be calculated. Second, it facilitates interactive multiple objective decision making. The basic principles of DEA and its use in production possibility estimation are discussed in Macmillan (1985). Space restrictions preclude anything more from being said here about these issues.

2. THE RELATIONSHIP BETWEEN THEORY AND COMPUTATION IN EQUILIBRIUM ANALYSIS

The problem to be examined in this paper, then, concerns the relationship between theory and computation. It focuses on the role of models, and addresses the question of how, if at all, should models be used to satisfy the theoretical and computational requirements of spatio-temporal equilibrium analysis. To answer this question it is desirable to begin by clarifying the concept of spatio-temporal equilibrium, and then to look at some example models.

In single location, single period models it may be assumed that if a stable equilibrium exists it will be realised by a process of successive adjustment of prices and allocations taking place 'within' a given time period. However, such an assumption is not necessary. What is required is a specification of the characteristics that a state of the system must possess if it is to be stable. The generalised market clearing condition is such a specification. For a multi-locational system, a spatial or inter-locational price equilibrium condition is needed as well. Once time is introduced, an inter-temporal price equilibrium condition also must be added (assuming temporary equilibria are to be avoided).

It may be objected immediately that real spatial economies do not exhibit the degree of stability that the satisfaction of these conditions would imply. Indeed it might be argued that many spatial systems are inherently unstable. Undoubtedly there is some merit in these criticisms. But this does not imply that work on equilibria is misguided. On the contrary, a proper understanding of the theoretical and computational features of equilibrium models promises to provide a sound basis for the examination of stability questions.

3. SOME SPECIMEN EQUILIBRIUM MODELS

Theoretically, the aim of the model builder is to produce a representation of the collective circumstances and behaviour of a set of economic agents, which is consistent with the existence of an equilibrium. Computationally, the aim is to devise a method for identifying an equilibrium consistent with the theory. Frequently, a single mathematical programming model is employed for these two tasks, the objective of which is interpreted in a way that is supposed to give the model its behavioural content. The circumstances of the agents--typically the production possibilities available to producers--are represented by a subset of the constraints of the programming problem (if they are represented at all). The other equilibrium conditions also appear as constraints, or as first order solution conditions. Just how this is done in practice will be shown with the aid of the complete range of inter-temporal spatial price equilibrium models presented by Takayama and Judge (1971). The structures of these models are as follows:

Supply Function Problems

Single Period Storage

Problem (3.1) (integrable case)

$$\text{MAX:} \quad K + \underline{\theta}'\underline{\Sigma}\underline{X} - (1/2)\underline{X}'\underline{R}\underline{\Sigma}\underline{X}$$

$$\text{s. t.:} \quad \underline{B}\underline{X} \geq \underline{L}$$

$$\underline{X} \geq \underline{0}$$

Problem (3.2) (non-integrable case)

$$\text{MAX:} \quad \underline{\theta}'\underline{\Sigma}\underline{X} - \underline{X}'\underline{R}\underline{\Sigma}\underline{X} + (\underline{\Psi}_c\underline{\rho})'\underline{L}$$

$$\text{s. t.:} \quad \underline{\Psi}_c\underline{B}\underline{X} \geq \underline{\Psi}_c\underline{L} \quad \text{or} \quad \underline{B}\underline{X} \geq \underline{L}$$

$$\underline{\Sigma}\underline{\theta} - \underline{\Sigma}\underline{R}\underline{X} + \underline{B}\underline{\Psi}_c\underline{\rho} \leq \underline{0}$$

$$\underline{\rho} \geq \underline{0} \;, \; \underline{X} \geq \underline{0}$$

Multi-period storage ($1 \leq g \leq \gamma$)

Problem (3.3) (integrable case)

$$\text{MAX:} \quad K_1 + \underline{\theta}_1'\underline{\Sigma}_1\underline{X}_1 - (1/2)\underline{X}_1'\underline{R}_1\underline{\Sigma}_1\underline{X}_1$$

$$\text{s. t.:} \quad \underline{B}_1\underline{X}_1 \geq \underline{L}_1$$

$$\underline{X}_1 \geq \underline{0}$$

Problem (3.4) (non-integrable case)

$$\text{MAX:} \quad \underline{\theta}_1'\underline{\Sigma}_1\underline{X}_1 - \underline{X}_1'\underline{R}_1\underline{\Sigma}_1\underline{X}_1 + (\underline{\Psi}_c\underline{\rho}_1)'\underline{L}_1$$

$$\text{s. t.:} \quad \underline{\Psi}_c\underline{B}_1\underline{X}_1 \geq \underline{\Psi}_c\underline{L}_1$$

$$\underline{\Sigma}_1\underline{\theta}_1 - \underline{\Sigma}_1\underline{R}_1\underline{X}_1 + \underline{B}_1'\underline{\Psi}_c\underline{\rho}_1 \leq \underline{0}$$

$$\underline{\rho}_1 \geq \underline{0} \;, \; \underline{X}_1 \geq \underline{0}$$

Multi-period storage ($1 > \gamma$)

Problem (3.5) (integrable case)

$$\text{MAX:} \quad K_{1(2)} + \underline{\theta}_{1(2)}'\underline{\Sigma}_{1(2)}\underline{X}_{1(2)} - (1/2)\underline{X}_{1(2)}'\underline{R}_{1(2)}\underline{\Sigma}_{1(2)}\underline{X}_{1(2)}$$

$$\text{s. t.:} \quad \underline{B}_{1(2)}\underline{X}_{1(2)} \geq \underline{L}_{1(2)}$$

$$\underline{X}_{1(2)} \geq \underline{0}$$

Problem (3.6) (non-integrable case)

$$\text{MAX:} \quad \underline{\theta}_{1(2)}'\underline{\Sigma}_{1(2)}\underline{X}_{1(2)} - \underline{X}_{1(2)}'\underline{R}_{1(2)}\underline{\Sigma}_{1(2)}\underline{X}_{1(2)} + (\underline{\Psi}_c\underline{\rho}_{1(2)})'\underline{L}_{1(2)}$$

$$\text{s. t.:} \quad \underline{\Psi}_c\underline{B}_{1(2)}\underline{X}_{1(2)} \geq \underline{\Psi}_c\underline{L}_{1(2)}$$

$$\underline{\Sigma}_{1(2)}\underline{\theta}_{1(2)} - \underline{\Sigma}_{1(2)}\underline{R}_{1(2)}\underline{X}_{1(2)} + \underline{B}_{1(2)}'\underline{\Psi}_c\underline{\rho}_{1(2)} \leq \underline{0}$$

$$\underline{\rho}_{1(2)} \geq \underline{0} \;, \; \underline{X}_{1(2)} \geq \underline{0}$$

Activity Analysis Problems

Problem (3.7) (integrable case)

$$\text{MAX:} \quad \underline{\theta}'\underline{\Sigma}_4\underline{X} - (1/2)\underline{X}'\underline{M}\underline{\Sigma}_4\underline{X}$$

Problem (3.8) (non-integrable case)

$$\text{MAX:} \quad \underline{\theta}'\underline{\Sigma}_4\underline{X} - \underline{X}'\underline{M}\underline{\Sigma}_4\underline{X} - (\underline{\Psi}\underline{\rho})'(\underline{S}+\underline{L})$$

$$\text{s. t.:}\quad \underline{\beta X} \leq \underline{S} + \underline{L} \qquad\qquad\qquad \text{s. t.:}\quad \underline{\Psi\beta X} \leq \underline{\Psi}(\underline{S}+\underline{L})$$

$$\underline{X} \geq \underline{0} \qquad\qquad\qquad\qquad \underline{\beta'\Psi\rho} + \Sigma_4\underline{M'X} \geq \Sigma_4\underline{\theta}$$

$$\underline{\rho} \geq \underline{0} \quad , \quad \underline{X} \geq \underline{0}$$

Problems (3.1) to (3.6) inclusive are supply function problems. Problem (3.1) is the key formulation. The other five are variations that are supposed to be required to cover circumstances in which (3.1) is inapplicable. Problems (3.7) and (3.8) are presented as activity analysis counterparts of (3.1) and (3.2). That is, they are presented as single period storage models with activity analysis production possibilities instead of the earlier problem's supply functions. The structure and interpretation of problem (3.1) will be explained first, then the nature of the variations represented by problems (3.2) thru (3.6) will be outlined. The two activity analysis problems will be considered in some detail subsequently.

3.1. Supply Function Formulations

The objective function of problem (3.1) may be written as

$$K + \sum_{\tau=1}^{\gamma} \sigma^{\tau-1} \int_0^{\underline{Y}(\tau)} \sum_i [\underline{p}_i(\tau)\, d\eta_i(\tau) - \underline{p}^i(\tau)\, d\xi^i(\tau)]$$
$$- \sum_{\tau=1}^{\gamma} \sigma^{\tau-1} \sum_i \sum_j \sum_k t_{ij}^k(\tau)\, x_{ij}^k(\tau) - \sum_{\tau=1}^{\gamma-1} \sigma^{\tau} \sum_i \sum_k b_i^k(\tau,\tau+1)\, x_i^k(\tau,\tau+1) \quad , \tag{3.1.1}$$

where $\underline{p}_i(\tau)$ and $\underline{p}^i(\tau)$ are m-by-1, and are given by

$$\underline{p}_i(\tau) = \underline{\lambda}_i(\tau) - \underline{\Omega}_i(\tau)\, \underline{y}_i(\tau) \quad , \text{ for all } \tau, \tag{3.1.2}$$

$$\underline{p}^i(\tau) = \underline{v}_i(\tau) + \underline{H}_i(\tau)\, \underline{x}^i(\tau) \quad , \text{ for all } \tau, \text{ and} \tag{3.1.3}$$

$$\underline{Y}(\tau) = \{y_i^1(\tau),\, x_i^1(\tau);\, \ldots;\, y_i^m(\tau),\, x_i^m(\tau);\ i=1\ 2,\ldots,n\} \quad , \text{ for all } \tau.$$

On integration, expression (3.1.1) becomes

$$K + \sum_{\tau=1}^{\gamma} \sigma^{\tau-1} \sum_i \sum_k \Big[y_i^k(\tau)\{\lambda_i^k(\tau) - \sum_\delta [\omega_i^{k\delta}(\tau)\, y_i^\delta(\tau)]/2\}$$
$$-x_i^k(\tau)\{v_i^k(\tau) + \sum_\delta [\eta_i^{k\delta}(\tau)\, x_i^\delta(\tau)]/2\}\Big] - \sum_{\tau=1}^{\gamma} \sigma^{\tau-1} \sum_i \sum_j \sum_k t_{ij}^k(\tau)\, x_{ij}^k(\tau)$$

$$- \sum_{\tau=1}^{\gamma-1} \sigma^{\tau} \sum_{i} \sum_{k} b_i^k(\tau,\tau+1)\ x_i^k(\tau,\tau+1) \quad . \qquad (3.1.4)$$

The terms in these expressions are defined (for $i,j=1,\ldots,n$; $k=1,\ldots,m$; and, $\tau=1,\ldots,\gamma$) as follows: $y_i^k(\tau)$ and $x_i^k(\tau)$ are, respectively, the amounts demanded and supplied of commodity k in region i in time period τ, $p_i^k(\tau)$ and $p^{ik}(\tau)$ are the corresponding demand and supply prices, $x_{ij}^k(\tau)$ is the amount of commodity k transported from location i to location j during time period τ, $t_{ij}^k(\tau)$ is the corresponding unit transport cost, $x_i^k(\tau,\tau+1)$ is the carry-over of k in i from τ to $\tau+1$ (i. e., the amount stored), $b_i^k(\tau,\tau+1)$ is the corresponding storage cost, σ is a time discount factor, K is a constant, and the remaining terms are parameters.

The objective function is formulated in the way shown in expression (3.1.1) so that it may be interpreted as a 'total discounted net quasi-welfare or ... total discounted net benefit function.' Quasi-welfare is defined, for each region and period, as the difference between the consumer and producer surplus, where these surpluses are the integrals under the multi-commodity demand and supply curves, respectively [the curves being defined by expressions (3.1.2) and (3.1.3)]. After forming the sum over all regions and the discounted sum over all time periods, total discounted transportation and storage costs are subtracted from the total discounted quasi-welfare to give a 'net' benefit expression.

The first point to note about this objective function is that the integrability of the quasi-welfare term is conditional on the symmetry of the matrices $\underline{\Omega}_i$ and $\underline{H}_i$. The implications of this condition will be taken up shortly. The second point is the ambiguity implicit in the designation of the objective function as a quasi-welfare function. On the one hand, this designation is a way of asserting that the agents in the economy behave (collectively) in a particular way. On the other hand, it suggests that any solution to the programming problem will be optimal in some social welfare sense. These two views are not in themselves inconsistent. Indeed, it is well known that a competitive equilibrium is a Pareto optimum, so the identification of an equilibrium through a process of optimisation (where that optimisation is said to represent social behaviour) seems unexceptionable. Yet it is this procedure which causes the theoretical and computational difficulties that are to be considered in detail once all eight of the example problems have been introduced.

Two tasks remain in connection with problem (3.1). One is to present the scalar form of the constraints. The other is to consider the associated equilibrium conditions. The constraint set is as follows:

$$\sum_i x_{ij}^k(\tau) \;-\; y_j^k(\tau) \;+\; [x_j^k(\tau-1,\tau)+x_j^{k*}(\tau-1,\tau)] \;\geq\; 0 \quad , \quad \text{for all } j,\ k \text{ and } \tau, \qquad (3.1.5)$$

$$x_i^k(\tau) \;-\; [x_i^k(\tau,\tau+1)+x_i^{k*}(\tau,\tau+1)] \;-\; \sum_j x_{ij}^k(\tau) \;\geq\; 0 \quad ,$$

for all i, k and τ, and (3.1.6)

$$\left.\begin{array}{ll} y_j^k(\tau),\ x_i^k(\tau),\ x_{ij}^k(\tau) \geq 0 & \text{, for all i, j, k and } \tau\text{, and} \\ x_i^k(\tau,\tau+1) \geq 0 & \text{, for all i, k and for } \tau=1,2,\ldots,(\gamma-1), \end{array}\right\} \quad (3.1.7)$$

where the terms with asterisks are parameters representing 'predetermined' storage quantities. The first constraint ensures that demand is at least satisfied for each commodity at each location. The second ensures that supply is at least adequate to meet distributional and storage requirements.

The equilibrium conditions rely on a definition of the term 'state.' A state of the economy is a set of quantities $\{y_i^k(\tau), x_i^k(\tau), x_{ij}^k(\tau), x_i^k(\tau,\tau+1), p_i^k(\tau), p^{ik}(\tau);$ for $i,j=1,\ldots,n,\ k=1,\ldots,\ m\}$. The economy is said to be in an inter-temporal spatial price equilibrium if a series of states for times $\tau=1,\ldots,\gamma$ satisfy the following conditions:

(a) homogeneity and uniqueness of the market price of each commodity in each region and time period,

(b) no excess demand and efficient market pricing, such that for all j, k and τ,

$$e_j^k(\tau) = \sum_i x_{ij}^k(\tau) - y_j^k(\tau) + [x_j^k(\tau-1,\tau)+x_j^{k*}(\tau-1,\tau)] \geq 0 \quad ,$$

and $$e_j^k(\tau)\ p_j^k(\tau) = 0 \quad ,$$

(c) excess supply possibility and efficient market pricing, such that for all i, k and τ,

$$e^{ik}(\tau) = x_i^k(\tau) - [x_i^k(\tau,\tau+1)+x_i^{k*}(\tau,\tau+1)] - \sum_j x_{ij}^k(\tau) \geq 0 \quad ,$$

and $$e^{ik}(\tau)\ p^{ik}(\tau) = 0 \quad ,$$

(d) inter-temporal price equilibrium, such that for all i, k and for $\tau=1,\ldots\gamma-1$,

$$e_i^k(\tau,\tau+1) = \sigma p_i^k(\tau+1) - p^{ik}(\tau) - \sigma b_i^k(\tau,\tau+1) \leq 0 \quad ,$$

and $$e_i^k(\tau,\tau+1)\ x_i^k(\tau,\tau+1) = 0 \quad \text{, and}$$

(e) spatial price equilibrium, such that for all i, j, k and τ,

$$e_{ij}^k(\tau) = p_j^k(\tau) - p_i^k(\tau) - t_{ij}^k(\tau) \leq 0 \quad ,$$

and $$e_{ij}^k(\tau)\ x_{ij}^k(\tau) = 0 \quad .$$

It can be shown that by deriving the Kuhn-Tucker (first-order solution) conditions for problem (3.1), the solution to this problem satisfies these above conditions.

If the matrices $\underline{\Omega}$ and $\underline{H}$ are not symmetrical the integration in expression (3.1.1) cannot be performed, so that the expression as a whole cannot be used as it is stated in problem (3.1). But if (3.1.1) is lost, then so too are the quasi-welfare and behavioural arguments that go with it. This situation cannot be avoided simply by employing (3.1.4) directly, since the partial differentiation, which must be performed to produce the Kuhn-Tucker conditions, would generate symmetrical supply and demand functions rather than reproducing the required non-symmetric ones. Thus, whatever the solution to problem (3.1) would be in these circumstances, it would not be a series of equilibrium states for the economy under investigation (with its non-symmetrical supply and demand functions).

To circumvent this difficulty, problem (3.2) is introduced. It is argued that this problem will produce a solution that will satisfy the equilibrium conditions when there is a lack of symmetry, and also will permit a behavioural interpretation, albeit a somewhat different one from that given to problem (3.1). The objective function of problem (3.2) is properly interpreted as a revenue expression, meaning the behaviour apparently represented by the model is that of revenue maximisation. The reference to consumer behaviour in the interpretation of problem (3.1) is lost.

The remaining supply-function models--problems (3.3) to (3.6), inclusive--can be dealt with rapidly. All four are designed to cope with multi-period storage, which is beyond the scope of problems (3.1) and (3.2). Problems (3.3) and (3.4) are intended to be used when the maximum storage time, g, is less than γ (the total number of time periods covered by the model). In problem (3.3), and in the associated equilibrium conditions, the demand and supply quantity relationships (3.1.5) and (3.1.6) are replaced by the following conditions:

$$\sum_i x_{ij}^k(\tau) - y_j^k(\tau) + \sum_{\varepsilon=1}^{g} x_j^k(\tau-\varepsilon,\tau) + \sum_{\varepsilon=1}^{g} x_j^{k*}(\tau-\varepsilon,\tau) \geq 0 \quad ,$$

for all j, k and τ, and (3.1.8)

$$x_i^k(\tau) - \sum_{\varepsilon=1}^{g} x_i^k(\tau,\tau+\varepsilon) + \sum_{\varepsilon=1}^{g} x_i^{k*}(\tau,\tau+\varepsilon) - \sum_j x_{ij}^k(\tau) \geq 0 \quad ,$$

for all i, k and τ, (3.1.9)

where $x_i^k(\tau,\tau+\varepsilon) = 0$ if $\tau+\varepsilon \geq \gamma+1$, and $x_i^k(\tau-\varepsilon,\tau) = 0$ if $\tau-\varepsilon \leq 0$ for all i, j, k, τ and ε, and where the asterisked terms are fixed storage quantities. This modification to the constraints requires consequent modification to the objective function. Otherwise problem (3.3) mirrors problem (3.1) in its structure. Similarly, problem (3.4) mirrors problem (3.2)--it is the 'non-integrable' counterpart of problem (3.3), just as problem (3.2) is the non-integrable counterpart of problem (3.1). Problems (3.5) and (3.6) have a similar relationship. They are designed to deal with situations in which the maximum storage time exceeds the total number of time periods covered by the model. The only other comment that needs to be made about these two problems is that their constraints resemble equation (3.1.8) and (3.1.9) with modified horizon conditions.

3.2. Activity Analysis Models

The last pair of examples to be examined are the activity analysis models, namely problems (3.7) and (3.8). Once this has been done, critical comments will be made on all eight examples. The activity analysis format is introduced to improve the description of the supply side of the economy. This goal is achieved by replacing the supply function by a system of inequalities that defines available production possibilities. The demand side still is described by a linear demand function. Problem (3.7) is intended to be used when this function is integrable (i. e., when its coefficient matrix is symmetrical). Its scalar form is as follows:

$$\text{MAX:} \Big[\sum_{\tau} \sum_{i} \sum_{\mu(1)} \sigma^{\tau-1} y_i^{\mu(1)}(\tau) \{ \lambda_i^{\mu(1)}(\tau) - (1/2) \sum_{\delta=1}^{k} [\omega_i^{\mu(1)\delta}(\tau)\, y_i^{\delta}(\tau)]\} - \sum_{\tau} \sum_{\xi} \sum_{\theta(\xi)} \sum_{i} \sum_{j} \sigma^{\tau-1} t_{ij}^{\theta(\xi)}(\tau)\, x_{ij}^{\theta(\xi)}(\tau) - \sum_{\tau} \sum_{\mu} \sum_{i} \sigma^{\tau} b_i^{\mu}(\tau,\tau+1)\, x_i^{\mu}(\tau,\tau+1) \Big]$$

$$\text{s. t.:} \quad y_i^{\mu(1)}(\tau) - \sum_{j} \sum_{\theta[\mu(1)]} x_{ji}^{\theta[\mu(1)]}(\tau) - x_i^{\mu(1)}(\tau-1,\tau) + x_i^{\mu(1)}(\tau,\tau+1) \leq s_i^{\mu(1)}(\tau) + x_i^{\mu*(1)}(\tau-1,\tau) - x_i^{\mu*(1)}(\tau,\tau+1) \quad \text{for all } i,\ \mu(1) \text{ and } \tau, \qquad (3.2.1)$$

where $x_i^{\mu(1)}(0,1) = 0$ and $x_i^{\mu(1)}(\gamma,\gamma+1) = 0$, for all i and $\mu(1)$

$$\sum_{\mu(1)} \sum_{j} \sum_{\theta[\mu(1)]} a_i^{\mu(2)\theta[\mu(1)]}(\tau)\, x_{ij}^{\theta[\mu(1)]}(\tau) - \sum_{j} \sum_{\theta[\mu(2)]} x_{ji}^{\theta[\mu(2)]}(\tau) - x_i^{\mu(2)}(\tau-1,\tau) + x_i^{\mu(2)}(\tau,\tau+1) \leq s_i^{\mu(2)}(\tau) + x_i^{\mu*(2)}(\tau-1,\tau) - x_i^{\mu*(2)}(\tau,\tau+1) \quad , \qquad (3.2.2)$$

with $x_i^{\mu(2)}(0,1) = 0$ and $x_i^{\mu(2)}(\gamma,\gamma+1) = 0$ for all i, $\mu(2)$ and τ,

$$\sum_{\mu(1)} \sum_{j} \sum_{\theta[\mu(1)]} a_i^{\mu(3)\theta[\mu(1)]}(\tau)\, x_{ij}^{\theta[\mu(1)]}(\tau) + \sum_{\mu(2)} \sum_{j} \sum_{\theta[\mu(2)]} a_i^{\mu(3)\theta[\mu(2)]}(\tau)\, x_{ij}^{\theta[\mu(2)]}(\tau) + \sum_{j} \sum_{\theta[\mu(3)]} [x_{ij}^{\theta[\mu(3)]}(\tau) - x_{ji}^{\theta[\mu(3)]}(\tau)] - x_i^{\mu(3)}(\tau-1,\tau) + x_i^{\mu(3)}(\tau,\tau+1) \leq s_i^{\mu(3)}(\tau) + x_i^{\mu*(3)}(\tau-1,\tau) - x_i^{\mu*(3)}(\tau,\tau+1) \quad , \qquad (3.2.3)$$

with $x_i^{\mu(3)}(0,1) = 0$ and $x_i^{\mu(3)}(\gamma,\gamma+1) = 0$ for all i, $\mu(3)$ and τ,

$$\sum_{\mu(1)} \sum_{j} \sum_{\Theta[\mu(1)]} a_i^{\mu(4)\Theta[\mu(1)]}(\tau)\, x_{ij}^{\Theta[\mu(1)]}(\tau) + \sum_{\mu(2)} \sum_{j} \sum_{\Theta[\mu(2)]} a_i^{\mu(4)\Theta[\mu(2)]}(\tau)\, x_{ij}^{\Theta[\mu(2)]}(\tau) - x_i^{\mu(4)}(\tau-1,\tau) + x_i^{\mu(4)}(\tau,\tau+1) \le s_i^{\mu(4)}(\tau) + x_i^{\mu^*(4)}(\tau-1,\tau) - x_i^{\mu^*(4)}(\tau,\tau+1) \quad , \qquad (3.2.4)$$

with $x_i^{\mu(4)}(0,1) = 0$ and $x_i^{\mu(4)}(\gamma,\gamma+1) = 0$ for all i, $\mu(4)$ and τ,

$y_i^{\mu(1)}(\tau)$, $x_{ij}^{\Theta(\xi)}(\tau)$, $x_i^{\mu}(\tau,\tau+1) \ge 0$,
for all i, j, $\mu(1)$, $\Theta(\xi)$, ξ, μ and τ ,

where $\mu(1)$, $\mu(2)$, $\mu(3)$ and $\mu(4)$ are indices for final, intermediate, mobile primary and immobile primary commodities, respectively, $x_{ij}^{\Theta(\xi)}$ represents the amount of output from the production or flow process $\Theta(\xi)$ that is transported from location i to location j during time τ [where $\xi = \mu(1)$, $\mu(2)$ or $\mu(3)$], $a_i^{v\Theta(\eta)}(\tau)$ is the quantity of input v required for one unit of output of η from process Θ^{η} in location i at time τ [where $v = \mu(2)$, $\mu(3)$ or $\mu(4)$ and $\eta = \mu(1)$ or $\mu(2)$], and $s_i^{\mu}(\tau)$ is the initial endowment of μ in location i at time τ [where $\mu = \mu(1)$, $\mu(2)$, $\mu(3)$ or $\mu(4)$].

The optimality and inter-temporal spatial equilibrium conditions associated with this problem are obtained from the Lagrangian

$$L(\underline{X},\underline{\rho}) = \underline{\theta}'\underline{\Sigma}_4\underline{X} - (1/2)\underline{X}'\underline{M}\underline{\Sigma}_4\underline{X} + (\underline{I}^{\varphi}\underline{\rho})'(\underline{S} + \underline{L} - \underline{\beta X}) \quad . \qquad (3.2.5)$$

The partial differential of this expression, with respect to $\underline{X}$, yields conditions that take the following scalar form:

$$\partial L/\partial y_i^{\mu(1)}(\tau) = \sigma^{\tau-1}[\lambda_i^{\mu(1)}(\tau) - \sum_{\delta=1}^{k} \omega_i^{\mu(1)\delta}(\tau)\, \bar{y}_i^{\delta}(\tau)] - \sigma^{\tau-1}\, \bar{\rho}_i^{\mu(1)}(\tau) \le 0$$

$$\partial L/\partial x_{ij}^{\Theta[\mu(1)]}(\tau) = \sigma^{\tau-1}\, \{\bar{\rho}_j^{\mu(1)}(\tau) - \sum_{v} \bar{\rho}_i^{v}(\tau)\, a_i^{v\Theta[\mu(1)]}(\tau) - t_{ij}^{\Theta[\mu(1)]}(\tau)\} \le 0$$

$$\partial L/\partial x_{ij}^{\Theta[\mu(2)]}(\tau) = \sigma^{\tau-1}\, \{\bar{\rho}_j^{\mu(2)}(\tau) - \sum_{z} \bar{\rho}_i^{z}(\tau)\, a_i^{z\Theta[\mu(2)]}(\tau) - t_{ij}^{\Theta[\mu(2)]}(\tau)\} \le 0$$

$$\partial L/\partial x_{ij}^{\Theta[\mu(3)]}(\tau) = \sigma^{\tau-1}\, \{\bar{\rho}_j^{\mu(3)}(\tau) - \bar{\rho}_i^{\mu(3)}(\tau) - t_{ij}^{\Theta[\mu(3)]}(\tau)\} \le 0$$

$$\partial/\partial x_i^{\mu(1)}(\tau,\tau+1) = \sigma^{\tau}\, \bar{\rho}_i^{\mu(1)}(\tau+1) - \sigma^{\tau-1}\, \rho_i^{\mu(1)}(\tau) - \sigma^{\tau} b_i^{\mu(1)}(\tau,\tau+1) \le 0$$

$$\partial/\partial x_i^{v}(\tau,\tau+1) = \sigma^{\tau}\, \bar{\rho}_i^{v}(\tau+1) - \sigma^{\tau-1}\, \bar{\rho}_i^{v}(\tau) - \sigma^{\tau} b_i^{v}(\tau,\tau+1) \le 0$$

and $[\partial L/\partial y_i^{\mu(1)}(\tau)]\bar{y}_i^{\mu(1)}(\tau) = 0$, $[\partial L/\partial x_{ij}^{\Theta[\mu(1)]}(\tau)]\bar{x}_{ij}^{\Theta[\mu(1)]}(\tau) = 0$,

$[\partial L/\partial x_{ij}^{\Theta[\mu(2)]}(\tau)]\bar{x}_{ij}^{\Theta[\mu(2)]}(\tau) = 0$, $[\partial L/\partial x_{ij}^{\Theta[\mu(3)]}(\tau)]\bar{x}_{ij}^{\Theta[\mu(3)]}(\tau) = 0$,

$[\partial L/\partial x_i^{\mu(1)}(\tau,\tau+1)]\bar{x}_i^{\mu(1)}(\tau,\tau+1) = 0$, $[\partial L/\partial x_i^{v}(\tau,\tau+1)]\bar{x}_i^{v}(\tau,\tau+1) = 0$,

for all i, j, $\mu(1)$, $\mu(2)$, $\mu(3)$, $\mu(4)$, v and τ. Together these conditions may be expressed in matrix terms as

$$\underline{\theta}'\underline{\Sigma}_4 - \underline{X}'\underline{M\Sigma} - ({}^{\psi}\underline{I}\underline{\rho})'\underline{\beta} \leq \underline{0} \quad \text{, and} \tag{3.2.6}$$

$$\underline{\theta}'\underline{\Sigma}_4\underline{X} - \underline{X}'\underline{M\Sigma X} - ({}^{\psi}\underline{I}\underline{\rho})'\underline{\beta X} = 0 \quad . \tag{3.2.7}$$

The partial differential of equation (3.2.5) with respect to ρ yields the second set of conditions, which in matrix form are

$${}^{\psi}\underline{I}(\underline{S} + \underline{L} - \underline{\beta X}) \geq \underline{0} \quad \text{, and} \tag{3.2.8}$$

$$\underline{\rho}'\,{}^{\psi}\underline{I}(\underline{S} + \underline{L} - \underline{\beta X}) = 0 \quad . \tag{3.2.9}$$

The remaining conditions are the inequalities $\underline{X} \geq \underline{0}$ and $\underline{\rho} \geq \underline{0}$.

The important point to notice about this collection of conditions is that problem (3.8) may be regarded as being constructed from them. The objective function of (3.8) is the left-hand side of equation (3.2.7) with (3.2.9) substituted into it. The constraints are equations (3.2.6), (3.2.8) and the non-negativity conditions. Problem (3.8) is intended to be used when $\underline{M}$, the demand function matrix, is not symmetric (that is, when the integrability condition is not satisfied).

It also should be noted that this relationship between the 'non-integrable case' problem and the Kuhn-Tucker conditions of the 'integrable case' problem applies to the supply function formulations as well.

4. SOME CRITICISMS

Four pairs of examples of spatio-temporal equilibrium models have been reviewed here in varying amounts of detail. Three major criticisms may be levelled at these models. The modifications arising out of the first of these criticisms enables each of the pairs of models to be replaced by a single formulation, whilst those arising from the second and third allow the three resulting supply function problems to be condensed into one. Thus, the eight examples may be reduced to two.

4.1. The Symmetry Problem

One should note that problem (3.8), on its own, provides no guarantee of obtaining non-negative prices for all consumer goods, and includes no mention of decentralised maximisation of profits (i. e., of the existence and profit maximising behaviour of individual producers). Following the argument of Macmillan (1980), both of these deficiencies may be remedied by a separation of the computing task of finding, inter alia, non-negative prices, and the theoretical task of describing the circumstances and behaviour of a set of economic agents. This separation requires the abandonment of (or at least renders redundant) the principal product of

Takayama and Judge's method—the model. The theoretical task of describing the behaviour of agents and the constraints under which they are operating over space and time is better performed by a set of axioms. These axioms must, of course, be particular rather than general. That is, they must describe each producer's production possibility set, not simply as being a closed and convex subset of n-dimensional Euclidean space, but as being of the activity analysis form with particular numerical values for each of the activity parameters. Provided they are properly formulated, it is possible to derive from such axioms a theorem concerning the properties that an equilibrium will possess. Given this axiomatic framework, the computational task can be seen for what it is—a means for identifying a set of allocations and prices over space and time, which together exhibit the required equilibrium properties. In the case of the theory implied by problem (3.8), these properties are specified by the conditions:

$$\underline{\Psi}\underline{\beta}\underline{X} \leq \underline{\Psi}(\underline{S} + \underline{L}) \quad , \quad \underline{\Sigma}_4'\underline{\theta} - \underline{\Sigma}_4\underline{M}\underline{X} - \underline{\beta}\underline{\Psi}\underline{\rho} \leq \underline{0} \quad , \quad \underline{\rho} \geq \underline{0} \quad , \quad \underline{X} \geq \underline{0} \tag{4.1.1}$$

$$\underline{\theta}'\underline{\Sigma}_4\underline{X} - \underline{X}'\underline{M}\underline{\Sigma}_4\underline{X} + (\underline{\Psi}\underline{\rho})'(\underline{S} + \underline{L}) = 0 \quad , \text{ and} \tag{4.1.2}$$

$$\underline{p}_i^{\mu(1)}(\tau) = \text{MAX}\{0, \lambda_i^{\mu(1)}(\tau) - \sum_{\delta=1}^{k} \omega^{\mu(1)\delta}(\tau)\, y_i^{\delta}(\tau)\} \quad , \quad \text{for all } \mu(1), i \text{ and } \tau \quad . \tag{4.1.3}$$

To find a set of prices and allocations that satisfies these conditions, it is possible to use a slightly extended version of Wolfe's simplex based algorithm for solving quadratic programming problems, notwithstanding the fact that we no longer have a quadratic programming problem to solve! Conditions (4.1.1) could be treated _as if_ they were the first-order Kuhn-Tucker conditions for a quadratic programming problem, and (4.1.2) could be treated as if it was the associated side condition. Conditions (4.1.3) then come into play as an end condition to ensure price non-negativity. In effect, Wolfe's algorithm treats the inequality components of a set of Kuhn-Tucker conditions as a linear programming problem without an objective. To convert these inequalities into the equation form needed for the application of the simplex (or revised simplex) method, it is necessary to introduce slack and artificial variables. The sum of the artificial variables then may be taken as a minimand, and computations equivalent to the first phase of the two-phase simplex procedure may be performed, subject to the added restriction imposed by the side condition (which has the effect of limiting the choice of the entering basic variable at each iteration).

It is interesting to note that if this approach was to be presented, despite what has been said above, in terms of a mathematical programming model, the appropriate form of that model would be to maximise nothing subject to conditions (4.1.1) and (4.1.2) [with (4.1.3) once more as an end condition].

Takayama and Judge prefer a specification in which an objective function is given that can be interpreted in such a way that the model is endowed with some behavioural content and/or some social planning significance. The objection to this procedure is not that it leads to inefficient computations (the primal-dual quadratic programming routine suggested by Takayama and Judge is essentially identical to that outlined above, although it lacks the necessary end condition for ensuring non-negative prices), but that it is at best unnecessary and at worst theoretically misleading. As a consequence, Takayama and Judge are led (quite unnecessarily) to abondon all reference to individual behaviour, and even to assume that in the aggregate it is revenue and not profit that is being maximised.

In the axiomatic approach all of the assertions made about the spatial economy are contained in the axioms and (tautologically) in any theorems derivable from the axioms. Consequently, there is much less risk of ambiguity and inconsistency than there is with a model having an interpretation supplemented by a commentary. With the axiomatic approach there is simply no need to specify and interpret an objective function in order to establish some behavioural proposition (that is done, and done better, by the axioms), or to suggest the optimality of the equilibrium state that is to be identified (this can be done, if it is required, by a theorem), or even to enable the calculation of the equilibrium state to be undertaken.

4.2. The Storage Location Problem

The second class of revisions is concerned with the constraint systems employed by Takayama and Judge. Again, it is sensible to start with the activity analysis problem. Constraints (3.2.1) in the scalar form of problem (3.7) allow units of output from storage to be consumed only at the location at which they have been stored. Constraints (3.2.2) and (3.2.3) are similarly restrictive. The following system of constraints imposes no such restrictions:

$$y_i^{\mu(1)}(\tau) - \sum_{\theta[\mu(1)]} x_i^{\theta[\mu(1)]}(\tau) + \sum_j x_{ij}^{\mu(1)}(\tau) - \sum_j x_{ji}^{\mu(1)}(\tau) - x_i^{\mu(1)}(\tau-1,\tau) + x_i^{\mu(1)}(\tau,\tau+1) \leq 0 \quad , \tag{4.2.1}$$

$$\sum_{\mu(1)} \sum_{\theta[\mu(1)]} a_i^{\mu(2)\theta[\mu(1)]}(\tau)\, x_i^{\theta[\mu(1)]}(\tau) - \sum_{\theta[\mu(2)]} x_i^{\theta[\mu(2)]}(\tau) + \sum_j x_{ij}^{\mu(2)}(\tau) - \sum_j x_{ji}^{\mu(2)}(\tau) - x_i^{\mu(2)}(\tau-1,\tau) + x_i^{\mu(2)}(\tau,\tau+1) \leq 0 \ , \tag{4.2.2}$$

$$\sum_{\mu(1)} \sum_{\theta[\mu(1)]} a_i^{\mu(3)\theta[\mu(1)]}(\tau)\, x_i^{\theta[\mu(1)]}(\tau) + \sum_{\mu(2)} \sum_{\theta[\mu(2)]} a_i^{\mu(3)\theta[\mu(2)]}(\tau)\, x_i^{\theta[\mu(2)]}(\tau) + \sum_j x_{ij}^{\mu(3)}(\tau) -$$

$$\sum_j x_{ji}^{\mu(3)}(\tau) \;-\; x_i^{\mu(3)}(\tau-1,\tau) \;+\; x_i^{\mu(3)}(\tau,\tau+1) \;\leq\; S_i^{\mu(3)}(\tau) \quad \text{, and} \tag{4.2.3}$$

$$\sum_{\mu(1)} \sum_{\theta[\mu(1)]} a_i^{\mu(4)\theta[\mu(1)]}(\tau)\, x_i^{\theta[\mu(1)]}(\tau) \;+\; \sum_{\mu(2)} \sum_{\theta[\mu(2)]} a_i^{\mu(4)\theta[\mu(2)]}(\tau)\, x_i^{\theta[\mu(2)]}(\tau) \;-\; x_i^{\mu(4)}(\tau-1,\tau) \;+\; x_i^{\mu(4)}(\tau,\tau+1) \;\leq\; S_i^{\mu(4)}(\tau) \quad , \tag{4.2.4}$$

for all i, $\mu(1)$, $\mu(2)$, $\mu(3)$, $\mu(4)$, and $\tau=1,2,\ldots,(\gamma-1)$,

where $x^{\theta(\eta)}_i(\tau)$ is the level of output from process $\theta(\eta)$ in i at τ, and $x^{\xi}_{ij}(\tau)$ is the amount of commodity ξ transported from i to j in τ. The $x_{ij}(\tau)$ terms that appear in the scalar form of problem (3.7) have now disappeared, as have both $s^{\eta}_i(\tau)$ terms (on the grounds that it is not particularly sensible to talk about future 'endowments' of a good that has to be produced), and the asterisked storage terms (on the grounds that these are strictly unnecessary for the purpose of the present argument).

Given the above changes in the constraints, the objective function of problem (3.7) would need to be amended to

$$\sum_\tau \sum_i \sum_{\mu(1)} \sigma^{\tau-1}\, y_i^{\mu(1)}(\tau)\, [\lambda_i^{\mu(1)}(\tau)-(1/2) \sum_{\delta=1}^{k} \omega_i^{\mu(1)\delta}(\tau)\, y_i^{\delta}(\tau)] \;-$$

$$\sum_\tau \sum_\xi \sum_i \sum_j \sigma^{\tau-1}\, t_{ij}^{\xi}(\tau)\, x_{ij}^{\xi}(\tau) \;-\; \sum_\tau \sum_\mu \sum_i \sigma^{\tau}\, b_i^{\mu}(\tau,\tau+1)\, x_i^{\mu}(\tau,\tau+1)$$

if the programming formulation is to be retained.

4.3. The Storage Period Problem

The third class of revisions is concerned with storage duration. As noted earlier, problem (3.1) cannot cope with multi-period storage [see constraints (3.1.5) and (3.1.6)]. Problem (3.3) [plus (3.4)] and problem (3.5) [plus (3.6)] were designed to overcome this difficulty. Both of these pairs of problems rely on the replacement of constraints (3.1.5) and (3.1.6) by constraints of the form (3.1.8) and (3.1.9).

Dealing with storage in this way is extremely cumbersome. It is also quite unnecessary. Instead of using (3.1.8) and (3.1.9), inequalities (3.1.5) and (3.1.6) may be replaced by the following single expression:

$$\sum_j x_{ji}^{k}(\tau) \;-\; \sum_j x_{ij}^{k}(\tau) \;+\; x_i^{k}(\tau-1,\tau) \;-\; x_i^{k}(\tau,\tau+1) \;+\; x_i^{k}(\tau) \;-\; y_i^{k}(\tau) \;\geq\; 0 \quad \text{, for all i, k, and } \tau, \tag{4.3.1}$$

where the asterisked terms have been omitted for the same reason as before.

It is interesting to compare this expression with the three-tier and two-tier systems of inequalities proposed by Guise (1979) and Takayama and Hashimoto (1979), respectively, as revisions for expressions (3.1.5) and (3.1.6). Using expression (4.3.1), there is obviously no need to have separate single and multiple storage period problems. Thus, the three revised versions of the supply function problems referred to earlier may be replaced by a general supply function problem consisting of a set of particularised axioms, and an equilibrium identification procedure in which the inequalities become

$$\sigma^{\tau-1}[\sum_j x_{ji}^k(\tau) - \sum_j x_{ij}^k(\tau) + x_i^k(\tau-1,\tau) - x_i^k(\tau,\tau+1) + x_i^k(\tau) - y_i^k(\tau)] \geq 0 \quad ,$$

$$\sigma^{\tau-1}[\lambda_i^k(\tau) - \sum_\delta \omega_i^{k\delta}\ y_i^\delta(\tau) - \rho_i^k(\tau)] \leq 0 \quad ,$$

$$\sigma^{\tau-1}[-v_i^k(\tau) - \sum_\delta \eta_i^{k\delta}(\tau)\ x_i^{k\delta}(\tau) + \rho_i^k(\tau)] \geq 0 \quad ,$$

$$\sigma^{\tau-1}[-t_{ij}^k(\tau) - \rho_i^k(\tau) + \rho_j^k(\tau)] \leq 0 \quad , \text{ and}$$

$$\sigma^{\tau-1}[-\sigma b_i^k(\tau,\tau+1) + \sigma\rho_i^k(\tau+1) - \rho_i^k(\tau)] \leq 0 \quad ,$$

for all i, k, and for $\tau=0,1,\ldots,(\gamma-1)$,

and the side condition becomes

$$\sum_i \sum_k \sum_\tau \Big[\sigma^{\tau-1}\{y_i^k(\tau)[\lambda_i^k(\tau) - \sum_\delta \omega_i^{k\delta}\ y_i^\delta(\tau)] - x_i^k(\tau)[v_i^k(\tau) + \sum_\delta \eta_i^{k\delta}(\tau)\ x_i^{k\delta}(\tau)]\}$$

$$- \sum_k \sum_i \sum_j \sigma^{\tau-1}\ t_{ij}^k(\tau)\ x_{ij}^k(\tau) - \sum_{\tau=1}^{\gamma-1} \sum_i \sum_k \sigma^\tau\ b_i^k(\tau,\tau+1)\ x_i^k(\tau,\tau+1)\Big] = 0 \ ,$$

and the same end condition as in the activity analysis problem [condition (4.1.3)] is used.

Returning to the so-called 'multi-region one storage period activity analysis model,' it is now worth asking what the corresponding multi-period storage problem might look like? The answer to this question is that the expression 'one-storage period' in connection with this model is a misnomer. Takayama and Judge's constraints (3.2.1) do not allow only single-period storage. Thus, the revised version of the activity analysis problem outlined in the preceding section may be regarded as the activity analysis equivalent of the above general supply function problem.

The eight formulations referred to earlier now have been reduced to two. It is easy to show these two allow the problems of negative consumer good prices and the absence of a decentralised profit maximisation assumption to be overcome. Moreover, the method that appears to be responsible for these shortcomings has been shown to be unnecessary. The answer to the question posed at the beginning of Section 2, which asked how mathematical programming models should be used to satisfy the theoretical and computational requirements of spatio-temporal equilibrium analysis, is that they should not be used at all. The associated question of the proper interpretation of the objective function in such a model is rendered meaningless.

5. CONCLUSIONS

This paper has had both a methodological and a practical purpose. An attempt has been made to clarify the relationship between theory and computation at a methodological level, and to show how the results obtained may be applied to the practical business of theory construction. The underlying objective has been to show that methodological and practical questions are intimately related, and that progress on one front should stimulate developments on the other.

6. REFERENCES

Alonso, W., 1964, Location and Land Use, Cambridge, Mass.: Harvard University Press.

Guise, J., 1979, An Expository Critique of the Takayama-Judge Models of Inter-regional and Inter-temporal Market Equilibrium, Regional Science and Urban Economics, 9: 83-95.

Herbert, J., and B. Stevens, 1960, A Model for the Distribution of Residential Activity in Urban Areas, Journal of Regional Science, 2: 21-36.

Macmillan, W., 1980, Some Comments on the Takayama and Judge 'Multi-region One-storage Period Inter-temporal Activity Analysis Models,' paper presented at the 20th European Congress of the Regional Science Association, Munich.

_____, 1985, The Estimation and Application of Multi-regional Economic Planning Models Using Data Envelopment Analysis, paper presented at the 25th European Congress of the Regional Science Association, Budapest, August 27-30.

Puryear, D., 1975, A Programming Model of Central Place Theory, Journal of Regional Science, 15: 307-316.

Schweizer, U., P. Varaiya and J. Hartwick, 1976, General Equilibrium and Location Theory, Journal of Urban Economics, 3: 285-303.

Takayama, T., and H. Hashimoto, 1979, An Expository Critique of the Takayama--Judge Models of Inter-regional and Inter-temporal Market

Equilibrium: a Rejoinder, Regional Science and Urban Economics, 9: 97-104.

_____, and G. Judge, 1971, Spatial and Temporal Price and Allocation Models, Amsterdam: North Holland.

von Thunen, J., 1826, Der Isolierte Staat in Beziehung auf Landwirtschaft und Nationalokonomie. translated by C. Wartenberg and edited by P. Hall as von Thunen's Isolated State (1966), London: Pergammon Press.

Weber, A., 1909, Uber den Standort der Industrien, translated by C. Freidrich as Theory of the Location of Industries (1929), Chicago: University of Chicago Press.

TRADE AS SPATIAL INTERACTION, AND CENTRAL PLACES

Leslie Curry
University of Toronto
Canada

1. INTRODUCTION

We have lived for so long under the naive morphological studies of commercial geography and the spaceless structural ideas of economics that it requires a real effort to envisage a genuine geographical study of trade. Written by economists for economists, the theory of trade rarely provides answers to specifically geographical questions; in particular, it has not provided any results concerning spatial arrangement. While criticism is easy (Curry, 1985a) it is much more difficult to provide the right sort of analytical apparatus.

Chipman's (1965, 1966) thorough analysis of the theory of international trade provides a convenient source of economists' criticisms of that theory. In treating specialisation of production and trade flows, the relative numbers of products and factors turns out to be critical. Since their definitions are essentially arbitrary, involving subtle differences in elasticities of substitution and of aggregation, this result must appear artificial. 'It may be that only a radical departure from conventional theory, in terms of a continuum of gradations in commodities and factors, will provide a satisfactory solution.' The notion of heterogeneous capital can perhaps be included here. Second, he points to the mistreatment of transport costs and the gap which exists between the theories of trade and location. 'Transport costs in international trade theory are either zero or infinite; there is nothing in between. So radical an idealization is this that it is a wonder that the theory has any bearing on reality at all.' Isard's early work is mentioned but no account was taken there of transport operations or freight costing. Third is the possibility of multiple equilibria and the partiality of Pareto optimality of equilibrium in such circumstances. The function of international exchange rates is raised here. If prices of goods are flexible, then exchange rates are redundant. At best they can substitute for any prices that are rigid, either for goods or factors, say wages. Alternatively where there are untraded domestic commodities such as housing then the exchange rate can be used to adjust for the relative prices of home and traded goods. Finally, Chipman discusses the controversies over external economies and their effect on trade, particularly the possibilities of multiple equilibria. Again questions of the classification of an industry are raised and the quandary of using only a static analysis. There have been many warnings against following the logic of modern theory that treats the complexities of world trade as a two country, two commodity, two factor problem. Unfortunately, the simplification introduced radically alters the situation to the point where its results are unlikely to apply in a larger case.

Some of these issues such as factor homogeneity are crucial to an adequate geographical treatment. Further criticisms from a geographical viewpoint are: (1) locational specialisation and inter-regional interchange are frequently mutually causative, (2) treating regions with their factor proportions and demand as given prevents tackling the essential spatial relationships involved, and (3) discussion of the paths of interdependence by which trade is balanced multilaterally is necessary. We shall be content here to delve into the spatial interactions of a trading system and defer a frontal assault on more general topics.

Conceiving trade purely as spatial interaction between buyers and sellers is an obvious yet little used approach and captures aspects of trade not vulnerable to other approaches. Essentially spatial price equilibrium (SPE) ideas are to be followed in examining the effects of/on spatial arrangements on/of multi-commodity exchanges, perhaps the most naive geographical approach possible. As a lead in, the effect of the geometrical configurations in which interaction occurs is demonstrated, with autocorrelation in resource distributions emphasised. Then an analogue to multi-commodity, multi-region trading is sketched on the lines of the SPE literature: this is not formalised because it proves sufficient to outline the issues of income and currency balancing as well as substitutability, accessibility and heterogeneity, which are discussed next. A very general solution is formulated using potentials. The constituent potential describing substitutability of commodities and the balancing potential specifying the equating of imports and exports are added, giving the intra-regional potential, which then is combined with the inter-regional potential from SPE notions. Direct as well as indirect flows of goods necessary for a region to be paid in goods for its exports (i. e., circularity of balancing is represented as various random walks and as an areal transfer function). The reaction-diffusion literature is examined in this context, first treating how the central place system could occur. Diffusion within the hierarchical structure then is examined.

Trade is composed of one-way commercial transactions. The goods of an area are bought by another area without, in the first instance, any thought being given to the direct or indirect reverse flow that must occur to make sufficient medium of exchange available to pay. The existence of export-import firms provides a stimulus to two way traffic as does the return of empty bottoms and politically arranged treaties. But the network of multilateral trade with its indirect method of paying for imports is clearly of major importance. It is easy enough to see why multilateral trade is to be preferred to bilateral trade: the chances of country A wanting to buy goods from B to the same value as B wants from A seems remote. It would be better to have an interconnected system of many countries exporting and importing without regard to bilateral balances, but with regard to individual and thus overall balances. Perhaps the cardinal feature of economic geography is that districts and people produce only a few commodities whereas they all consume many. Thus direct barter between two districts can be only a minor component of trade: complex trade networks are normal. In the final reckoning, each district must disburse the receipts from its exports to all the districts from which it obtains goods. Collection and distribution centres are no more than nodes on such nets of interconnections. The theory of trade does not demonstrate the paths of interdependence; presumably it should be possible to trace the multilateral network by which each dollar output of some market is sent out to a consumer and the path travelled before it eventually comes back to the producer (now consumer). Certainly theory should obtain the various complete circuits of commercial transactions by which obligations are met multilaterally.

2. GEOMETRY, AUTOCORRELATION AND INTERACTION

The geometry of the locations of possible partners must be important in affecting the amount of trade that occurs. To show this in simple fashion, and to use the work of Percus (1977), let the squares of a mesh be either a or b with only contiguous squares interacting. The strengths of interaction are λ_{aa}, λ_{ab} and λ_{bb} and the numbers of each type of contact are N_{aa}, N_{ab}, N_{bb} so that the total interaction is

$$E = N_{aa}\lambda_{aa} + N_{ab}\lambda_{ab} + N_{aa}\lambda_{bb}$$

The extent of E is limited by geometry. Given the total number of squares, N_a, N_b and that the number of neighbours of a single square is C,

$$C\,N_a = N_{ab} + 2N_{aa}$$

$$C\,N_b = N_{ab} + 2N_{bb}$$

$$E = 0.5C\,(\lambda_{aa}N_a + \lambda_{bb}N_b) - N_{ab}\,(0.5\,\lambda_{aa} + 0.5\,\lambda_{bb} - \lambda_{ab}) \quad .$$

The only significant interaction variable is N=a=b and its effect is determined by the average excess of interaction between likes over that between unlikes:

$$\mu_{ab} = 0.5\,(\lambda_{aa} + \lambda_{bb}) - \lambda_{ab} \quad .$$

If a and b are randomly distributed, the only factor affecting interaction is their relative densities:

$$\lambda_{aa} = \lambda_a^2; \quad \lambda_{ab} = \lambda_a\lambda_b; \quad \lambda_{bb} = \lambda_b^2$$

$$\mu_{ab} = 0.5\,(\lambda_a^2 + \lambda_b^2) - \lambda_a\lambda_b = 0.5\,(\lambda_a - \lambda_b)^2 \quad .$$

In the square mesh C = 4 so that $4N_i = 2N_{ii} + \sum_{j \neq i} N_{ij}$.

$$\text{For } E = \sum_{ij} \lambda_{ij} N_{ij}$$

$$= 2 \sum_i \lambda_{ii} N_i - \sum_{ij} \mu_{ij} N_{ij} ,$$

where $\mu_{ij} = 0.5 (\lambda_{ii} + \lambda_{jj}) - \lambda_{ij}$. In the case of only a and b

$$E = k - \mu N_{ab} .$$

If $\mu < 0$, Figure 1a results with max N_{ab};
$\mu = 0$, all patterns such as Figure 1b have equal interaction; and,
$\mu > 0$, N_{ab} has to be minimised, depending upon the relative numbers of a and b

Figure 1c is obtained if $N/4 < N_a < 3N/4$. Figure 1d results if $N_a < N/4$ or $N_b < N/4$. Thus it can be seen that total interaction (trade) will depend very much on the spatial configurations present. Percus demonstrates that the pair correlation function can be used to designate the differences between an independently random arrangement of (a,b) and one that is autocorrelated. Indeed the autocorrelation function is the natural measure for discussing the effect of configuration on interaction.

In specifying the autocorrelation function of resources, it should be remembered that the earth is finite. When the processes producing the distribution are of about the same order of magnitude as the space they occur in (i. e., they are planetary in scale in this case, then the autocorrelation function is not a good guide, while the variance spectrum is better. For example (Curry, 1967), if the space is large relative to the process concerned, we may assume $\rho = kd$; in a small space this becomes $\rho = k [1 - (d/D)]$, where D is the distance beyond which autocorrelation declines to zero. Both functions are linear. However in its inverse Fourier

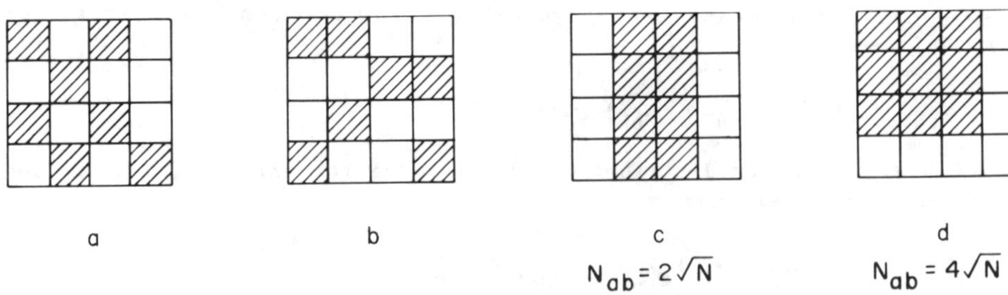

Figure 1: Possible locations of interacting partners (Perous, 1977).

form, while the former declines monotonically as the square of frequency, the latter has a periodicity superimposed on this (Figure 2). Consequently we are likely to find that the variance of global measures of climate has a periodic component because of the standing planetary waves and meridianal circulations. In the same way sedimentary basins that are associated with oil may well have a periodic component to their areal differentiation. Transport costs with distance argument can be replaced by a frequency filter that usually would be monotonically increasing as before. Identically, preference structures are likely to show a spatial ordering that can be represented by their autocovariance, reflecting the manner in which cultures have become differentiated.

As an example consider an autocorrelation function a as in Figure 3 describing the average degree of similarity in the occurrences of a resource at all places according to their distance separation. It may be seen that, in this case, nearby places are similar so that if the resource is present (or absent) here it is likely to be the same a short distance away. There is little correlation with distant places however. Nearby places then are similar so that it is unlikely they will trade with each other in this respect, whereas heterogeneity increases with distance and with it the desire for trade. The complement of this function will be the need for trade. Graph b displays a transport cost function for the resource in question; this could be a general friction of distance curve reflecting any or all of the many types of distance decay that can occur. Graph c combines a and b. On the one hand, when transport costs are low there is not much desire for trading and on the other, when need is high so also are transport costs. It is thus in the middle range then where trade is likely to be greatest: the intensity in this zone is shown by shading. Of course, this is the result of the particular resource distribution depicted--it is, however, very common.

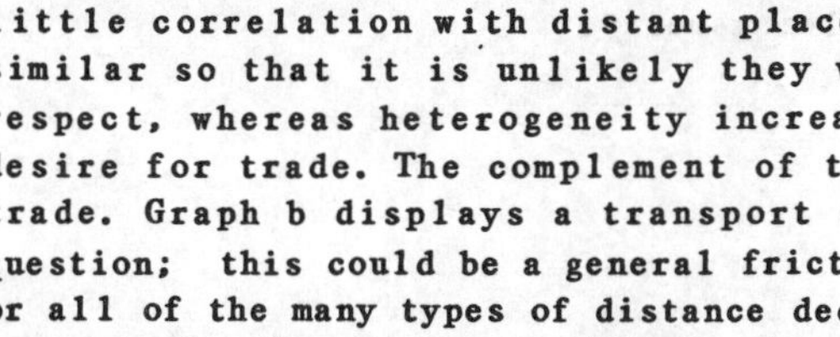

Figure 2: Spectral density for linear autocorrelation space (Curry, 1967).

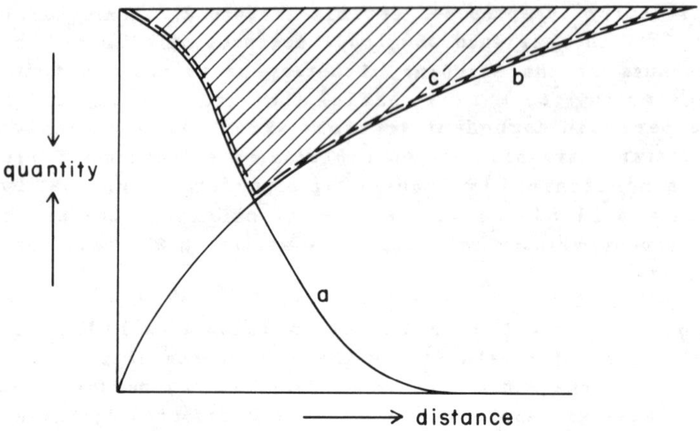

Figure 3: **Interaction and resource homogeneity.**

3. SUBSTITUTABILITY

If the demands for separate commodities were independent, the trade model envisaged would be little more difficult than for the single commodity. But this is not so. A pattern of inter-regional trade will reflect demands in different regions for a good, the degree to which it can be substituted for by other goods in various technical operations, its relative price elasticity and the relative prices of these other goods. Demand could go up for another good in another region, which is partly supplied by our region and occasion increased imports of raw materials and a whole series of shifts in technology, in supply routes, in consumption patterns. For any one region, a price is set for each of a number of commodities that will just balance supply and demand for each without any trade occurring, $V(x_1, \ldots, x_n)$. The vector of first order derivatives is $\underline{v}(x)$ and the matrix of second order derivatives is $\underline{V}$. The off-diagonal elements of $\underline{V}$,

$$v^{(1)}_{1,2}(x) = (\partial^2/\partial x_1 \partial x_2)\ \underline{V}(x_1, x_2 \mid x_3, \ldots, x_n) \quad ,$$

are measures of direct substitutability between x_1 and x_2 given a certain position in $(x_3, \ldots, x_n)$. Substitutability is given by a negative value and complementarity is positive with the dividing line of zero implying independence. However interactions with x_3 will provide indirect substitutability with x_3:

$$v^{(2)}_{1,2} = \sum_3 v_{13}\, v_{32} = \underline{V}^2 \quad .$$

The k-th order effect are obtained as indirect interactions work themselves out:

$$v_{1,2}^{(k)} = \underline{V}^k \quad .$$

To obtain the total effect, each of these subsidiary effects needs to be added:

$$u_{ij} = \sum_{k=0}^{\infty} v_{1,2}^{(k)} \rightarrow \underline{U} = \underline{I} + \underline{V}^1 + \underline{V}^2 + \underline{V}^3 + \ldots + \underline{V}^k + \ldots$$

$$= (\underline{I} - \underline{V})^{-1} \quad .$$

This prices matrix $\underline{U}$ is in equilibrium in that all substitutions have adjusted and thus can be regarded as autarkic potentials. This argument can be phrased in terms of stochastic processes, (i. e., by disturbing the system by random shocks). The covariances of errors are proportional to the total cross substitution effects, (i. e., the off-diagonal elements except for sign and a positive constant). Variances are proportional to the diagonal elements (Phlips, 1974).

Each region has a demand function that is an aggregate of heterogeneous individuals. Some activities simply will use less of a commodity as its price rises, others may easily substitute an alternative commodity if its relative price falls, while yet others may do so only if the preferred commodity becomes extremely expensive. Merchants will know the play of the market: in some cases they may automatically follow the price drop of a highly substitutable commodity. In other cases, where there is a degree of complementarity between goods, quantities will follow depending on the initial reaction to prices. Given that the demand curve sums all this up, it is unlikely that it is linear. However, we shall assume it is. While it is reasonable to accept demand curves as given in the single commodity spatial price equilibrium case, this practice is suspect when the many commodity regional trade model is considered. Indeed these curves are somewhat ambiguous here. In empirical terms, they can be established taking the whole trade set-up into account and especially that the books must be balanced regionally. Conceptually, they are analogous to their separate components, the marginal utilities of goods for an individual that are dependent on the amount of the particular good he possesses, then on the amounts of other goods he possesses, and finally on the amounts of goods possessed by other individuals. But if we are starting from scratch with none of the prices fixed that will determine the individual or collective baskets of goods consumed, then we cannot assume the environment to define the individual preferences to specify demand curves.

If prices are the integrals of interaction, what guides interaction? In a market economy, the motivation must still be prices, but presumably they

will be inefficient and incoherent. Opportunities will exist for making profits and slowly the efficient equilibrium prices will be approached and flows occur up potential gradients. With a changing environment affecting supply and demand, there will always be a searching-learning tatonnement process in progress, just as important as the equilibrium flows. This is the justification in substantive terms for using a stochastic process to describe trade flows. We may postulate flows first, just as we postulated interaction first, so that we are assuming short random price gradients. Through time, and depending on the constraints and dependencies introduced, a steady state will develop. Mean flows will move up potential gradients as before. If we do not emphasize the tatonnement process the deterministic and probabilistic formulations seem incompatible, at least looked at naively, for in the first you have prices producing flows, while in the second flows produce prices.

Previously we criticised the fact that regional demand and supply curves already take into account the inter-regional structure of trade they are used to derive. Again presumably a tatonnement process is occurring continually to control the flow of funds so that the supply and demand functions are historical products, always nearly in step with the overall conditions. We never have start-up conditions in reality--only theorists need to be concerned. However, violent shocks to the system such as the recent oil-price rise can approximate this situation. Given tatonnement here, is there an equivalent stochastic process? We could postulate flows of funds based on random excess supply functions. In the steady state each area would be in balance with the level of activity adjusted to the total structure.

4. SPE AND SUBSTITUTABILITY

In phrasing the spatial price equilibrium problem in terms of potentials the concern is for interactions between pairs of regions. Supply and demand functions that are the properties of the individual region are translated as 'elasticity for trading,' which is the responsiveness of regional imports and exports to local price changes and thus refers to both local absorption and inter-regional interaction. Sheppard and Curry (1982) represented the circumstances as an electrical circuit, as in Figure 4. ξ_i is the price (potential) in each region considered independently when supply equals demand. The resistances are the slopes of the excess supply curves relative to price and thus show how autarkic production can be absorbed in each region. 'Any of the boxes can be opened or closed; if they are closed flow can be in either direction with the appropriate transport cost inserted': they are labelled flap valves and compare the potentials at each end to determine the direction of flow and whether the difference is greater than transport costs so that the link is open. This system can be solved using potentials and without resort to mathematical programming (MP).

Equilibrium occurs when total power (revenue) is minimal corresponding

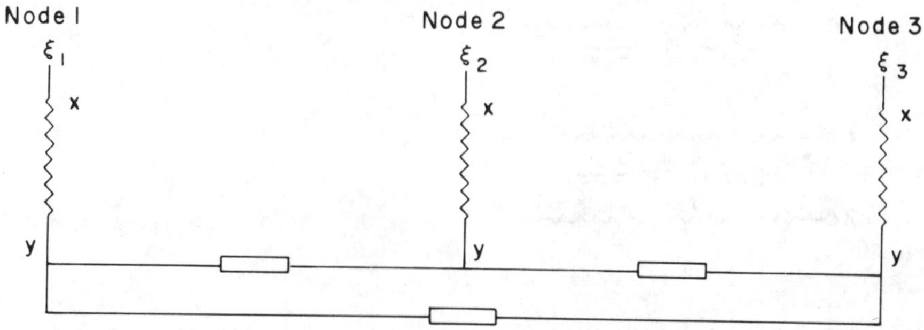

Figure 4: Spatial price equilibrium, the single commodity case.

to the flattest possible potential map where potentials are weighted by flows. Assume linear indifference and production functions, linear interactions, transitivity and additivity with spending equalling earnings at all levels. A unit export of Commodity 1 from Region A requires internal consumption of so much of 1 and import of and internal consumption of so much of 2, 3, . . . , n. The total costs of the quantities specified moving into A must equal the revenue earned by the single export of Commodity 1 if all relationships are specified unambiguously and quantities are correct.

We do not intend to model in such a sufficiently precise way that a MP solution could be obtained, or even present an electrical analogue sufficiently faithful to reproduce the system. Rather, we demonstrate the main principles involved. First the substitutability of Commodities 1 and 2 in Region A is taken up (see Figure 5a). The elasticities for trading resistors have a current flowing through them mirroring the degree of autonomous absorption with potentials for prices, as before. To introduce dependence between commodities, local absorption needs to change. In the case of substitutability, this is done most easily by joining the resistors in series and providing a variable tapping point joined to earth. Then, as the absorption of Commodity 1 is increased, that of Commodity 2 decreases as required. Where complementarity occurs so that greater use of 1 induces greater use of 2 and vice versa, the connections are made so that both resistors increase or decrease together (Figure 5b). The two figures may be combined as in 5c, where the angle represents the degree of substitutability while lateral motion with a vertical link represents complementarity. Finally if, say, three commodities are to be investigated, each pair would be treated as in Figure 5c, which we shall symbolise as in 5d. Figure 5e combines Figure 4 and 5 and thus combines the interaction of the commodities within each region and the interactions between regions for each commodity. Consequently each price, each flow, each region's production and

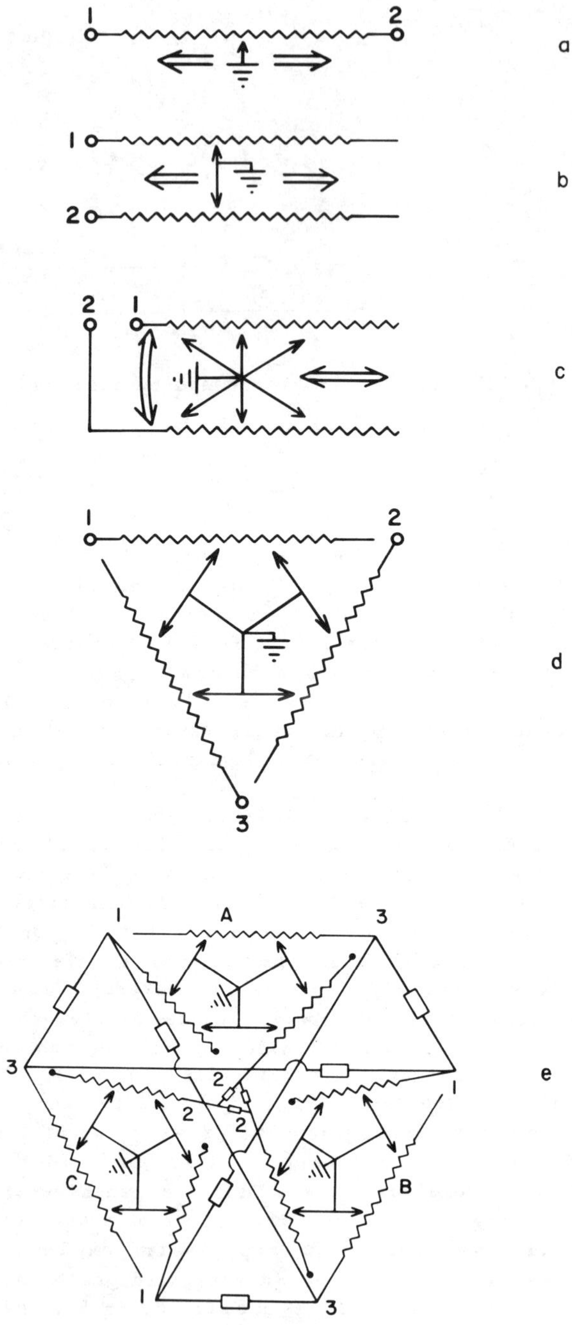

Figure 5: Spatial price equilibrium with commodity substitution.

consumption is dependent on every aspect of the total system. Presumably there will be an overall balance for the total space and a separate balance for each commodity in the minimum power (revenue) equilibrium. However, it is extremely unlikely that the MP solution to this problem would produce balanced import-exports in each region. Each MP problem could confer a region with a large negative excess supply in each commodity so that the aggregate balance is unrealistic (within our assumptions).

Sheppard and Curry pointed to the arbitrary nature of the territorial bounds over which SPE was established so that, given a loss of local autonomy under the overall objective function and a lack of compensatory payments, welfare of some regions could well decline with trade. How far do these criticisms apply to multi-commodity trade? Where there is a considerable degree of substitutability (i. e., commodities are competitive) so that the results of individual SPEs would disappear in the overall equilibrium, this reduction in welfare appears to be much less likely.

5. BALANCING OF PAYMENTS

It is an easy matter to write a model ensuring that each region in a trading system will have imports equalling exports. Essentially it is a variant of the model of substitutability of commodities used earlier. Write the export interaction $\underline{X} = (x_{ij})$, the transpose of which is $\underline{M} = (m_{ij})$, for the imports into the i-th place from j (Seninger, 1977). Propensities to import, $p_{ij} = m_{ij}/\sum_j m_{ij}$ become a matrix of transition probabilities, $\underline{P} = (p_{ij})$

$$P = P_{ij}(n,n+1) = P(\underline{X}_{n+1} = j \mid \underline{X}_n = i) = p_{ij} \quad .$$

This is a one step transition probability of a dollar going to j after n+1 transitions. The $\underline{P}$ matrix for an absorbing Markov chain can be written in canonical form as

$$P = \left(\begin{array}{c|c} I & 0 \\ \hline R & Q \end{array}\right) \quad ,$$

where $\underline{Q}$ and $\underline{R}$ are non-negative matrices representing transitional probabilities for non-absorbing states, and $\underline{I}$ is the identity matrix. Define an initial endowment vector at time zero by $V^0 = (V_2^{(0)}, V_3^{(0)}, \ldots)$, a first round redistribution vector by $\bar{V}^{(1)} = V^{(0)}\underline{Q}$ where income is redistributed according to the transitional probabilities in matrix $\underline{Q}$. A second round results in $\bar{V}^{(2)} = \bar{V}^{(1)}\underline{Q} = V^{(0)}\underline{Q}^2$, and by the n-th round $\bar{V}^{(n)} = V^{(0)}\underline{Q}^n$. Since $\underline{Q}^n$ goes to zero as n increases, then $V^{(0)}\underline{Q}^n$ also goes to zero. Allowing income generation in all places, the exogenous input can be written as $u = (u_2, u_3, \ldots)$. The fundamental matrix can be derived as $(\underline{I} - \underline{Q})^{-1}$. Combined with $\underline{u}$, a limiting distribution $\underline{y}$ can be obtained as

$$\underline{y} = \underline{u}(\underline{I} - \underline{Q})^{-1} \quad .$$

In this steady state incomes in equal incomes out for each region, which is what is needed. However, this simple Markovian formulation is quite inadequate as a description of spatial interaction.

6. EXCHANGE RATES AND EQUILIBRIUM

There appear to be two ways by which we could proceed further. The first is to reduce or increase the share of the home market in each region and, allowing the system to adjust, iterate to the required solution. This would be done by varying all the excess supply curves of a region to the same proportion so that either demand or supply or both could be said to be changed. Figure 6 gives a schematic interpretation of this situation in

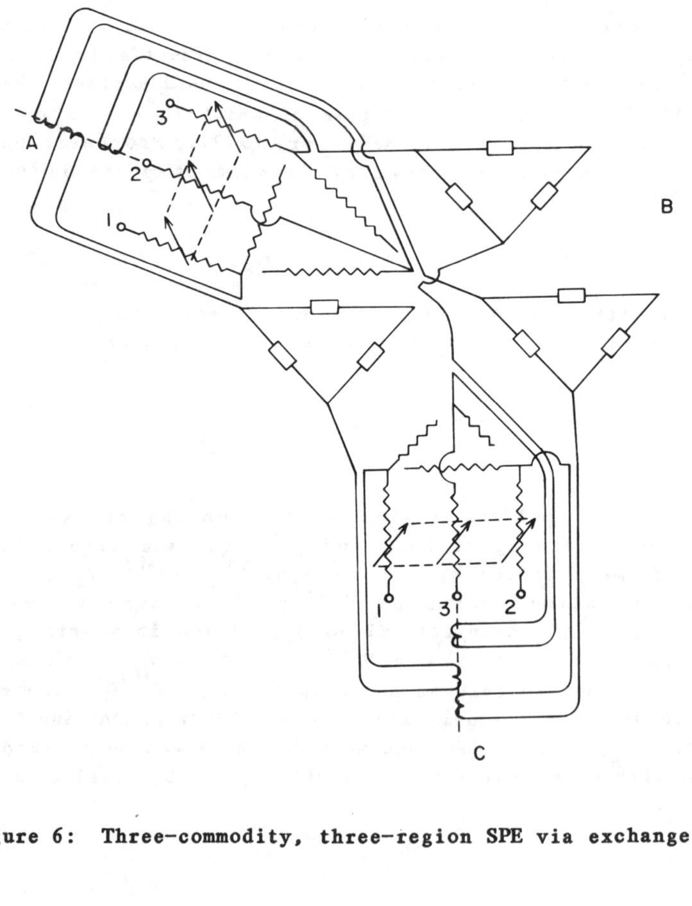

Figure 6: Three-commodity, three-region SPE via exchange rate adjustment.

which the revenue balances (flows x price) of each commodity in a region are summed and the total, positive or negative, automatically adjusts the resistors representing the properties to trade in identical proportions for each commodity. By inserting an ammeter, the direction of the current would show where there is a net flow of funds and potential could be adjusted automatically until there was no flow through the ammeter. By judicious choice of the total reaction in each region, each regional equilibrium is automatically maintained. This is similar to using rates of exchange as the adjustment mechanism provided individual commodity merchants do their own financial exchanges. For example, exporters could be more profitable and divert more to export. Imports would decline and the home market in both sectors would expand. We would need to postulate a fairly long-term adaptation by which, say, producers and importers (but not exporters) could not sell their outputs and consumers could not afford to buy as supply and demand curves suggest. There would be local deflation and the prices of factors would come down, relative to their previous ratios, against other regions. This would lower the supply curves, forming lower internal prices and opening the way to regional balance with a greater home market and lower imports. However, it is unlikely that all supply curves are simply lowered while maintaining their slopes. It is more likely that commodities with strong export components will change little while strong importers will change much so that regional structures will change. Kemeny and Snell (1972) have supplied a method of discounting the currency in different countries in order to obtain rates of exchange so that each country will be in equilibrium. Only trade payments are allowed so that the total supply of currency is constant.

Let there be n countries, $i = 1,..., n$, with $\underline{m} = (m_i)$ the currency presently in various countries, $\underline{f} = (f_i)$ the currency to be transferred to countries as an adjustment and $\underline{g} = (g_i)$ the currency distribution goal that will give equilibrium. Vectors $\underline{m}$ and $\underline{g}$ are positive, whereas $\underline{f}$ may be of either sign. Let p_{ij} be the function of currency in country i that ends up in country j at the end of one period, $0 < p_{ij} < 1$. For our assumptions, $P = (p_{ij})$ and $\sum_j p_{ij} = 1$ so that the matrix $\underline{P}$ is the transition matrix of an ergodic Markov chain. Since at least one and presumably every country retains some of its currency within its borders for a period, the positive p_{ii} diagonal entry assures that the chain is not cyclic. After one period of business, of the original amounts held, $\underline{m}$, $\sum_i m_i p_{ij}$ is at city j. Vector $\underline{mP}$ is the distribution of currency after the one period, $\underline{mP}^2$ after two periods, and so on. The adjustment shifts of currency will in like manner change this total by $\underline{fP}^n$ in n periods; a transfer in the first period causes a cumulative change by the n-th period of $\underline{fP}^{n-1}$. The total amount available after n periods is $\underline{mP}^n + \sum_{k=0} \underline{fP}^k$, which is required to be non-negative and to converge to vector $\underline{g}$. Let the first term tend to $\underline{m1}$,a where $\underline{m1}$ is the total currency in circulation at the outset so that the original sums redistribute themselves to amounts proportional to a. If the second term consisting of the series $\underline{f} + \underline{fP} + \underline{fP}^2$ is to converge, $\underline{fP}^n$ must tend to zero. Since $\underline{fP}^n$ tends to $\underline{f1}$a, with $a \neq 0$, convergence will occur when $\underline{f1} = 0$. Thus, net total currency adjustments must total zero and equilibrium is established by redistribution alone. Indeed convergence is assured since, if $\underline{f}\ \underline{1} = 0$ then $\underline{fA} = 0$ and $\underline{f} + \underline{fP} + \underline{fP}^2 + \ldots = \underline{f} + \underline{f} \sum_{n=1}^{\infty} (\underline{P}^n - \underline{A}) = \underline{fZ}$.

7. A 'THERMODYNAMIC' MODEL

Instead of having equally adjusted variable resistors on the lines of balancing the gravity model for known destinations and origins, it is preferable to have a separate control for each commodity (see Figure 7). Certainly it may be necessary to deflate the regional economy; but probably only some import sectors would need to be affected. Importers of raw materials for export firms should not suffer. Only importers catering directly and indirectly to the home market should feel the pinch. This result suggests reducing the home market; but what mechanism can be appealed to? What signals will summon it?

The answer lies in establishing that each individual and institution balance his/her/its books or that the value of goods moving into an area equals that of exports. Any arbitrary area can be chosen save only that workplaces and homes should not be separated. For given regions as in the analogue circuit, automatic adjustments in choice of commodities as well as in amounts and venues of imports and exports will not only allow intra-regional balance, but also an overall maximum value of exports. An operational model capable of handling a number of commodities and areas that

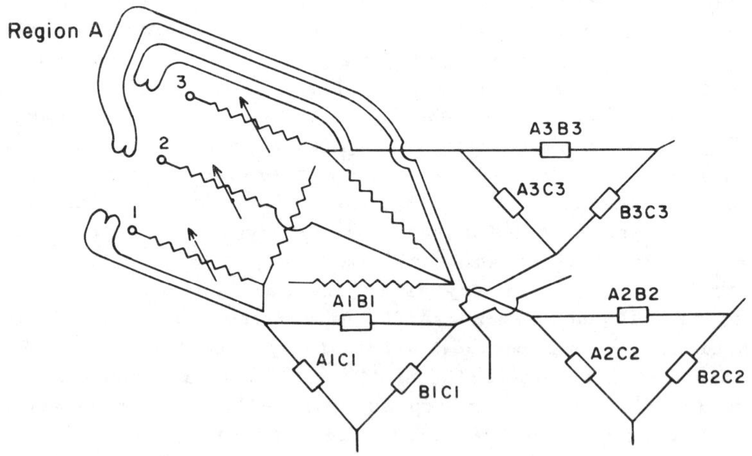

Figure 7: Three-commodity, three-region SPE via regional balancing.

would allow geographical principles to be established is what is wanted but is difficult to conceive. We particularly want to see how many commodity flows interact. We have categorised a region in terms of supply and demand functions plus its freight charges to other regions. The state of the system can be described by the spatial entropy corresponding to the minimum average potential (price), ξ, where the regional prices are weighted by the associated flows, ${}^{D}r$, and by the spatial entropy based on actual prices and flows, S, plus the correlation between the two maps of prices, using flow weights, T (Curry, 1984a). Differentiating yields

$$dU = T\,dS - \alpha\,dr \quad , \quad dE = T\,dS + r\,d\alpha \;,$$

$$dA = -S\,dT - \xi\,dr \quad , \text{ and } \quad dG = -S\,dT + r\,d\xi \quad ,$$

where dU is the increase in price gradients due to a lack of exploitation of opportunities, dE is the loss of flows due to opportunities not being exploited, dA represents wrong price gradients, and dG is the inefficient flows resulting from wrong price gradients. If each of a number of commodities are included and if they are independent of each other, values for r or for ST can be added and averaged, if desired, to give the joint information. To the extent that they are spatially correlated, the entropies cannot be added so that for n maps there will be (n-1)! correlation coefficients from which the extent to which entropies can be added is obtained. All of this can be repeated for derivatives and differences when changes in time are studied. However, dependence of the commodities in time is now involved; interactions are likely to cause changes in amounts and vice versa so that the composition of the commodity mix will change. Here we are concerned with changes in relative amounts due to substitutability.

Let different commodity matrices of substitution coefficients, μ_{ij}, describe each region so that, as the spatial price equilibria iterate towards their solutions for each commodity so also do the amounts substituted. While the substitutability of each pair of commodities can be linear for prices, this relation will be different for different commodities so that there will be considerable interplay between changes in the product mixes used and those supplied in different areas. In the final equilibrium, the overall change in reaction to a unit change in the amount may be termed the constituent potential since it has all the attributes of a potential measure:

$$\mu = \begin{cases} \partial A/\partial n_i & \text{, where } T,\ r,\ n_j \neq n_i \text{ given;} \\ \partial G/\partial n_i & \text{, where } T,\ \xi,\ n_j \neq n_i \text{ given;} \\ \partial U/\partial n_i & \text{, where } r,\ S,\ n_j \neq n_i \text{ given; and,} \\ \partial E/\partial n_i & \text{, where } S,\ \xi,\ n_j \neq n_i \text{ given.} \end{cases}$$

In equilibrium $\sum_i \mu_i \partial n_i = 0$ (Poland, 1978). The parameter μ_{ij} is the substitutability of commodity i for j in a region, and $\sum_j \mu_{ij}$ is the total substitutability of i in a region, the constituent potential. The differential ∂n_i is the unit change in the amount of commodity i so that $\sum_i (\sum_j \mu_{ij})\partial n_i$ is the total change in all commodity amounts in a region for a change in each.

Define a dissipation function β as income that could be generated but which is being lost in various inefficiencies of organisation. The supply of inefficiencies capable of being corrected for and so of being converted into income is due to innovations that define the efficient possible, and thus increases the gap between this level and realisation:

$$\partial U = T\,\partial S - \xi\partial r + \sum_i \sum_j \mu_{ij}\,\partial n_i; \quad \text{for constants } r,\ S.$$

Setting ∂r and ∂U to zero,

$$T\,\partial S = \sum_i \sum_j \mu_{ij}\,\partial n_i \quad .$$

Taking derivatives with respect to time,

$$\beta = T\dot{S} = -\sum_i \sum_j \mu_{ij}\,\dot{n}_i \quad .$$

The rate at which substitution may be achieved is

$$V = \dot{n}_i / v_i \quad ,$$

where v_i is the overall substitutability coefficient of the i-th commodity. Summing the product of this and the constituent potential over all commodities, the affinity is obtained:

$$F = -\sum_i (\sum_j \mu_{ij}\,v_i) \quad .$$

From this result we obtain β = VF (Caplen and Essig, 1983).

There is another aspatial influence between components that needs to be examined as well as the constituent potential. This is due to the fact that each region needs to balance its books. Let π be the balance of imports over exports for all commodities in a region. When the regional system is in equilibrium we may call this the balancing potential. If we multiply it by a term z_i related to the slope of the surplus demand/supply curve of commodity i, then we get the degree of diminution of imports, and augmentation of exports of this commodity in a region. Obviously, in equilibrium π is zero. Since potentials can be added, combining the balancing and constituent potentials per unit change, $\sum_i [(\sum_j \mu_{ij}) + z_i\pi]\partial n_i$, we may speak of the intra-regional potential. Averaged over all regions we have

$$\partial U = T\,\partial S - \xi\partial r + \sum_i [(\sum_j \mu_{ij}) + z_i\pi]\partial n_i \quad .$$

Instead of defining $\beta = VF$ we now have

$$\beta = \Delta[(\sum_j \mu_{ij}) + z_i \pi]\mathring{n}_i + VF \quad .$$

Consequently $\beta < VF$, which implies that the necessity for balancing has reduced the rate of dissipation; since it constrains trade in goods, it affects the rate at which the inefficiency gap defined on individual goods is reduced.

8. CIRCULARITY OF MULTILATERAL BALANCES

Multilateral trade is simply the aggregate version of interpersonal trade. When one gets down to the level of individuals, the multilateral interactions involved in settling the accounts between two persons are absolutely fantastic. Say the reporter on a local newspaper in Thailand buys an Eskimo carving from N.W.T. of Canada. Imagine the transactions involved in real terms before the return flow is closed. The variety of paths by which transactions are made before the circuit is complete is almost boundless. If one considers the vast number of goods and services that an individual uses compared with the one or two activities that he/she contributes, it is clear that the real trade circuit looks less like the path of a rocket and more like a continually star-bursting effect at a fireworks display. In money terms, the path is quite different since receipt of exports by B is likely to stimulate a direct return of payment to A almost immediately. However, these funds then will be spent by A in many places so that the return path to B will look like a diffusion process, first spreading out and then as a set of targets (the sources of B's imports) are hit this financial circuit is closed. We cannot distinguish the tracks of dependence categorically. When A sells to B who sells to C,D,E and F we can only give probabilities of the path being ABC, ABD, or the such. Even if we emit a marked dollar bill from a source the actual track it takes is only one of a large number of possible tracks it could have taken, so that again the appropriate phrasing is of random flights.

Consider a map of value of exports per person: perform cascaded averaging and differencing on it in order to determine its scale composition (Curry, 1969). This procedure also may be regarded as breaking up the map into slope components such that the Laplace equation is applicable at each scale. Assuming flows are proportional to slopes, we are assuring that gradient and thus flow in equals gradient and flow out locally at each scale and that the map is in equilibrium. Empty lattice points would also balance since flows would simply go in and out again. There could even be losses from the system, say payments of tribute abroad; then the Poisson is the operative elliptical equation and balances are maintained. The resulting function shows an 'equivalent' geographical make-up of trade partners. Although there need be no distance dependence in actual flows as everyone may be sending payment to distant partners, the scale composition function says something about the degree of complementarity of areas according to the degree of localness relative to the whole system. It describes the least possible scales required to make the unit payment of a local area return to itself. In this sense it describes minimum circularity of flows. The uniform spectrum of an independently random distribution means considerable return at all scales and goes with zero autocorrelation--say no distance weighting in the generating equation. With a large autocorrelation there is considerable return from a distance.

Another way of looking at the circularity of trade links is by setting up the problem as a symmetric random walk of a dollar bill in two dimensions (2D). It will always return to its starting point if we take a long enough period, so that completion will always occur, eventually. Take first a one dimension random walk of N steps, each λ long, with A forwards and B backwards and starting at x. What is the probable distance travelled from the origin (Tabor, 1969)?

$$A + B = N \quad ,$$

$$A\lambda - B\lambda = x \quad ,$$

$$A = 0.5\,[N + (x/\lambda)] \quad , \text{ and}$$

$$B = 0.5\,[N - (x/\lambda)] \quad .$$

The number of ways (W) that taking N steps with A forwards and B backwards will give x,

$$W = N!/(A!\,B!) \quad .$$

Using Stirling's approximation to the factorial and since $x << N\lambda$,

$$\ln(W) = N\ln(2) - x^2/(2N\lambda^2 \quad \text{or} \quad W = Q\exp(-B^2x^2) \quad ,$$

where $B_1 = 1/(2N\lambda^2)$. Meanwhile, the probability of distance x occurring is proportional to W, such that

$$P(x) = C\exp(-B^2x^2) \quad .$$

The constant C is determined by observing that

$$\sum_x = -N\lambda\, P(x) = 1 \simeq \int_{-\infty}^{\infty} P(x)\, d(x),$$

and so $C = B_1/\pi^{0.5}$. Now turning to 2D, with the origin at (0,0) and ending at (x,y), the probability of finding the bill there is given by

$$P(x,y) = (B^2/\pi)\exp[-B^2(x^2 + y^2)] \quad , \text{ with } B^2 = 1/(N\lambda^2) \ .$$

Only the exponential factor is of interest. The most probable length travelledis

$$P(r) = A \exp(-B^2 r^2) \tag{8.1}$$

The maximum of this density occurs at r=0 but this does not mean that the most probable distance between origin and termination is zero. To remove the directional effect, ask for the terminations occurring in an annulus of area $2\pi r\, dr$. The number of lengths, then, π, between σ and σdr is

$$\pi dr = A\, 2\pi r \exp(-B^2 r^2)\, dr \quad .$$

The maximum of this distribution occurs at

$$r_m = B^{-1} = \lambda N^{0.5} \quad .$$

From equation (8.1), the configurational entropy is

$$S = K \log(P) = \text{constant} - k B^2 r^2 \quad .$$

Let us assume that this is a description of an efficient system so that its average potential is minimum (i. e., the gradients driving the dollar bill are as flat as possible). The effect of reducing efficiency marginally, say, by incrementing the average potential difference positively is

$$\xi dr = d(TS) = dU - dA \quad .$$

In words, the increase in the price gradients due to lack of exploitation of opportunities (dU) minus the change in the wrong price gradients of the real world (i. e., transaction costs), dA, (if negative the two are added) equals the decrease in efficiency d(TS). Assume this latter is measured on the areal entropy of the real world map, dS, efficiency T assumed unchanged. Rewriting as

$$\xi dr = dA = d(U - TS) \quad ,$$

then

$$\xi = (dU/dr) - T(dS/dr) = (dU/dr) - 2\, k\, T\, B^2 r \quad , \text{and}$$

$$\xi \simeq 2\, k\, T\, B^2 r \quad .$$

Thus the changes are proportional to $B^2 = 1/(N\lambda^2)$ and therefore inversely proportional to the square of step length.

9. REACTION DIFFUSION AND CENTRAL PLACES

A considerable part of the flows of payments for exports is via purchases of consumer goods. We already have asserted that such return flows will be by diffusion so it is interesting to examine such a mechanism in detail for the simplest case of the central place system. We shall first show how central places develop from such flows and then how diffusion occurs within the hierarchical structure. Money, of course, moves from consumer to suppliers; but it is easier to think in terms of goods and thus of indebtedness, a, which diffuses from suppliers to consumers. The rate of change of a will depend on

1. production rate ρ,
2. decay rate $-\mu a$, and
3. its diffusion in space $D_a[(\partial^2 a/\partial x^2) + (\partial^2 a/\partial y^2)]$.

A central place appearing as activator a will appropriate consumers h, to give a market that will be at least large enough to allow a reasonable price to be set. Repeating previous arguments (Curry, 1960, 1967), those retailer's costs relevant to central place analysis are the holdings of inventories. As size of market goes up, the relative size of the upper tail of the probability density of weekly consumer demands decreases relative to the mean. The cost of holding inventories that allow only some small risk of not meeting requirements thus will go down non-linearly as market size increases. We may postulate

4. an autocatalytic process affecting the production rate $k\,a^2$,
5. the inhibitor can hinder the autocatalytic rate ka^2/h [since autocatalysis must be greater than decay, the former has to be non-linear, and the net effects then are

   $$\partial a/\partial t = \rho + (ka^2/h) - \mu a + D_a \Delta_a],$$

6. the inhibitor will have a production rate, in this case taken to be based on the activator, ca^2,
7. it will have a decay rate of $-vh$, and
8. it can diffuse $D_h[(\partial^2 h/\partial x^2) + (\partial^2 h/\partial y^2)]$, where $\partial h/\partial t = ca^2 - vh + D_h \Delta h$.

What we are doing is using Meinhardt's (1977) particular example of the general reaction-diffusion model in order to generate a pattern of the central place system. Even with fairly homogeneous conditions, small deviations have a strong positive feedback such that they reinforce their own growth. To get a stable pattern this autoregressive growth must concomitantly have an antagonistic process surrounding it. This may result either from depletion of material necessary for the autocatalytic centre, the absorption of purchasing power and time as in our case, or by lateral inhibition produced as a by-product of the self-reinforcing growth. In either case, the growth center is isolated by its neutralised surrounds. A local increase of a cannot be compensated for by the increasing concentration of h because it diffuses more rapidly into the surrounds, increasing inhibition there but at the centre the advantage of a increases.

A limit will be set when the loss by diffusion is equal to the net production of a. h should equilibrate rapidly and, to inhibit sufficiently, it should have a higher diffusion rate than a.

10. GAS DIFFUSION AND CENTRAL PLACES

The black box nature of the simulation generally used to solve reaction-diffusion models hides the specifics of the development of regularity. Some idea of the equilibrating mechanism can be seen in a short note by von Neumann (1963) when he discusses two dimensional bubbles and the changes in their geometry due to diffusion. The diffusion of gas across a 'wall' will depend on the pressure differences across it, and he shows that the pressure in a polygonal bubble will depend on its shape. Specifically, the rate of loss of gas is proportional to 6-n, n being the number of sides, or alternatively the gain is proportional to n-6. 'Thus every hexagonal bubble (irrespective of further details of shape!) has a constant-gas-content, every pentagonal bubble loses gas at the same rate; every heptagonal bubble gains at the same rate as the pentagonal ones lose it; every tetragonal (octagonal) bubble loses (gains) gas at twice the rate at which pentagonal (heptagonal) bubbles lose (gain) it etc.' In case this result appears magical, von Neumann's argument is in terms of polygons formed of circular arcs with of course a side bulging out being of a cell of higher pressure, and with the side bulging in belonging to the lower pressure cell. The arcs then are replaced by their chords and the results hold. It should be realised that, as diffusion occurs across the boundary, the boundary in fact moves.

We can translate pressure as potential, ξ, and therefore as price and the gas as customers with the diffusive flow r being the change in numbers attracted to a region. In this case, pressure can be given a potential interpretation because the boundaries are always in balance with the pressure configuration. Flows will of course move away from high prices and towards low ones. Equilibrium can presumably only occur with a hexagonal lattice, which then will have a potential surface as flat as possible--indeed it will be without relief--and thus be of minimum average potential. This will be a system of maximum efficiency, T = 1, in which the areal entropies of the 'real' map and the theoretical perfectly efficient one are identical:

$$\xi r = S \quad ; \quad T = 1 \quad .$$

Perfect hierarchical nesting is an ideal rarely met in practice: geometrically it requires conditions such as uniformity, or rather isotropy, of demand, an exact matching of centres that could only occur by sequential spreading from some initial point, and the such. It is likely that there will not be an exact integer multiple between orders so that the empirical spectrum will show harmonic periodicity only at the very local level. Further, the allocation of incomes between different levels will be an extremely complex function integrating past accessibilities, present profitability, anticipated future specialisation and so forth, which are likely to vary widely in space. It is possible to define an efficient system at any time using current technologies and living standards, anything less efficient involving extra costs and/or higher prices and thus lower incomes.

11. TURBULENCE AND EFFICIENT CENTRAL PLACES

We shall now take up the question of diffusion of purchasing power within a hierarchical central place system. Nearly twenty years ago, I pursued a suggestion of Richardson (1926):

> The problem (of atmospheric turbulence) is rather like that of finding some simple specification of the extent to which the population is divided between cities, towns, villages and isolated homes without making a map; far in the atmosphere a map of all eddies would be too bewilderingly complicated.

In fact, I attempted to use what I knew of turbulence theory within the inertial subrange and of its use of dimensional analysis in order to try to understand central places (Curry, 1967). Basically I was exploiting the common feature of scaling in both these phenomenon as Mandelbrot (1977, 1982) would now call it. The main results were that allocation of personal income was an inverse linear function of distance and the spectral density of consumers' behaviour varied directly as the square of wavelength (i. e., of scale). In a limited area the spectrum developed periodic waves of decreasing amplitude as shown in Figure 8. It should have been pointed out that while this represented flows, the spectrum of the attractive forces of places was complementary to this. The argument here is that in a conservative system, the waves would not diminish and since they represent flows, their complement would be market prices, (i. e., the forces causing the flows). To the extent that some degree of conservatism remains, this relationship would be present (Curry, 1984b). Diffusion of purchasing power increased in efficiency upwards through the central place hierarchy as the 1.5 power of distance, downwards as the two-thirds power. Some indirect evidence was presented to justify these results. Ergodicity was hypothesised in which the operation of consumers' purchases and retailers' inventories in time had its exact counterpart in consumers' shopping trips and retailers' locations in space. One or the other aspect can be looked at as appropriate.

The central place hierarchy, conceived as a distributive system for convenience goods is noteworthy in that specialised functions and the resulting relative settlement sizes appear to be allocated orthogonally. Successive band-stop filters are applied which, because of the ergodicity of the signal, apply in both the areal and periodic domains--the signal is

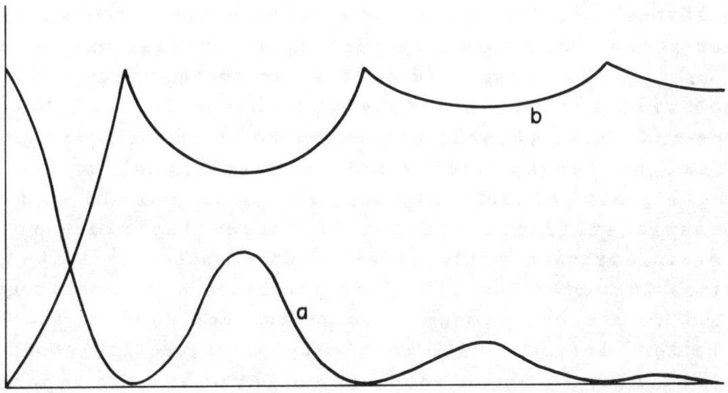

Figure 8: Spatial spectra of flows (a), and of forces and prices (b).

demand for goods. There are no continuing non-linear interactions in these domains; the divisions are neutral, they do not produce effects, and the geography is inert. Because of the geometrical nesting that occurs and thus the uniform ratios between n-th and (n+1)-th orders of centre, the aggregate amounts of buying and selling in each order form a linear progression. Individual commodities are offered independently. Perhaps the band of goods offered has to be above a certain width and perhaps the collection of separate goods is not independent, but this need not affect linearity. Again, the shopping problem can be non-linear if phrased in terms of spatial behaviour of consumers, especially economy of travel. But conceived as a mapping of product availability among scales of center it is linear. It is in this sense that interaction between scales is inert; in most other geographical circumstances the allocation of functions would interact non-linearly with population numbers and thus with locations of certain sizes to change both.

Considering the central place system as within the trading system for which payments in equals payments out in any part is the rule, then we can expect equilibrium flows that exhibit temporal stability. The central place system has adjusted into quasi-equilibrium, of which the numbers of functions in the different orders of settlement is a manifestation. Changes in preferences, in products, in incomes, and the such, will affect the balance of payments between the centres, causing a reallocation of functions and of settlement sizes to a new equilibrium. The system is inert in this condition.

In reality, the hierarchical arrangement of central places is the result of a highly complex and little understood relationship between the demands at various stores. At the upper end of each range of inventories kept by an n-th order store, holding costs just equal average net revenue minus a traditional profit. The lower end is limited by the upper end of n-1 order stores which will have a locational advantage for all but the nearest of customers and those already attending to buy n-th order goods. There is a great deal of leeway here since 'a traditional profit' is necessary only over all goods collectively and many goods could be sold at a loss in order to ensure sufficient customers attend the store to buy profitable items. Again, pricing policies can induce customers to buy and retain large quantities in home storage so that purchases are less frequent and switches from n-1 to n order stores result for such goods. The same result occurs when better storing, and more slowly deteriorating goods are provided. This flexibility in where goods are offered is most important because it allows a strong ordering principle in central places. If each good had its own order of store strictly based on frequency of purchases there would be no reason to expect that the bandwidth of goods offered from a centre would be enough to allow sufficient profit to be earned to provide a traditional income. Thus there would be a dichotomy between a nested hierarchical arrangement and a minimisation of travel for n-th order goods. It is because of competition between orders of center, competition allowed because order of centre is not strictly established for many goods, that a reconciliation develops. Considered from a static viewpoint, an equilibrium can be envisaged in which there is a constant multiple between n and n+1 order stores that allows a nested, but not necessarily perfectly nested, hierarchy. The multiple is established by allocating goods between order of stores so that normal incomes are earned in each order. It is assumed that the practices producing flexibility are available to all orders of store.

There is an interesting corollary to these assertions, particularly if we take a hierarchical central place system as given historically, although this is not essential. The multiple allocating goods to orders and concomitantly the allocation of travel to orders is not fixed. Consequently, the finding that the spectral density of consumers' behaviour varies as the square of wavelength (scale) is suspect. It would appear that this relationship applies only to perfect nesting. When we have centres that should have n+1 order services but stop at n order, when we have centers with two or more identical stores, then we depart from perfect time and space periodicity. There is a reduction in efficiency, basically because of a lack of competition perhaps resulting from historical accidents in sequencing locations and routeways, perhaps from an inhomogeneous terrain, perhaps lack of information, or the such.

In a subsequent paper discussing town existence (Curry, 1978), it was pointed out that while zero points in the spectrum were adduced in the 1967 paper and thus distinct levels of activity with characteristic spacing allowed, there was not the sharp concentration of facilities at virtual points and thus what should be a virtual line spectrum locally did not occur. Wiener's non-linear conservative system was appealed to in which random noise representing a continuum of goods ranked on their initial differentiation of availability was fed to a Hamiltonian dynamical, system which then oscillated in discrete harmonic lines by a process of entrainment. From a continuum of goods each of equal amplitude at its spatial scale but without overall spatial pattern, we see that these goods all adapt to a single scale by increasing or decreasing the density of production, and a pattern results. High frequency items of standard amplitude change to medium scales of higher intensities. Large scale features of the same amplitude increase their frequency, while reducing their intensity. The scales are interacting because of non-linearities. This pattern would have a minimum configurational entropy in Prigogine terms since covariance is maximum. Clearly there is a possibility of operating on the harmonics with a hierarchy of central place functions, but this does not occur in the model. This discrete non-overlapping hierarchical arrangement with no blanks must be maximally efficient, $T = 1$. The extent to which the hierarchy has elements in any scale missing, or the whole is not neatly compartmentalised, will give $T < 1$. If we still assume a structure in locally isotropic space and without external influences, the relations between the n-th and (n+1)-th scales are likely to be the same as the (n-1)-th and n-th, as before. If we write $f(\mu) = T_{max}/T_{real}$, we specify the extent of non-conformity.

12. TURBULENCE AND INEFFICIENT CENTRAL PLACES

Let us first write down some definitions (Frisch, Sulem and Nelkin, 1978).

l_n = the scale (distance apart) of n order centres,
x_n = the rate of sales at this scale,
$t_n \sim l_n/x_n$ = the turnover time of inventories since l_n will be proportionate to inventory size, and
$\varepsilon_n \sim x_n/t_n$ = the rate of allocation of purchases between scales of center by a consumer.
$\varepsilon_n \sim x_n^2/l_n$; $x \sim (\varepsilon l)^{0.5}$.

Taking the Fourier transform yields the variance spectrum of sales by scale:

$$x_n^2 \sim \varepsilon w_n^{-2} \text{ in terms of wavelength, and } \varepsilon f_n^2 \text{ in terms of frequency.}$$

This is the result previously obtained much more laboriously for the average generation of sales by scale, G (Curry, 1967). Now introduce an inefficiency term so that

$$G \sim \varepsilon w_n^{-2} \, l_n^{-\mu} \quad ,$$

where $\mu > 0$ is the degree of non-conformity to a perfect hierarchical arrangement. The mean number of centers at the lower scale relative to the next higher scale is

$$N = \alpha^{D} \quad , \quad D \leq 2 \quad .$$

In the case of no inefficiency, D = 2 so that the fraction of space serviced efficiently by a centre is

$$N/\alpha^{2} = \alpha^{-\mu} \equiv \beta \quad , \quad \mu \equiv 2 - D > 0 \quad .$$

At the n-th scale, assuming the highest scale is 0, only a fraction β^{n} of space will be efficiently served:

$$\beta_{n} = (l_{n}/l_{o})^{D-2} \quad .$$

This gives

$$x_{n} \sim \varepsilon^{0.5}\, l_{n}^{0.5}\, (l_{n}/l_{o})^{-0.5(2-D)} \quad ,$$

$$t_{n} \sim \varepsilon^{-0.5}\, l_{n}\, (l_{n}/l_{o})^{0.5(2-D)} \quad ,$$

$$G(w) \sim \varepsilon^{0.5}\, w^{-2}\, (l_{n}/l_{o})^{0.5(2-D)} \quad , \text{ and}$$

$$\varepsilon_{n} \sim x_{n}^{2}\, \beta_{n} \quad .$$

It should be noted that n-order centres are defined as containing all functions below n so that the largest scale of activity is regarded as completely space filling. Higher order centers may usurp more of the lower order functions than they are 'entitled' to so that increased travel is necessary and lower order centers cover less of the available space than efficiency requires.

Fujisaka and Mori (1979), Mori and Fujisaka (1979) essayed a maximum entropy hypothesis to evaluate μ, N and D. They used a physical interpretation of entropy referring to the probability of whether one of the n-order spatially randomly distributed eddies occupies a given volume element f or not (1-f). We prefer to use entropy in an inferential context in which we maximise our ignorance of the probability distribution of D. The information entropy of the fractal D is

$$H(D) = -\sum_{n} [f_{n} \ln(f_{n})] = -f \ln(f) - (1-f)\ln(1-f) \quad ,$$

which is maximum for $f(D_{max}) = 0.5$. Translating their arithmetic into two dimensions,

$$N_{n} = \alpha^{Dn}\, (t_{n}/t_{o})\, N_{o}$$

yields $N_{n}/N_{o} = (l_{n}/l_{o})^{-D+(1+\mu)/2}$, and $O_{n}/O_{o} = (l_{n}/l_{o}) = \alpha^{-nz}$,

where $z = (3\mu + 1)/2 \quad ,$

$$\mu^{*} = \{2/[3 \log(\alpha)]\} \log(2\alpha^{0.5} - 1) - 1/3 \quad , \text{ and}$$

$$N^* = \alpha^{5/3} (2\alpha^{0.5} - 1)^{-2/3} \quad .$$

Figure 9 shows the nesting ratios, N* for various degrees of inefficiency.

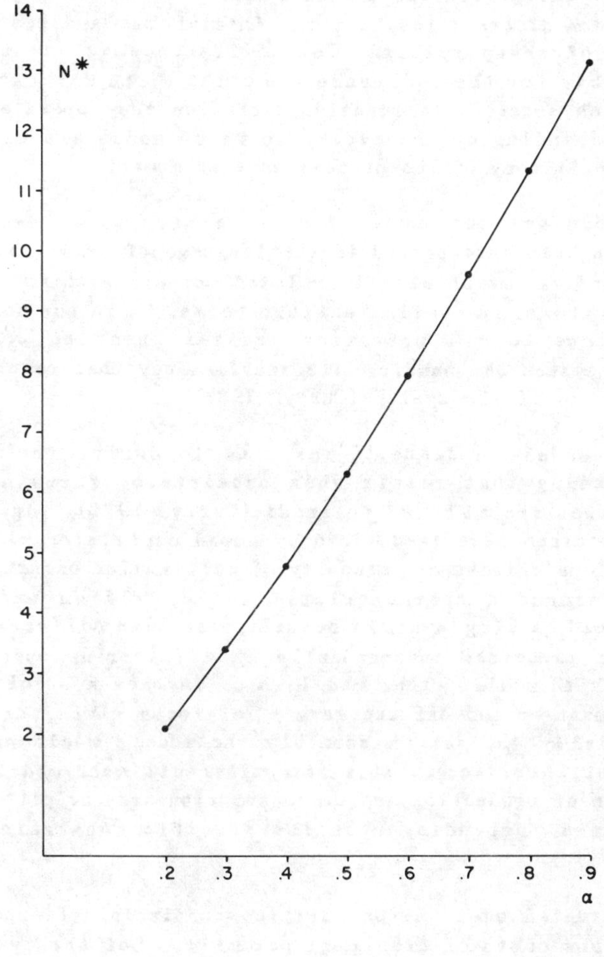

Figure 9: Maximum entropy estimates of the nesting ratio of centre numbers, N*, for various degrees of inefficiency, α.

13. COMPARISONS

The results of some of the previous specific models may best be appreciated by comparing them to somewhat similar models. For example, the use here of a diffusive mechanism for activation and inhibition is very like Whittle's model of growth in which both promoting and retarding influences decline with distance. His results are analytic from probabilistic arguments rather than from simulation so that comparison is difficult. Simple diffusion would give linearly declining influence, but his interactions are more complex. Negative autocovariance would encourage growth, positive would not. Inhibition results from competition and is therefore proportional to the mean density and is maximal for identical densities (Curry, 1977). In the context of settlements, there is a general tendency to a more uniform distribution than the initial one and certainly not to the production of sharp spikes. This difference is due to the different paths available for the influences to work. With Whittle they operate in rays from each source. In reaction-diffusion they operate on a lattice, their paths depending on the extent to which nodes are already occupied and thus on the history of the process to that moment.

The much quoted 'Brusselator' model from chemistry is a reaction-diffusion model that has been interpreted in the language of trade to yield relatively autarkic subsystems of closely related members with only weak links between the subsystems. A spatial pattern forms, as in the central place case, but here due to a fluctuation greater than the control mechanisms of the trade system can handle. The inefficiency that results is to be found in the break-up of the system (Curry, 1983).

A rather similar model to Wiener's was used to derive the areal specialisation and trading that result when subsistence farms having different man/land ratios are allowed to trade (Curry, 1970). Again an initial flat spatial spectrum here feeds into a damped oscillator with the result that a texture of patchiness of intensity of cultivation occurs, with the spectrum peaking around a characteristic scale. If an undamped oscillator is postulated, a single scale occurs. The main difference is that each wavelength is processed independently by a filtering operation. Wiener's is more general than this other model which invokes a specialised operator comparing movement on and off the farm. Yet for a given variance, the results are essentially the same; presumably the models would produce different values for total variance. This harmonic oscillator model does stress that localisation of production and of consumption and the patterning of trade go hand in hand, depending upon the specific constraints on behaviour considered.

Take an economy situated on a uniform lattice consisting of exporters having either high or low costs of transport per mile. Let the types be independently randomly distributed so they exhibit a flat spectrum of areal differentiation. There is a diffusion of payment between lattice points representing their imports. But high cost points are limited in the distance they can send exports so that it is unlikely that a steady state can be reached. To maintain equal accessibility to imports it is necessary for high cost shippers to become more regularly distributed, emphasising a range of the shorter wavelengths of the spectrum. Low cost shippers need to relocate to fill the resulting gaps in order to preserve the uniform lattice, emphasising long wavelengths. From an initial zero autocorrelation in shipper categories, there are probably two peaks in the locational spectrum (see Figure 10), one in the short- to mid-distances and one in the long-distances. Movement of funds between these categories will exhibit a complementary spectrum.

The play of trade through time has not proved interesting in our models; we have been assuming a monotonic multiplier argument heading to a steady state. But in a trading network are there actual or possible links that provide large multiplier effects? Presumably these are bottlenecks where small and perhaps expensive additions would pay off handsomely. Is such a feature possible or are all links basically the same? Clearly links can only be evaluated for given networks. There are some mathematical results that appear to support such a notion. The theory of random graphs (Erdos and Renyi, 1961), for example, postulates separate but spatially unrelated points and two of them are randomly chosen and connected. This operation is continued sequentially and the connectedness of the resulting graphs examined. In fact at some stage the probability of high connectivity, of a single system, occurs abruptly suggesting that the last link is extremely important. Connectivity then is not a linear function of number of links.

14. CONCLUSION

In commodity trade, labour markets and similar exchanges there is need to handle both substitution between 'species' within a region and

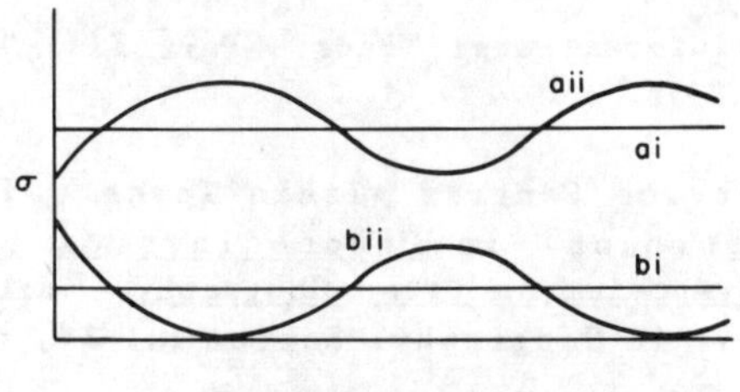

Figure 10: Spatial spectra of (a) locations of shippers (i) originally, and (ii) after readjustment, and (b) associated flows of funds (i) originally, and (ii) after readjustment.

interactions between regions. There does not appear to be an ideal way to handle these topics. In SPE there are distinct homogeneous regions with flows allowed between them subject to transport costs; inefficiencies of pricing are introduced via the thermodynamic approach. This can be extended to handle many substitutable commodities within each separate region, but reliance on the bounded area is highly constricting for further analysis, say relating to labour markets. In labour markets, intra-regional substitutions between occupations can be made with or without regard to prices via the predator-prey equations (Curry, 1981a). Unfortunately all these yield are statements as to system stability. Spatial considerations can be brought in to some extent via diffusion equations, but the type of results remain the same. More heterogeneous space may be considered using the competitive Lotka-Volterra equations (Curry, 1981b), but not, I believe, for the case of substitution. Heterogeneity is an ambiguous concept in the many species case. Autocorrelation or spectral density functions can best be used to describe areal differences for one or the ratio of two species. In these models spatial interactions and substitution interactions become one via the impulse response function (Green function) or the transfer function. What is needed is a multiple cross spectrum or similar measure, such as a scale component function for each species with the corresponding transfer function and mutual and joint information statistics describing the relation between each pair of species. Then if one could manipulate a whole 'matrix' of such features, the dichotomy between intra- and inter-regional interactions would disappear. But this appears to be a horrendous task even for linear relations, and when non-linear constraints are introduced it becomes impossible.

15. REFERENCES

Caplan, S. and A. Essig, 1983, Bioenergetics and Linear Nonequilibrium Thermodynamics: The Steady State, Cambridge, Mass.: Harvard University Press.

Chipman, J., 1965, A Survey of the Theory of International Trade: Part I, The Classical Theory, and Part II, The Neo-classical Theory, Econometrica, 33: 477-519, 685-760.

_____, 1966, A Survey of the Theory of International Trade: Part III, The Modern Theory, Econometrica, 34: 18-76.

Curry, L., 1960, The Geography of Service Centres within Towns: The Elements of an Operational Approach, in Proceedings of the International Geographical Union Symposium in Urban Geography, edited by K. Norborg, Lund: Lund Studies in Geography, Series B., 24, pp. 31-53.

_____, 1967, Central Places in the Random Spatial Economy, Journal of Regional Science, 7 (supplement): 217-238.

______, 1969, Univariate Spatial Forecasting, Economic Geography, 46 (supplement): 241-58.

______, 1970, Geographical Specialisation and Trade, Studies in Regional Science, 2: 85-95.

______, 1977, Stochastic Spatial Distributions in Equilibrium: Settlement Theory, in Man, Culture and Settlement: Essays in Honor of Professor R. J. Singh, edited by R. Eidt, et. al., New Delhi: Kalyani, pp. 228-237.

______, 1978, Position, Flow and Person in Theoretical Economic Geography, in Time and Regional Dynamics, vol. 3, Timing-Space and Spacing Time, edited by T. Carlstein, D. Parkes and N. Thrift, London: Edward Arnold, pp. 35-50.

______, 1981a, Macro-ecology of Vacancy Chains, in Dynamic Spatial Models, edited by D. Griffith and R. MacKinnon, Alphen aan den Rijn: Sijthoff and Noordhoff, pp. 159-85.

______, 1981b, Division of Labor from Geographical Competition, Annals, Association of American Geographers, 71: 133-65.

______, 1983, Inefficiency and Instability of Trade Patterns, in Evolving Geographical Structures, edited by D. Griffith and A. Lea, The Hague: Martinus Nijhoff, pp. 278-92.

______, 1984a, Inefficiency of Spatial Prices Using The Thermodynamic Formalism, Environment and Planning A, 16: 5-16.

______, 1984b, A Trade-Localisation Operator: The Conservative Case, Canadian Geographer, 28: 373-6.

______, 1985a, Factor Returns and Geography, Geografiska Annaler, Series B, forthcoming.

______, 1985b, Formalised Behaviour in Agents Markets, unpublished manuscript, Department of Geography, University of Toronto.

Erdos, P. and A. Renyi, 1961, On the Evolution of Random Graphs, Bull. de l'inst. Int. de Statistique, 38: 343-347.

Frisch, U., P.-L. Sulem, and M. Nelkin, 1978, A Simple Dynamical Model of Intermittent Fully Developed Turbulence, Journal of Fluid Mechanics, 87: 719-36.

Fujisaka, H. and H. Mori, 1979, Maximum Principle for Determining the Intermittency Exponent of Fully Developed Steady Turbulence, Progress *in Theoretical Physics*, 62: 1.

Kemeny, J. and J. Snell, 1972, *Mathematical Models in the Social Sciences*, Cambridge, Mass.: M. I. T. Press.

Mandelbrot, B., 1977, *The Fractal Geometry of Nature*, San Francisco: Freeman.

______, 1982, The Many Faces of Scaling: Fractals, Geometry of Nature, in *Self-Organization and Dissipative Structures*, edited by W. Schieve and A. Allen, Austin: University of Texas Press, pp. 91-109.

Meinhardt, H., 1977, The Spatial Control of Cell Differentiation by Autocatalysis and Lateral Inhibition, in *Synergetics*, edited by H. Haken, Berlin: Springer-Verlag, pp. 214-223.

Mori, H. and H. Fujisaka, 1979, A Simplified Theory of Intermittent Fully-Developed Turbulence, in *Pattern Formation by Dynamic Systems and Pattern Recognition*, edited by H. Haken, Berlin: Springer-Verlag, pp. 82-89.

Percus, J., 1977, *Combinatorial Methods in Developmental Biology*, New York: Courant Institute of Mathematical Science, New York University.

Phlips, L., 1974, *Applied Consumption Analysis*, Amsterdam: North-Holland.

Poland, D. 1978, *Co-operative Equilibria in Physical Biochemistry*, London: Clarendon Press.

Richardson, L., 1926, Atmospheric Diffusion on a Distance-neighbour Graph, *Proceedings of the Royal Society of London* Series A, 110: 709-37.

Seninger, S., 1977, Expenditure Diffusion in Central Place Hierarchies: Regional Policy and Planning Aspects, *Journal of Regional Science*, 18: 243-61.

Sheppard, E. and L. Curry, 1982, Spatial Price Equilibria, *Geographical Analysis*, 14: 279-304.

Tabor, D., 1969, *Gases, Liquids and Solids*, Harmondsworth: Penguin Books.

von Neumann, J., 1963, Discussion-Shape of Metal Grains, in *Collected Works*, vols. V and VI, edited by A. Taub, New York: Pergamon Press.

INCOME DIFFUSION AND REGIONAL ECONOMICS

Robert Haining
University of Sheffield
England

1. INTRODUCTION

Space often is important in considering economic events. At the San Miniato NATO Institute, entitled 'Evolving Geographical Structures,' an example was given where competitive interactions between spatially distributed markets had implications for spatial price variation (Haining, 1983). There was evidence to suggest that retailers engaged in local or neighbourhood price comparisons, and that the spatial distribution of markets had implications for certain properties of the macro or regional price distribution, most notably overall price variability (Haining, 1985). This line of work has since focused on two models of spatial pricing with different mechanisms for binding together the space economy (Haining, 1984). In one model the mechanisms were retailer competitiveness and 'local price awareness,' and in the other they were consumer price sensitivity and inter-site demand elasticities. Such mechanisms, it was suggested, underlie the existence of interpretable patterns in the spatial distribution of prices.

The theme of this paper also is the importance of space in economic events, but dealing here situations where the evolution of spatial pattern arises from direct spatial transfers of income. Local scale inter-regional diffusion processes generate a wide range of questions of considerable interest in the field of regional economic analysis. Indeed it is perhaps the potential strength of such transfers and their consequences at a variety of scales that contribute much to the distinctive flavour of regional economic analysis. This paper will attempt to review and explore certain aspects of inter-regional economic variation that may arise from such income transfers.

2. FLOWS, PATTERN AND STRUCTURE IN THE REGIONAL ECONOMY

A number of systems of geographical interest involve the simultaneous treatment of flows, pattern, and structure. Spatial flow simply refers to the physical transfer of commodities, people, information, or the such. The important distinction for present purposes is between spatial pattern and spatial structure. The term structure relates to the background geography (the 'built form'), whilst pattern relates to more volatile or changeable levels of spatial regularity that may be imposed on or with respect to the more permanent structure. These three elements may be closely dependent, although in any particular context one or two of the elements may be held constant in order to study the other(s). Two examples may help to exemplify these ideas, while Table 1 provides a summary of them.

2.1. Retailing Systems

The size and distribution of retail sites within a city might be considered to form the structural level whereas such variables as price levels and commodity mix form the pattern level. The aggregate movement behaviour of consumers in terms of both volume and direction of movement represents the flow level. Although one could argue about the interdependencies between each of these three attribute levels of a retailing system, in practice linking them together appears to be difficult.

In the pricing problem mentioned above, the structure is fixed (the size and distribution of sites) and the movement behaviour of individuals is also an exogenous input into the models. Intersite demand elasticities and 'local price awareness' underlie spatial price variation, and these in turn

TABLE 1

FLOWS, PATTERNS AND STRUCTURE FOR RETAILING MODELING AND SPATIAL INCOME MODELING

Component	Retail System	Spatial Income Model
flow	consumer movements; income expenditure flows	expenditure flows
pattern	price distribution; commodity mix	distribution of wages
pattern parameters	local price awareness; inter-site demand elasticities	propensity to consume non-locally
structure	size and distribution of retail sites	distribution of population and economic activity
structure parameters	friction of distance; scale economies	transport costs

presumably depend upon characteristics of the movement behaviour of consumers (Haining, 1984), and perhaps also the actual size and siting of the retail centres (Haining, 1985). One attempt to simultaneously link two of the elements (structure and movement) is provided by the work at Leeds University. Here the size and distribution of shopping centres as well as the matrix of inter-area income flows are determined in a simultaneous manner. Friction of distance and economies of scale parameters determine the equilibrium structures that can be generated (Clarke and Wilson, 1983).

2.2. Spatial Income Models

In this case movement or flow relates to income transfers between regions, whereas pattern relates to the distribution of that income and regional variations in income levels. The structural level includes the distribution of economic activity and population. It is this system that will be considered in more detail in this paper. Important parameters underlying the formation of spatial income variation are the propensities to consume local and non-local goods.

The study of these types of interdependency raises some complications. First, the three elements involve different time scales of reaction. While adjustments between pattern and flow may take place over quite short time spans, structural change, for example, usually involves longer term adjustment. The question of how to tie together short period change to longer term structural adjustment appears to be an open question. The approach suggested by Wilson in studies of retail change introduces bifurcation theory to handle structural shifts. This may be one way to move forward, as might be the study of phase transitions (Bennett, Haining and Wilson, 1985). Second, the systems are highly non-recursive at two levels, first in terms of the feedbacks involved between the three elements themselves, and second at the spatial level. Inter-regional transfers involve simultaneous flows from place to place that generate spatial multiplier effects. In the study of these sorts of systems one is interested in both the temporal trajectory of the system, and the properties of the equilibrium state.

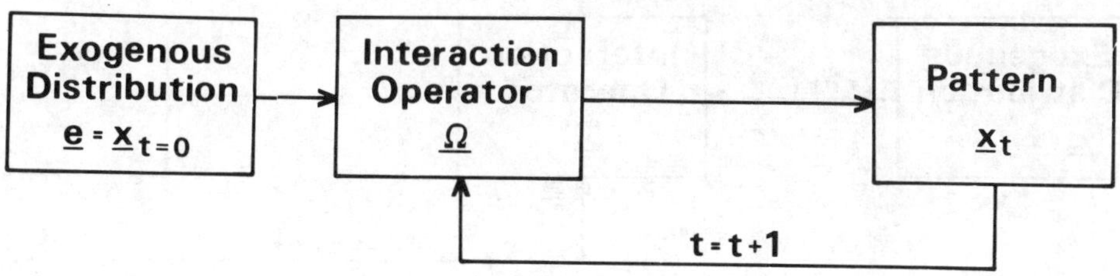

Figure 1: Pattern-flow system (Type 1).

2.3. Representations of Pattern-flow Relationships

Without attempting an exhaustive discussion, three cases will be introduced here that may have general significance. In the first case (see Figure 1), an initial exogenously given distribution is transformed through a time invariant interaction operation given by $\underline{\Omega}$. So, for instance, the vector $\underline{e}$ represents the initial distribution of wage income across a set of n regions. The matrix operator $\underline{\Omega}$ specifies the spatial form of the redistribution process (i. e., the spatial pattern of wage expenditure). Hence,

$$\underline{X}_o = \underline{e} \quad \text{, and}$$

$$\underline{X}_T = \sum_{t=0}^{T} \underline{\Omega}^t \underline{e} \quad , T = 1, 2, 3, \ldots ,$$

where $\underline{X}_T$ represents wage income across the n regions by time T. If $\underline{X}$ is equal to total income across the n regions, then

$$\underline{X} = \sum_{i=0}^{\infty} \underline{\Omega}^i \underline{e} = (\underline{I} - \underline{\Omega})^{-1} \underline{e} \quad ,$$

assuming row sums are less than unity.

The second case (see Figure 2) again assumes an exogenously given

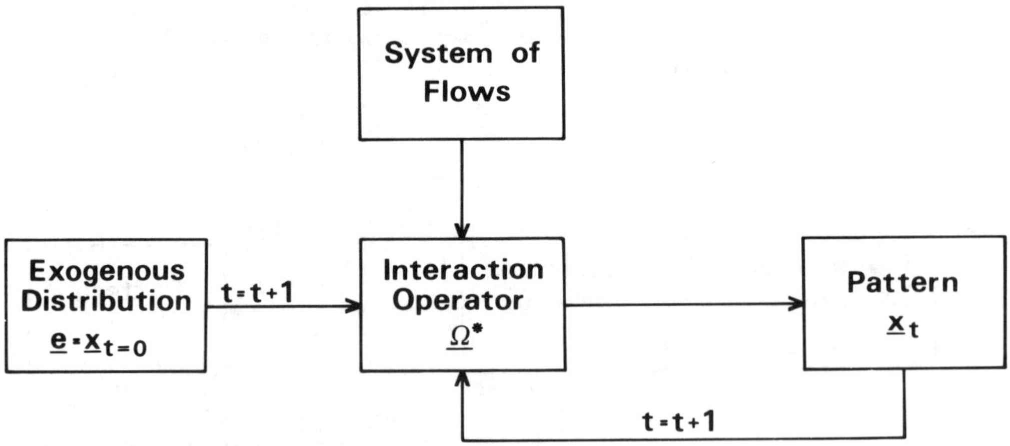

Figure 2: Pattern-flow system (Type 2).

distribution transformed by a time invariant operator $\underline{\Omega}^*$. In this case the structure of the operator is determined by the background system of flows. For example, if $\underline{X}$ denotes the equilibrium price level over a set of n markets, then

$$\underline{X} = \underline{\Omega}^* \underline{X} + \underline{e} \quad ,$$

so that

$$\underline{X} = (\underline{I} - \underline{\Omega}^*)^{-1} \underline{e} \quad .$$

The form of $\underline{\Omega}^*$, and hence the spatial distribution of the reference group adopted by each retailer, is assumed to be influenced by the background system of consumer flows.

The third case (see Figure 3) is where the system of flows arises from the distribution $\underline{X}_T$, which in turn influences the interaction operator by which $\underline{X}_T$ is transformed to $\underline{X}_{T+1}$. Here

$$\underline{X}_0 = \underline{e}$$

$$\underline{X}_T = [1 + \sum_{t=0}^{T-1} \prod_{j=0}^{t} \underline{\Omega}(\underline{X}_j)]\underline{e} \quad , T = 1, 2, \ldots,$$

where $\underline{\Omega}(\underline{X}_j)$ implies that $\underline{\Omega}$ is functionally dependent on $\underline{X}_j$. An example of this type of economic system might be the case in which capital flows are responding to interest rate differentials (pattern), and interest rate

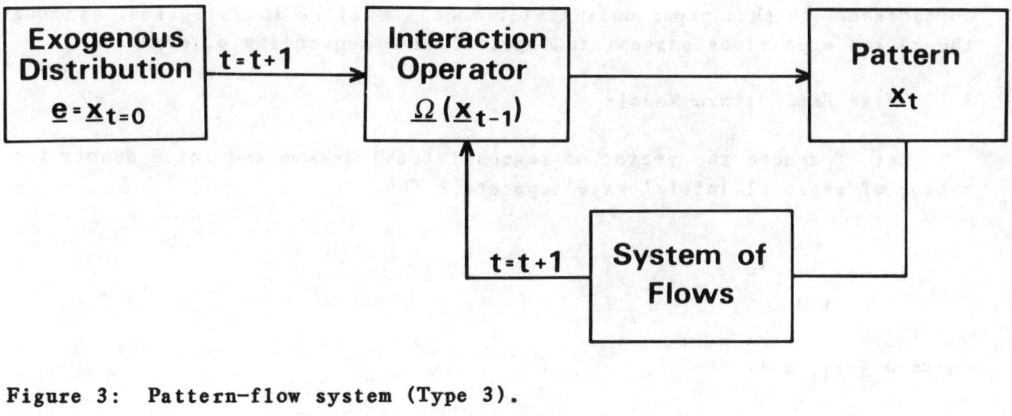

Figure 3: Pattern-flow system (Type 3).

levels in turn depend upon net capital flows. It would appear that such systems are likely to be very unstable, a property not characteristic of the previous two cases.

2.4. Mathematical Descriptions of Pattern-flow Relationships

This paper will address models of the form

$$\underline{X} = \underline{\Omega X} + \underline{e} \qquad \text{or} \qquad \underline{X} = (\underline{I} - \underline{\Omega})^{-1}\underline{e} \quad .$$

Several features of these types of equations are noteworthy. First, as has been noted on several occasions (Curry, 1978; Sheppard, 1979), the term $(\underline{I} - \underline{\Omega})^{-1}$ is called the potential. The i-th row of the matrix defines the total spatial impact of a unit impulse at site i ($e_j = 1$ if $j = i$, and 0 otherwise). This value can be thought of as identifying the influence of the impulse at location i, or the total flows induced by the event that takes place at location i. For continuous spatial systems the Green function plays the same role.

For stochastic systems, where $\underline{e}$ is a random variable, the term

$$[(\underline{I} - \underline{\Omega}^T)(\underline{I} - \underline{\Omega})]^{-1}$$

is proportional to the autocorrelation matrix, and thus provides a description of the second-order or pattern properties of $\underline{X}$. A more technical description of the links between the Green, potential and autocorrelation functions is given in the Appendix.

3. INTER-REGIONAL WAGE EXPENDITURE AND INCOME MODELS

Two types of models will be distinguished here, namely inter-regional wage expenditure models and regional per capita income models. There are a number of important distinctions between them, not least of which is the empirical issues they raise; but first some theoretical issues will be considered. In this paper only static models will be investigated, although the series expansions suggest an implicit time sequencing of events.

3.1. Wage Expenditure Models

Let $\underline{Y}$ denote the vector of regional total income and let $\underline{E}$ denote the vector of regional initial wage payments. Then

$$\underline{Y} = \sum_{i=0}^{\infty} \underline{\Gamma}^i \underline{E} \quad ,$$

where $0 \leq \gamma_{ij} \leq 1$, and

γ_{ii} = the propensity to spend (wage) income in i that was initially received in i itself, for all i,

γ_{ij} = the propensity to spend (wage) income in i that was initially received in j, for all i and j,

and $0 \leq \sum_{i=1}^{n} \gamma_{ij} \leq 1$. So

$$\underline{Y} = (\underline{I} - \underline{\Gamma})^{-1}\underline{E} = \underline{E} + \underline{\Gamma Y} \quad , \tag{1}$$

which divides the total impact into initial ($\underline{E}$) and direct plus indirect ($\underline{\Gamma Y}$) impacts. Of course $\underline{E}$ can be treated as either a fixed variable or a stochastic variable, but may be determined by the underlying regional economic structure reflecting the distribution of economic activity and the distribution of the working population with respect to that activity.

The matrix $(\underline{I} - \underline{\Gamma})^{-1}$ is a matrix of wage impact multipliers and is also a wage income potential. The matrix $[(\underline{I} - \underline{\Gamma}^T)(\underline{I} - \underline{\Gamma})]^{-1}$ describes the spatial pattern of the wage income surface. If one assumes that $\gamma_{ij} < \gamma_{ik}$ if region k is closer to i than to j, then this matrix generates a potential function that declines with distance. Further, if the initial wage distribution is random across the regions, then the net effect of the transfer process is to create a distribution $\underline{Y}$ with a pattern that is smoothly varying and shows positive spatial autocorrelation, but has no periodicities in the surface. This result follows directly from equation (1), which in the case of an independent random $\underline{E}$ is a simultaneous autoregressive (SAR) model.

Now let

$$\gamma_{ii} = (c - d) \quad , \text{and}$$

$$\gamma_{ij} = \begin{cases} m \ , & \text{if } i \varepsilon N_1(j) \\ 0 \ , & \text{otherwise} \end{cases} ,$$

where c is the propensity to spend and (c-d) represents the propensity for a unit of spending to accrue as local income, so that d is a measure of income leakage for the region. Then $\gamma_{ij} = m$ is the propensity for a unit of income in region j to accrue as income in i, and the only assumption is that local trading exists, so that $N_1(j)$ denotes the first-order spatial neighbours of region j. The regional system is not closed. Rather, income accrues either in the sub-region of receipt, its nearest neighbour, or leaves the region completely. Thus $d \geq |N_1(i)|m$, where $|N_1(i)|$ denotes the number of first-order neighbours of i.

Assuming a rectangular lattice of regions, the continuous analogue of equation (1) with the above assumptions is

$$\left[\partial^2/\partial s_1^2 + \partial^2/\partial s_2^2 + \{1 - [(c - d) + 4m]\}/m\right]y(s_1,s_2) = \varepsilon(s_1,s_2) \quad ,$$

where s_1 and s_2 denote the spatial co-ordinate system. Let

$$\eta = \left[\{1 - [(c - d) + 4m]\}/m\right]^{1/2} \quad ,$$

and assuming that $\varepsilon(s_1,s_2)$ is an independent (i. e., white noise) random variable, then following results in Whittle (1954) the autocorrelation function depends only on distance (i. e., r) and is given by

$$\rho(r) = \eta r K_1(\eta r) \quad ,$$

where $K_1(\eta r)$ is a modified Bessel function of the first order. This indicates a very strong local spatial income correlation before the occurrence of long distance tailing off typical of a distance decay or negative exponential function. The Green function is given by

$$G(r) = (2\pi)^{-1}K_0(\eta r) \quad ,$$

where K_0 is a modified Bessel function of order zero (Heine, 1955). Spatial income correlation increases as (c–d) and m increase (some numerical examples are given in Table 2). This result is in accordance with those for the discrete space case (for values of the Green function, see Table 3). The shapes of the spatial correlation and Green functions are shown, for

TABLE 2

INTEGER DISTANCE CORRELATIONS $\rho(r)$ FOR THE STATIC TWO-DIMENSIONAL INCOME-EXPENDITURE MODEL

c	d	m	Distance r						
			0	1	2	3	4	5	6
0.8	0.35	0.08	1.0	0.35	0.08	0.01	0.00	0.00	0.00
0.8	0.70	0.08	1.0	0.15	0.01	0.00	0.00	0.00	0.00
0.8	0.70	0.15	1.0	0.44	0.14	0.03	0.00	0.00	0.00
0.8	0.42	0.13	1.0	0.64	0.34	0.16	0.07	0.03	0.00

TABLE 3

GREEN FUNCTION VALUES FOR THE STATIC TWO-DIMENSIONAL INCOME-EXPENDITURE MODEL

c	d	m	Distance r					
			1/2	1	2	3	4	5
0.8	0.35	0.08	0.085	0.026	0.003	0.000	0.000	0.000
0.8	0.70	0.08	0.042	0.008	0.000	0.000	0.000	0.000
0.8	0.70	0.15	0.104	0.038	0.006	0.001	0.000	0.000
0.8	0.42	0.13	0.170	0.079	0.024	0.008	0.003	0.001

different parameter values, in Figures 4 and 5. Note that for small values of x, $K_0(x)$ behaves as $-\ln(x)$.

The results presented here on the influence of the transfer function on the spatial patterning of the vector $\underline{Y}$ depend upon an initial random configuration of wage income. Where that is not the case, arising say from some initial structure of income receipt and spending opportunities, then not only will the results be more complex but the entries in the $\underline{\Gamma}$ matrix, for instance, will tend to mirror that underlying structure. For example, a central place structure might generate a pseudo-periodic income surface, but the problems of introducing heterogeneity of structure into these models are severe. Nonetheless one would argue that this framework does shed light on the role of inter-regional income transfers, as well as the role of certain economic parameters in the development of local economic surfaces and inter-regional impacts.

3.2. Regional Income Models

Although very similar in general structure, this class of models starts from a Keynesian accounting identity, and indeed ten Raa (1984) has referred to these regional models as 'Keynes spatialized.' Thus one begins with the expression

$$\underline{Y} = \underline{I}^* + \underline{C} \quad , \tag{2}$$

where $\underline{Y}$ is now a vector of regional incomes (for n regions) and $\underline{I}^*$ and $\underline{C}$

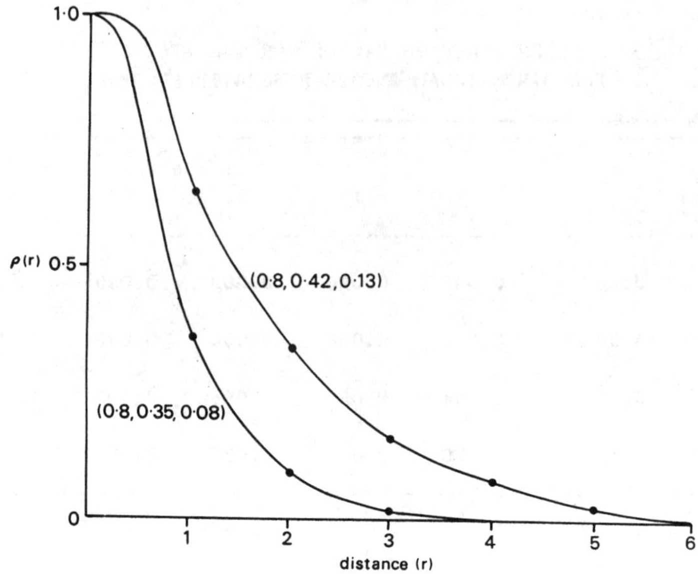

Figure 4: **Autocorrelation funtion ρ(r) for the static two-dimensional wage expenditure model.**

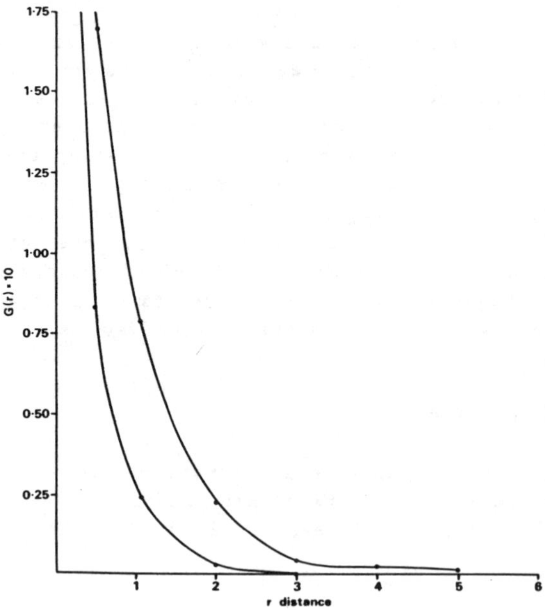

Figure 5: **Green function G(r) for the static two-dimensional wage expenditure model**

denote regional investments or exogenous expenditures and regional consumption, respectively. Transfers occur through the consumption term, and adopting a modified consumption function results in

$$\underline{C} = \underline{\Omega Y} \quad , \tag{3}$$

where $\underline{\Omega} = \{\omega_{ij}\}$ and ω_{ij} is the proportion of income received in region j and used for consumption expenditure in region i. As before, $0 \leq \omega_{ij} \leq 1$.

Putting equations (2) and (3) together yields the familiar expression

$$\underline{Y} = \underline{I}^* + \underline{\Omega Y} \quad , \tag{4}$$

or

$$\underline{Y} = (\underline{I} - \underline{\Omega})^{-1}\underline{I}^* \quad . \tag{5}$$

This static framework could be developed for a dynamic system using both consumption and accelerator coefficients for $\underline{C}$ and $\underline{I}^*$, respectively, yielding

$$\underline{Y}_t = \underline{\Omega Y}_{t-1} + \underline{\xi}(\underline{Y}_{t-1} - \underline{Y}_{t-2}) + \underline{I}^*_{a,t} \quad ,$$

where $\underline{I}^*_{a,t}$ is autonomous expenditure. Although $\underline{\Omega}$ usually consists of non-zero off-diagonal entries, in most regional models investment decisions are treated as not involving inter-regional comparisons, so that ξ is diagonal. Gertler (1984) however has suggested that regional growth rates, relative to other regions, are important so that ξ need not be treated as diagonal.

While ξ is treated as diagonal, the pattern properties of the income surface will not be affected by assumptions on the elements of this matrix; this will not be the case if ξ is not diagonal. The main point, however, is that the static theory outlined in the previous section can be carried over if one is prepared to place the same sorts of assumptions on $\underline{I}^*$ that were placed on $\underline{E}$. The same sorts of criticisms also could be levelled though. Models of the type described by equation (4) have been used to study spatial variation in per capita income levels ($\underline{Y}$), and in the next section attention will be turned from the present theoretical approach to empirical analyses of local scale spatial variations in per capita income and wage income diffusion.

4. STATISTICAL ESTIMATION

The wage expenditure models involve direct estimation of $\underline{\Gamma}$, preferably from survey data of wage receipts and patterns of wage expenditure. Currently, at the University of Sheffield, Ms. Shabnam Bhatara is undertaking a wage income expenditure survey of workers at the Bhilai steel plant (Madhya Pradesh, India), as part of a study of the production and consumption impacts of a large regional investment project. Part of the work involves estimating the elements of $\underline{\Gamma}$. Employees are distinguished by place of residence (among other attributes), and the spatial pattern of their income expenditure is recorded. Interest has focused on linking variations in inter-urban propensities to spend, $\{\gamma_{ij}\}$, with attributes of the urban places. This work is similar in spirit to, for example, that of Erickson (1977).

The regional per capita income models, on the other hand, involve a different type of statistical estimation problem. On the basis of the preceding arguments one might specify a model of the form

$$\underline{Y} = \underline{C} + \underline{I}^* \quad ,$$

$$\underline{C} = \underline{RY} + \underline{e} \quad , \text{ and}$$

$$\underline{I}^* = \underline{Zb} \quad ,$$

where $\underline{Y}$ refers to the set of n regional per capita incomes and $\underline{e}$ is a vector of iid random variables; that is, $E(\underline{ee}^T) = \sigma_e^2\underline{I}$. $\underline{R}$ is a matrix of inter-regional dependencies, regional investment is accounted for by a set of explanatory variables, say $\underline{Z}$, and $\underline{b}$ is a vector of associated parameters. Accordingly,

$$\underline{Y} = \underline{RY} + \underline{Zb} + \underline{e} \quad . \qquad (6)$$

For estimation purposes, $\underline{R}$ and $\underline{\Omega}$ are not the same since $\underline{\Omega}$ will contain diagonal elements and $\underline{R}$ usually must consist of zeroes in the diagonal cells. Note, however, that equation (4) can be rewritten as

$$\underline{Y} = \underline{I}^* + \text{diag}(\underline{\Omega})\underline{Y} + \overline{\text{diag}}(\underline{\Omega})\underline{Y} \quad , \qquad (7)$$

where diag and $\overline{\text{diag}}$ refer to the diagonal and off-diagonal decomposition of $\underline{\Omega}$. Thus,

$$\underline{Y} = [\underline{I} - \text{diag}(\underline{\Omega})]^{-1}\underline{I}^* + \underline{QY} \quad , \qquad (8)$$

where $\underline{Q} = \{q_{ij}\}$, and

$$q_{ij} = \begin{cases} \omega_{ij}/(1 - \omega_{ii}) \text{ , if } i \neq j \\ 0, \text{ otherwise} \end{cases} \quad .$$

$\underline{R}$ is related to $\underline{Q}$ rather than to $\underline{\Omega}$.

The correlation between $\underline{Y}$ and $\underline{e}$ in equation (6) means that least squares estimators of the parameters of the model are biased and inconsistent. Two cases are of interest.

4.1. Case 1: An Endogenous Investment Function

Investment may be treated as an endogenous function of regional income levels, so that $\underline{I}^* = \underline{SY}$ ($\underline{S}$ may be diagonal or otherwise). This case produces a model of the general form

$$\underline{Y} = (\underline{I} - \underline{R}^*)^{-1}\underline{e} \quad , \tag{9}$$

where $\underline{R}^* = (\underline{R} + \underline{S})$. Then equation (9) can be written as

$$\underline{Y} = \underline{u} \quad ,$$

where $E(\underline{uu}^T) = \sigma_e^2[(\underline{I} - \underline{R}^{*T})(\underline{I} - \underline{R}^*)]^{-1} = \sigma_e^2\underline{V}$. Ord (1975) has discussed a maximum likelihood procedure for estimating the unknown parameters of $\underline{R}^*$ and σ_e^2. The map of regional per capita income variation specified by equation (9) has a constant level of first-order (or mean) variation. One way to allow for spatial variation in the level of the mean would be to include trend surface coefficients. The model (9) becomes

$$\underline{Y} = \underline{X\beta} + \underline{u} \quad , \tag{10}$$

where $\underline{u}$ is as before and $\underline{X\beta}$ is a trend surface of some given order. $\underline{X}$ is the n-by-k matrix of location co-ordinates for each county (with 1s in the first column), and $\underline{\beta}$ is a k-by-1 vector of unknown trend surface parameters. For first-order trend surfaces, k=3, but clearly higher order surfaces can be estimated by introducing higher order terms into $\underline{X}$ and more parameters into $\underline{\beta}$.

The parameters of the model defined by equation (10) can be estimated by generalized least squares if $\underline{R}^*$ is known. Let $\underline{V} = \underline{LL}^T$; then transforming equation (10) yields

$$\underline{L}^{-1}\underline{Y} = \underline{L}^{-1}\underline{X\beta} + \underline{u}^* \quad , \tag{11}$$

where $\underline{u}^* = \underline{L}^{-1}\underline{u}$ and $E(\underline{u}^*\underline{u}^{*T}) = \sigma_e^2\underline{I}$. Three components of surface variation can be identified from this result, namely

(a) trend: $\underline{X\beta}$;

(b) noise (random variation): $\underline{L}^{-1}(\underline{Y} - \underline{X\beta}) = \underline{u}^*$; and,

(c) 'signal': $\underline{Y} - \underline{X\beta} - \underline{u}^*$.

In general, $\underline{R}^*$ is not known so that estimation proceeds by a form of iterative generalized least squares. The empirical analysis in the next section uses a modification of the Cochran-Orcutt method, which is described in detail in Ord (1975). A first estimate of $\underline{\beta}$ is obtained by setting $\underline{R}^* = \underline{0}$ so that an ordinary least squares estimate ($\hat{\underline{\beta}}$) is obtained. The residuals from this model, namely

$$\underline{e}^* = \underline{Y} - \underline{X}\hat{\beta} \quad ,$$

are obtained and the parameters of $\underline{R}^*$ are estimated by a maximum likelihood procedure. The trend surface coefficients are re-estimated in equation (10) using the transformation $\underline{L}^{-1}$ defined by equation (11), where the elements of $\underline{L}$ are specified by the estimates obtained for $\underline{R}^*$. A new set of residuals are obtained, the parameters of $\underline{R}^*$ are re-estimated and the cycle is repeated until convergence occurs.

4.2. Case 2: An Exogenous Investment Function

Investment ($\underline{I}^*$) also may be modelled in terms of exogenous explanatory variables, as is indicated in equation (6). If these other explanatory variables are represented as trend surface coefficients, then $\underline{Zb} = \underline{X\beta}$ in equation (6). In either case, however, equations (6) and (10) require different estimation procedures (Ord, 1975).

The estimation procedure for equation (6) has been discussed by Ord (1975), who has shown that

$$\hat{\underline{b}} = (\underline{Z}^T\underline{Z})^{-1}\underline{Z}^T\underline{Y}^* \quad , \text{ and}$$

$$\sigma^2_e = (1/n)\ \underline{Y}^{*T}\ \underline{M}\ \underline{Y}^* \quad ,$$

where n is the sample size, $\underline{Y}^* = (\underline{I} - \hat{\underline{R}})\underline{Y}$ and $\underline{M} = [\underline{I} - \underline{Z}(\underline{Z}^T\underline{Z})^{-1}\underline{Z}^T]$. $\underline{R}$ is obtained by a maximum likelihood procedure applied to the variable $\underline{Y}$ (rather than the residuals in Case 1, for $\underline{R}^*$), and once this estimate is obtained, it can be introduced directly into $\underline{b}$ and σ^2_e so that no iterative procedure is required. In both Cases 1 and 2, approximate significance tests can be constructed from the asymptotic variance-covariance matrices for the estimators of the model parameters.

These models are close in spirit to the model in Paelinck and Klaassen (1979), who, however, imposed additional closure constraints, presumably since these researchers were dealing with a system of regions that together comprised a national economy. Such a constraint does not seem necessary when the system of regions forms a sub-region of a national economy, so that there will be substantial extra-regional trade flows. Finally, whether estimating a model such as equation (6) or equation (10), considerable simplification of the matrix $\underline{R}$ (or $\underline{R}^*$) is necessary. Spatial interactions usually are represented in terms of a small number of parameters, a point that will be exemplified in the next section.

5. AN EMPIRICAL STUDY OF SPATIAL INCOME VARIATION: REGIONAL PER CAPITA INCOME IN PENNSYLVANIA COUNTIES, 1970

The transfer mechanisms identified in Section 3 and associated with the consumption element of regional income can, at best, operate over relatively small distances, so that whilst interaction effects would not tend to be observable between, say, the states of the United States, they might be observable at the county scale. As a consequence, models such as equation (6) or (10) might provide appropriate descriptions of county level per capita income variation. For example, average distances between the mid-points of the 67 Pennsylvania counties are of the order of 40 to 50 miles. Such distances mean that commuting fields will be large enough to permit significant levels of inter-county interaction; that is, levels of interaction might show up in the form of spatially correlated per capita income levels.

Per capita income levels for the 67 counties of Pennsylvania were extracted from the 1970 Census of Population (Table 124), and the model

$$\left.\begin{aligned} \underline{Y} &= \underline{X\beta} + \underline{u} \\ \underline{u} &= (\underline{I} - \rho\underline{W})^{-1}\underline{e} \end{aligned}\right\} \qquad (12)$$

was fitted to these data. $\underline{Y}$ refers to county average per capita income. $\underline{X}$ is a 67-by-k matrix of trend surface co-ordinates, where k is determined by the order of the model (k = 1 if it is a zero-order model, k = 3 if it is a first-order model, k = 6 if it is a second order model, and so on), and where $\underline{\beta}$ is the k-by-1 vector of trend surface coefficients. The 67-by-67 matrix $\underline{W}$ defines the join structure of the counties, such that

$$w_{ij} \begin{cases} > 0, & \text{if i and j are first-order neighbours,} \\ = 0, & \text{if } i = j, \\ = 0, & \text{otherwise, and} \end{cases}$$

$$\sum_{i=1}^{i=n} w_{ij} = 1 \quad .$$

Counties were neighbours if they shared a common border. The parameter ρ is a measure of inter-county interaction and must lie in the interval (-1,1). Note that by substitution, equation (12) can be rewritten as

$$\underline{Y} = \rho\underline{W}\,\underline{Y} + (\underline{I} - \rho\underline{W})\underline{X}\,\underline{\beta} + \underline{e} \quad ,$$

so that the model explicitly includes inter-county correlation in the $\underline{Y}$ variable.

Table 4 summarizes the estimation of the model for a first-order (k = 3) trend surface model. Assuming that $\rho = 0$, the model reduces to a conventional trend surface model and can be estimated by ordinary least squares. The β_1 coefficient refers to the North-South axis and the β_2 coefficient to the East-West axis. The trend surface coefficients are strongly significant, but the generalized Moran coefficient (see Cliff and

TABLE 4

FIRST-ORDER TREND SURFACE VERSION OF EQUATION (10)

Ordinary Least Squares Estimation (assuming $\rho = 0$)		
$\hat{\beta}_0 = 2.684$	$\hat{\beta}_1 = -0.053$ $t = 3.9$	$\hat{\beta}_1 = 0.028$ $t = 4.4$
$GMC = 0.459$	$\mu_{GMC} = -0.04$	$\sigma_{GMC} = 0.07$
$R^2 = 0.254$		
Iterative Estimation (assuming $\rho \neq 0$)		
$\hat{\beta}_0 = 2.619$	$\hat{\beta}_1 = -0.036$ $t = 1.14$	$\hat{\beta}_1 = 0.025$ $t = 1.29$
$\hat{\rho} = 0.71$	$LR = 0.643$	
R^2(full model) = 0.571		R^2(on trend component) = 0.240

NOTE: GMC denotes the generalized Moran coefficient, and LR denotes the likelihood ratio statistic.

Ord, 1981) indicates the presence of spatial autocorrelation. Re-estimating the full model ($\rho \neq 0$) results in the trend surface coefficients ceasing to be significant, while the interaction parameter ρ is strongly significant, as indicated by the likelihood ratio statistic, whose approximate critical value is 0.95.

Higher order trend surfaces were estimated, too. The R^2 values for a $k = 6$, second-order surface was .598, and for a $k = 10$, third-order surface was 0.619. Apart from one weakly significant second-order term, none of these trend surface coefficients were significant. It was concluded that there was no significant systematic variation in the data, and a zero-order model was estimated, with the results being summarized here in Table 5. The presence of significant autocorrelation in $\underline{Y}$ was evidenced by the generalized Moran coefficient test, and a large interaction coefficient (ρ) was obtained.

TABLE 5

ZERO-ORDER TREND SURFACE VERSION OF EQUATION (10)

Iterative Estimation (assuming $\rho \neq 0$)

$\hat{\beta}_0 = 2.716$

GMC = 0.600 $\mu_{GMC} = -0.04$ $\sigma_{GMC} = 0.07$

$\hat{\rho} = 0.786$ LR = 0.507

R^2(full model) = 0.570

NOTE: GMC denotes the generalized Moran coefficient, and LR denotes the likelihood ratio statistic.

TABLE 6

ZERO-ORDER TREND SURFACE VERSION OF EQUATION (10):
NON-SMSA LINKS

Iterative Estimation (assuming $\rho \neq 0$)

$\hat{\beta}_0 = 2.716$

GMC = - 0.027 $\mu_{GMC} = -0.028$ $\sigma_{GMC} = 0.092$

$\hat{\rho} = -0.058$ LR = 0.99

R^2(full model) < 0.01

NOTE: GMC denotes the generalized Moran coefficient, and LR denotes the likelihood ratio statistic.

TABLE 7

ZERO-ORDER TREND SURFACE VERSION OF EQUATION (10):
SMSA-SMSA LINKS

Iterative Estimation (assuming $\rho \neq 0$)

$\hat{\beta}_0 = 2.716$

GMC = 0.534 $\mu_{GMC} = -0.017$ $\sigma_{GMC} = 0.059$

$\hat{\rho} = 0.766$ LR = 0.599

R^2(full model) = 0.462

NOTE: GMC denotes the generalized Moran coefficient, and LR denotes the likelihood ratio statistic.

TABLE 8

ZERO-ORDER TREND SURFACE VERSION OF EQUATION (10):
NON-SMSA-NON-SMSA LINKS

Iterative Estimation (assuming $\rho \neq 0$)

$\hat{\beta}_0 = 2.716$

GMC = 0.189 $\mu_{GMC} = -0.026$ $\sigma_{GMC} = 0.068$

$\hat{\rho} = 0.497$ LR = 0.910

R^2(full model) = 0.133

NOTE: GMC denotes the generalized Moran coefficient, and LR denotes the likelihood ratio statistic.

The presence of strong positive correlation between neighbouring county per capita income levels seems to be confirmed by these results, and could be the product of income transfers arising from commuting flows and inter-county purchasing by residents. However, a further question arises that asks whether or not all inter-county connections are equally important. Twenty-four of the Pennsylvania counties are designated SMSAs, and 43 are designated non-SMSAs. In Tables 6-8, results are presented for estimating equation (12) with different $\underline{W}$ matrices. The definitions of $\underline{W}$ are as follows:

Table 6, w_{ij} { > 0, if i and j are neighbours and one is in an SMSA and the other is not,
= 0, otherwise;

Table 7, w_{ij} { > 0, if i and j are neighbours and both are in SMSAs,
= 0, otherwise;

Table 8, w_{ij} { > 0, if i and j are neighbours and both are not in SMSAs,
= 0, otherwise.

The evidence reported in these tables suggests that the strongest links are between the contiguous SMSA counties, followed by the non-SMSA counties.

TABLE 9

ZERO-ORDER TREND SURFACE VERSION OF EQUATION (10): MEAN ADJUSTED DATA FOR SMSAs AND NON-SMSAs

Iterative Estimation (assuming $\rho \neq 0$)

$\hat{\beta}_0 = 0.000$

GMC = 0.212 $\mu_{GMC} = -0.028$ $\sigma_{GMC} = 0.093$

$\hat{\rho} = 0.267$ LR = 0.942

R^2(full model) = 0.075

NOTE: GMC denotes the generalized Moran coefficient, and LR denotes the likelihood ratio statistic.

Table 6, on the other hand, reveals no significant autocorrelation between SMSA counties and their neighbouring non-SMSA counties. This finding might suggest the presence of a certain discontinuity between the 'rural' and 'urban' areas in terms of income transfer. However, the mean level of per capita income differs between SMSA and non-SMSA counties. If per capita income levels are adjusted by subtracting the appropriate mean for each county (depending upon whether or not the county is an SMSA), then Table 9 shows the results obtained using the same linkage rule as for Table 6. The implication here is that there may be spillover effects between SMSA and non-SMSA counties, although even with these adjustments the transfers are not strong.

6. CONCLUSIONS

Spatial pattern in economic variables may arise from a variety of mechanisms, including spatial transfers. Several types of transfer mechanisms, as well as some detailed models for wage expenditure diffusion and regional income, have been considered in this paper. Relationships between the form of the surface and economic impulses coupled with certain economic parameters were considered in Section 3. The last part of the paper has dealt with empirical issues, including first the construction of statistical models for estimating inter-regional income variation, and second the discussion of an empirical study. In an aggregate static framework it clearly is not possible to associate the patterns in the Pennsylvania county data with the mechanisms reviewed in earlier parts of this paper. There are many factors that could underly the observed spatial variation, most importantly continuities in the economic structure of, for example, the clusters of SMSA counties. Development of this research area requires both aggregate income studies (as reported here) and micro-level studies of actual wage expenditures over space.

7. REFERENCES

Bennett, R., R. Haining and A. Wilson, 1985, Spatial Structure, Spatial Interaction and Their Integration: A Review of Alternative Models, Environment and Planning A, 17: 625-45.

Clarke, M. and A. Wilson, 1983, Dynamics of Urban Spatial Structure: Progress and Problems, Journal of Regional Science, 23: 1-18.

Cliff, A. and J. Ord, 1981, Spatial Processes, London: Pion.

Curry, L., 1978, Demand in the Spatial Economy: II, Homo-stochasticus, Geographical Analysis, 10: 309-44.

Erickson, R., 1977, Sub-regional Impact Multipliers: Income Spread Effects From a Major Defense Installation, Economic Geography, 53: 283-94.

Gertler, M., 1984, Regional Capital Theory, Progress in Human Geography, 8: 50-81.

Haining, R., 1983, Spatial Structure and Spatial-temporal Processes, in Evolving Geographical Structures, edited by D. Griffith and A. Lea, The Hague: Martinus Nijhoff, pp. 323-46.

_____, 1984, Testing a Spatial Interacting Markets Hypothesis, Review of Economics and Statistics, 66: 475-483.

_____, 1985, The Spatial Structure of Competition and Equilibrium Price Dispersion, Geographical Analysis, 17: 231-242.

Heine, V., 1955, Models for Two-dimensional Stationary Stochastic Processes, Biometrika, 42: 170-8.

Ord, J., 1975, Estimation Methods for Models of Spatial Interaction, Journal of the American Statistical Association, 70: 120-6.

Paelinck, J. and L. Klaassen, 1979, Spatial Econometrics, Farnborough, Hants: Saxon House.

Sheppard, E., 1979, Geographic Potentials, Annals, Association of American Geogaphers, 69: 438-47.

ten Raa, T., 1984, The Distribution Approach to Spatial Economics, Journal of Regional Science, 24: 105-17.

United States Department of Commerce, 1970, Census of Population: General Social and Economic Characteristics for Pennsylvania, Washington, D. C.: U. S. Government Printing Office.

Whittle, P., 1954, On Stationary Processes in the Plane, Biometrika, 41: 343-9.

8. APPENDIX

If the set of regions forms a rectangular lattice, then the model

$$(\underline{I} - \underline{\Omega})\underline{X} = \underline{e} \qquad \text{(A1)}$$

can be written as

$$L(T_s, T_t)\, X_{s,t} = e_{s,t} \quad ,$$

where (s,t) defines a site on the lattice, and T_s and T_t are lag operators in the two spatial directions. Let

$$L(T_s,T_t) = \sum_j \sum_k a_{j,k} T_s^j T_t^k \quad .$$

Therefore the potential is

$$[L(T_s,T_t)]^{-1} = [\sum_j \sum_k a_{j,k} T_s^j T_t^k]^{-1} \quad .$$

The continuous analogue can be written as

$$L(\partial/\partial x_1,\partial/\partial x_2)\; X(x_1,x_2) = \varepsilon(x_1,x_2) \quad , \tag{A2}$$

where x_1 and x_2 are the spatial co-ordinates. Formally, solutions to equations of the type (A2) can be written as

$$X(x_1,x_2) = \iint_{-\infty}^{\infty} G(x_1 - u,x_2 - v)\; \varepsilon(u,v)\; du\; dv \quad , \tag{A3}$$

where $G(x_1,x_2)$, the Green function, satisfies

$$L(\partial/\partial x_1,\partial/\partial x_2)\; G(x_1,x_2) = \delta(x_1)\; \delta(x_2) \quad ,$$

and where δ denotes the Dirac delta function. The similarity between the role of the Green function in equation (A3) and the potential function for equation (A1) can be seen. Finally, one should note that if $\rho(s,t)$ denotes the autocorrelation function for X at lag (s,t) [so that ε is a stochastic variable], then

$$\rho(s,t) = R(s,t)/R(0,0) \quad ,$$

where

$$R(s,t) = \iint_{-\infty}^{\infty} G(u,v)\; G(u - s,v - t)\; du\; dv \quad .$$

Again the link with the autocorrelation function for equation (A1), assuming e is stochastic, can be followed through.

TRANSPORTATION FLOWS WITHIN CENTRAL-PLACE SYSTEMS

Michael Sonis
Bar-Ilan University
Israel

1. INTRODUCTION

It is quite understandable that the actual hierarchy of settlements puts strong restrictions on the type of optimal (minimal cost) transportation flows between settlements. In turn, the spatial and temporal stability of transportation flows may be an essential factor of growth or decline of an individual settlement in the hierarchy. Moreover, usually the optimal transportation flow does not cover all linkages of the transportation network between settlements. Therefore, the existence of structurally stable optimal transportation flows can result in a change in the transportation network itself, at the expense of non-used links.

It is generally accepted among regional scientists that central place theory constitutes a theoretical basis for the analysis of the hierarchical structure of settlement networks. In this paper an attempt is made to provide the theoretical analysis of all possible types of structurally stable optimal transportation flows associated with transportation networks within the original Christaller central place hierarchy, and its generalization--the Beckmann-McPherson central place system. The spatial description of the transportation network of the original Christaller central place model is based on three generic geometrical properties of market areas associated with this central place system. The first is that the market areas of all central places at the same hierarchical level form a hexagonal covering of the plane. The second is that if the hierarchical level increases from the lowest tier through to the highest one, then the sizes of market areas increase from the smallest size through to the largest by a constant nesting factor K, which expresses one of the three Christallerian optimization principles, namely marketing (K = 3), transportation (K = 4), or administration (K = 7). The third distinctive geometric property is that the center of a market area of a given size is also the center of market areas of each smaller size (Christaller, 1933).

The set of hierarchical models proposed by Beckmann and McPherson (1970) differ from the original Christaller model by variable nesting factors and by using the Loeschian idea of all possible coverings of the plane by hexagons, whose centers are the vertices of the initial Christallerian triangular lattice, with the sizes of market areas being integers (Loesch, 1940). In this paper we will consider the bounded

Beckmann-McPherson model with a single largest central place and a fixed finite number of hierarchical levels. In this model central places are arranged in a triangular lattice within a hexagon that presents (with possible small deviations) the transportation network within a bounded Beckmann-McPherson central place model (see Figures 1 and 2).

The question that naturally arises asks whether or not the topological structure of the above mentioned transportation network corresponds to the optimal minimal cost flow. The answer is surprising; only the topological structure of the transportation network within the Christaller bounded two-tier central place models (with one central town, six dependent villages, and the nesting factor of $K = 3$, $K = 4$ or $K = 7$) corresponds to the minimal cost flow [see Figures 1(a) and 1(b)]. This property is based on the following theorem:

> The arc density theorem: Consider a connected graph with v vertices and a arcs. If this graph represents a topological structure of the minimal cost flow of homogeneous production, then its arc density $a + 1/v$ has a range of
>
> $$1 \leq a + 1/v \leq 2 \quad . \tag{1}$$
>
> Proof: Each vertex of the graph can be an origin for supply, a destination for demand, or an origin-destination for supply and demand, simultaneously. Let $x, y, z \geq 0$ be the numbers of origins, destinations and origin-destinations, respectively. Then $x + y + z = v$, the number of suppliers is $m = x + z$, and the number of demanders is $n = y + z$. As is known from the theory of Linear Programming (Dantzig, 1963, Ch. 15, 17), the topological structure of the optimal flow between m suppliers and n demanders is the maximal tree (i. e., the maximal connected subgraph of the transportation network without cycles, which includes exactly $m + n - 1$ arc). In our case the number of arcs is
>
> $$a = m + n - 1 = x + y + 2z - 1 = v + z - 1 \quad .$$
>
> Therefore,
>
> $$\left.\begin{array}{lll} 1 + a - v = z \geq 0 & \text{, or} & a + 1 \geq v \text{, and} \\ 2v - a - 1 = x + y \geq 0 & \text{, or} & a + 1 \leq 2v \quad . \end{array}\right\} \tag{2}$$
>
> Consequently, $1 \leq a + 1/v \leq 2$.

Let us now apply conditions (1) and (2) to the triangular network organized into a hexagon whose sides include p arcs [see Figure 2(a)]. It

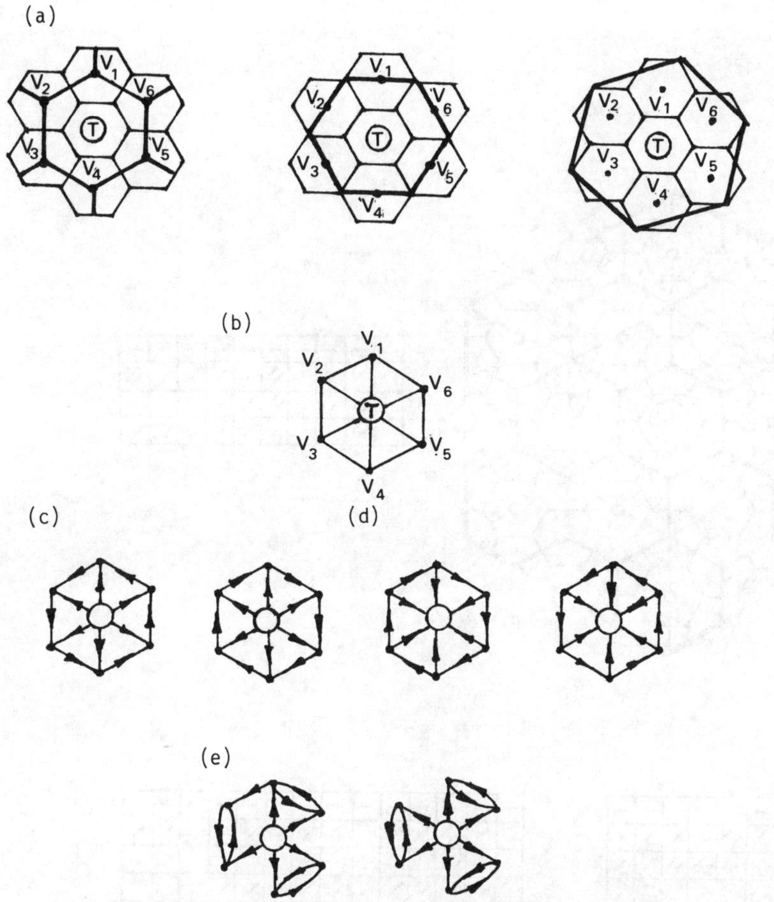

Figure 1: The topological structure of the transportation network within the two-tier Christaller bounded central place systems, which corresponds to the minimal cost flow; (a) shows the hierarchical models, (b) the transportation network, (c) and (d) the structures of the 'top-down' and 'bottom-up' structurally stable optimal flows, and (e) the examples of optimal structurally unstable flows.

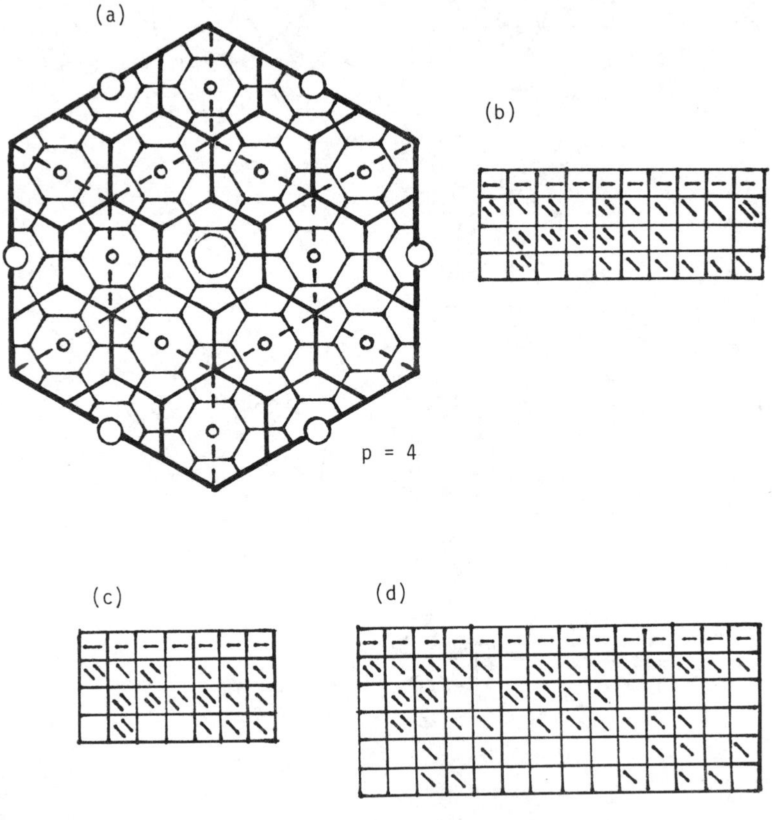

Figure 2: Aggregated schemes for the four-tier Beckmann-McPherson models; (a) shows the four-tier spatial organization with nesting factors $K_1 = 4$, $K_2 = 4$, and $K_3 = 3$, and (b), (c) and (d) show the aggregated schemes for models with the following combinations of nesting factors: 4,4,3; 3,4,3; and, 7,4,3.

is easy to see by mathematical induction that the numbers of vertices and arcs of this hexagonal graph are

$$v = 3p^2 + 3p + 1 \quad , \quad a = 9p^2 + 3p \quad ,$$

and from condition (2) that

$$z = 6p^2 \quad , \quad x + y = -3p^2 + 3p + 1 \quad .$$

Therefore the arc density

$$a + 1/v = (9p^2 + 3p + 1)/(3p^2 + 3p + 1)$$

satisfies condition (1) only if $p = 1$, and $v = 7$, $a = 12$, $z = 6$ and $x + y = 1$ (i. e., the hexagon must include only one pure supplier or only one pure demander and six suppliers-demanders).

By putting the pure supplier or pure demander into the center of the hexagon we obtain only four topologically possible structures of optimal flows, namely the pair of optimal 'top-down' flows [Figure 1(c)], and the pair of optimal 'bottom-up' flows [Figure 1(d)]. Obviously these four optimal flows are invariant with respect to rotations preserving the hexagonal transportation network. The most important property of such optimal flows is that they are structurally stable (i. e., they are unchangeable under admissible perturbations of supply, demand and transportation costs). As an interesting aside, comparing this property with the spatial organization of the structurally stable minimal cost flows within a continuous two-dimensional transportation field rules out the spatial hexagonal tessellation of Christaller and Loesch (Puu, 1980). The second important property of these four flows is that each 'top-down' or 'bottom-up' transportation flow in a two-tier Christaller model, which is invariant with respect to rotations, can be presented as a weighted superposition--convex combination--of these structurally stable flows.

This paper extends the preceding description of the structurally stable 'top-down' and 'bottom-up' optimal transportation flows to the case of flows in an arbitrarily bounded Beckmann-McPherson hierarchical model, under the assumption that in the case of 'top-down' flows each central place supplies homogeneous production to each dependent central place within its own market area and exchanges the production with the closest central places of the same hierarchical tier. Further, the supply and demand of each central place is not necessarily equal, and self-demands are mutually reducible. The second assumption is that the system of transportation costs is invariant with respect to rotations. Under these assumptions structurally stable transportation flows usually are associated with non-planar graphs, and the structure of all rotationally invariant flows is ascertained, each flow, which is invariant with respect to rotations, is decomposed into the convex superposition of optimal structurally stable flows.

2. DOMAINS OF STRUCTURAL STABILITY FOR OPTIMAL SOLUTIONS OF THE LINEAR PROGRAMMING PROBLEM

The following theorem gives a brief unification of the main results of Linear Programming:

Theorem: Consider a primal Linear Programming problem and its dual problem, such that

$$\begin{array}{llll} \text{MIN:} & \underline{cX} & \text{MAX:} & \underline{yb} \\ \text{s.t.:} & \underline{Ax} = \underline{b} & \text{s.t.:} & \underline{yA} \leq \underline{c} \\ & \underline{x} \geq \underline{0} & & \end{array}$$

Let $\underline{B}$ be an invertible submatrix of the matrix $\underline{A}$ ($\underline{B}$ is the basis matrix) with the properties

$$\underline{B}^{-1}\underline{b} \geq 0 \quad \text{, and} \quad \underline{c}_0\underline{B}^{-1}\underline{A} \leq \underline{c} \quad ,$$

where the coordinates of the vector $\underline{c}_0$ correspond to the columns of matrix $\underline{A}$, which are the columns of matrix $\underline{B}$. Then the primal problem has the optimal solution $\underline{X}$ (with zero non-basis components) whose basis components are $\underline{X}_0 = \underline{B}^{-1}\underline{b}$, and the dual problem has the optimal solution $\underline{Y} = \underline{c}_0\underline{B}^{-1}$.

Proof: The vectors $\underline{X}$ and $\underline{Y}$ are the admissible solutions for the primal and dual problems, respectively, because $\underline{X} \geq \underline{0}$, and

$$\begin{array}{l} \underline{AX} = \underline{BX}_0 = \underline{BB}^{-1}\underline{b} = \underline{b} \quad \text{, and} \\ \underline{YA} = \underline{c}_0\underline{B}^{-1}\underline{A} \leq \underline{c} \quad . \end{array}$$

Furthermore, $\underline{X}$ and $\underline{Y}$ are the optimal solutions, because if $\underline{X}_1$ and $\underline{Y}_1$ are the other admissible solutions, then

$$\underline{cX} = \underline{c}_0\underline{X}_0 = \underline{c}_0\underline{B}^{-1}\underline{b} = \underline{c}_0\underline{B}^{-1}\underline{AX}_1 = \underline{YAX}_1 \leq \underline{cX}_1$$

and

$$\underline{Y}_1\underline{b} = \underline{Y}_1\underline{AX} \leq \underline{cX} = \underline{c}_0\underline{X}_0 = \underline{c}_0\underline{B}^{-1}\underline{b} = \underline{Yb} \quad .$$

This theorem also gives a complete description of the domains for structural stability of the optimal solutions for primal and dual Linear Programming problems. Within the domains for structural stability are the domains of the permissible changes of the production costs $\underline{c}$ and the permissible fluctuations of the resources $\underline{b}$ under which optimal solutions correspond to the same basis $\underline{B}$ (i. e., the optimal assortment of production is preserved). Indeed, if the vectors $\underline{b}$ and $\underline{c}$ are changed, a cone

$$\underline{C} = (\underline{b}^*: \underline{B}^{-1}\underline{b}^* \geq \underline{0})$$

in the space of resources, and a wedge

$$\underline{W} = (\underline{c}^*: \underline{c}_0^*\underline{B}^{-1}\underline{A} \leq \underline{c}^*)$$

in the space of costs are obtained. This cone and wedge determine the domains of stability of the basis $\underline{B}$ for the optimal solutions of the primal and dual problems. Thus, the Cartesian product $\underline{C}x\underline{W}$ defines the domains of structural stability of the optimal solutions for the primal and dual problems. The construction of this Cartesian product for each given optimal solution is simple, since the last tableau of the Simplex algorithm contains the components of matrix $\underline{B}^{-1}\underline{A}$ (Dantzig, 1963, Ch. 6). Therefore, to obtain inequalities determining the domain of structural stability, we need only the components of the last Simplex tableau.

Moreover, the optimal solutions $\underline{X}$ and $\underline{Y}$ associated with the basis matrix $\underline{B}$ are the extreme points--vertices--of the corresponding convex polyhedrons of admissible solutions for the primal and dual problems. Since the matrix $\underline{A}$ contains a finite number of invertible submatrices, the space of resources and the space of costs are decomposable into a finite number of domains, namely

$$\underline{C}_1x\underline{W}_1 \;, \; \underline{C}_2x\underline{W}_2 \;, \; \ldots \;, \; \underline{C}_rx\underline{W}_r \quad ,$$

so that each of these domains corresponds to the preservation of some invertible basis submatrix of the matrix $\underline{A}$ (i. e., to the preservation of some optimal assortment of production).

Now we will apply the aforementioned results to the optimal solutions for the transportation problem of Linear Programming. Consider the minimal cost flow problem on the transportation network, which includes m suppliers with a_i units of supply for the i-th supplier (i=1,2,...,m), and n demanders with b_j units of demand for the j-th demander (j=1,2,...,n), such that the total supply is equal to the total demand. In addition, let c_{ij} be the cost of transportation of one unit of production from the i-th supplier to the j-th demander. For a reconstruction of the cone and the wedge of structural stability, we will use the well-known method of potentials--the MODI-method of Dantzig (Dantzig, 1951). First, a computing table is constructed that includes m rows corresponding to supply a_i, n columns corresponding to demand b_j, and mn transportation costs c_{ij}. The optimal solution for the transportation problem occupies m + n - 1 cells of the computing matrix. The occupied cells (basis cells) show the coordinates of the cost-vector c_0. The set of basis cells corresponds to the columns of the invertible basis matrix $\underline{B}$. Topologically, this set of cells defines the maximally connected graph of the transportation network without cycles, which includes exactly m + n - 1 arcs. Each such maximal tree corresponds to the topological structure of some optimal (minimal cost) flow.

The solution for the dual problem is the set of potentials (shadow prices) for suppliers, $u_1=0$, u_2, ..., u_m, and for demanders, v_1, v_2, ..., v_n, such that for each basis cell (i,j) the differences between these potentials equal the transportation costs in question:

$$v_j - u_i = c^0_{ij} \quad , \; i=1,\, 2,\, \dots,\, m \text{ and } j=1,\, 2,\, \dots,\, n \quad .$$

This system of linear equations means that each potential is the alternating sum of basis transportation costs c^0_{ij}. For the non-basis cells (r,s),

$$u_r + c_{rs} - v_s \geq 0 \quad .$$

For the changeable transportation costs c^*_{rs} this system of linear inequalities describes the wedge W̲ of structural stability for a given optimal flow in the space of transportation costs.

For the construction of the cone of structural stability we will use the following observation. For any optimal solution of the transportation problem the computing table contains a row or a column that includes only one basis cell. This means that we can insert into this cell the amount of supply or demand in such a way that the size of the computing matrix is reduced. After a finite number of such steps, we will fill each basis cell with the alternating sums of supply-demand amounts. The cone C̲ of structural stability in the space of supply-demand is defined by the requirement of the non-negativity of each such alternating sum.

It is possible to present this foregoing procedure for the construction of the inequalities, defining the domain of structural stability C̲xW̲, in the form of the vectorial generalization of the method of potentials (Sonis, 1982a).

Structural change in the topological structure of the optimal transportation flow is linked with the fact that one or more inequalities defining the cone and the wedge of structural stability are not satisfied. The domains of the structural change are the faces of the domain of the structural stability C̲xW̲ that are the closed hyperplanes in the supply-demand space or in the space of transportation costs. On the face of the cone C̲ the flow is degenerate; it divides into a few independent subflows that are the optimal solutions of smaller size problems. If one moves out of the cone C̲, then an admissible flow with a given topological structure does not exist, and a new flow structure must be constructed. If one moves out of the wedge W̲, then there is an admissible flow with a previous topological structure, but the condition of optimality of the transportation flow fails to hold, and the structure of flow must be changed by substituting one arc of the topological structure for another. For example, consider the construction of the domains of structural stability of the optimal transportation flow within a two-tier, bounded Christaller hierarchical system (see Figure 1). Assume that the central town, T, supplies a_0 units of a homogeneous product to the dependent villages V_1, each village is b_1 and its supply is a_1. Assume also that the cost of transportation from the central town is equal to c_1, and to the village is equal to c_2. Figure 3 presents the computing tables for the construction of the domains of structural stability for optimal flows, with this structure being described in Figures 1(c) and 1(e). These structures are marked in Figures 3(a) and 3(b) with the help of arcs. The flow with that structure described in Figure 1(e) is unstable, since the non-negativity of indices of non-basis cells $c_2 - c_1$, $2c_1 - c_2$, and $c_1 - c_2$ means the equality $c_1 = c_2$ will appear in the inequalities defining the wedge of structural stability. This implies that each small perturbation of transportation costs, breaking this equality, converts the optimal flow into a non-optimal one. The flow with the structure 1(c) is a structurally stable 'top-down' optimal flow that is invariant under rotations. This flow illustrates structurally stable flows within general hierarchical models.

3. STRUCTURALLY STABLE OPTIMAL FLOWS WITHIN A BECKMANN-MCPHERSON BOUNDED CENTRAL PLACE SYSTEM

Consider a bounded Beckmann-McPherson hierarchical model with a finite number of tiers. The central places of such a model are arranged in a triangular lattice that represents the transportation network within a given central place model. Due to the geometrical symmetry of this network, central places from the same hierarchical level and with the same topological location are supposed to have the same volumes of supply or demand. Moreover, the transportation costs of supplied or demanded production are supposed to be identical (supply is not necessarily equal to demand for the same central place, and the transportation costs depend on direction). Hence the distribution of supply, the distribution of demand, and also the distribution of transportation costs are invariant with respect to all possible rotations preserving the transportation network. In contradistinction, the topological structure of an optimal flow—the maximal tree—can be non-symmetrical [see Figure 1(e)]. But if an optimal flow is structurally stable, then its maximal tree must be rotationally invariant; if not, then there is an admissible rotation of the transportation network that transforms the given maximal tree into a different one, and these two different maximal trees correspond to transportation flows with the same minimal transportation cost. Therefore, the solution for the transportation problem is not unique, and the flows are on the face of the cone of structural stability. For the same reason, the distribution of volumes of flow on the arcs must be rotationally invariant, too.

The problem of enumerating maximal rotationally invariant trees is purely combinatorial and hence formidable, cumbersome and tedious. Its solution for each given hierarchical model can be found with the help of aggregated schemes of computing tables of the type described in Figure 2. Each cell of the aggregated scheme corresponds to a description of the possible rotationally invariant parts of transportation flows between the subset of suppliers and the subset of demanders, so that each subset belongs to the same hierarchical level and has the same relative location. This means that each subset is rotationally invariant, and the set of all central places from the same hierarchical level is divided into rotationally invariant subsets. Moreover, if the sets of suppliers and demanders include, respectively, M and N rotationally invariant subsets, then the maximal rotationally invariant tree will include 6(M + N − 1) linkages. Furthermore, each cell of the aggregated scheme includes one or two arcs that present the set of all possible linkages between central places. For the construction of the rotationally invariant optimal flow we must consider the set of all possible choices of M + N − 1 lines, and for each choice we

(a)

Demand

Supply	V_1 b_1	V_2 b_1	V_3 b_1	V_4 b_1	V_5 b_1	V_6 b_1	u_i
T a_0	c_0 / b_1-a_1	c_0 / c_2-c_1	c_0 / b_1-a_1	c_0 / b_1-a_1	c_0 / b_1-a_1	c_0 / b_1-a_1	0
V_1 a_1		c_1 / b_1-a_1				c_2 / a_1	c_0-c_2
V_2 a_1	c_2 / c_2-c_1		c_1 / a_1				c_0-c_1
V_3 a_1		c_2 / a_1		c_1 / $2(c_1-c_2)$			$c_0+c_1-2c_2$
V_4 a_1			c_2 / c_2-c_1		c_1 / a_1		c_0-c_1
V_5 a_1				c_2 / a_1		c_1 / c_1-c_2	c_0-c_2
V_6 a_1	c_1 / a_1				c_2 / c_2-c_1		c_0-c_1
v_j	c_0	$c_0+c_1-c_2$	c_0	c_0	c_0	c_0	

$C:\ b_1 \geq a_1$; $W:\ c_1 = c_2$

Figure 3: Analysis of the structural stability of minimal cost flows in the Christaller two-tier central place system with nesting factors $K_1 = 3,4,7$; (a) shows the computing table for the optimal unstable flow with a structure from Figure 1(e), and (b) shows the computing table for the optimal structurally stable flow with a structure from Figure 1(c).

(b)

Demand

Supply	V_1 b_1	V_2 b_1	V_3 b_1	V_4 b_1	V_5 b_1	V_6 b_1	u_i
T a_0	c_0 b_1-a_0	c_0 b_1-a_0	c_0 b_1-a_0	c_0 b_1-a_0	c_0 b_1-a_0	c_0 b_1-a_0	0
V_1 a_1		c_1 a_1				c_2 c_2-c_1	c_0-c_1
V_2 a_1	c_2 c_2-c_1		c_1 a_1				c_0-c_1
V_3 a_1		c_2 c_2-c_1		c_1 a_1			c_0-c_1
V_4 a_1			c_2 c_2-c_1		c_1 a_1		c_0-c_1
V_5 a_1				c_2 c_2-c_1		c_1 a_1	c_0-c1
V_6 a_1	c_1 a_1				c_2 c_2-c_1		c_0-c_1
v_j	c_0	c_0	c_0	c_0	c_0	c_0	

C: $b_1 \geq a_1$; W: $c_1 \leq c_2$

Figure 3: Continued.

need to check whether or not the corresponding subgraph will give the maximal tree.

Figure 2 presents the aggregated schemes for four-tier Beckmann-McPherson models, and emphasizes the fact that the problem of total enumeration of the set of possible structurally stable flows can be quite complicated. Now we will present the solution for the enumeration problem for all three-tier models whose nesting factors are associated with Christallerian optimization principles. To start with, consider the three-tier model with nesting factors $K_1 = K_2 = 3$. Figure 4 includes the geometrical description of this model, and the construction of the corresponding computing table and its aggregated scheme. For this model there are two rotationally invariant subsets of suppliers—the central city C, and the towns T_1, ..., T_6—and two rotationally invariant subsets of demanders—the towns T_1, ..., T_6, and the villages V_1, ..., V_6. Therefore,

(a)

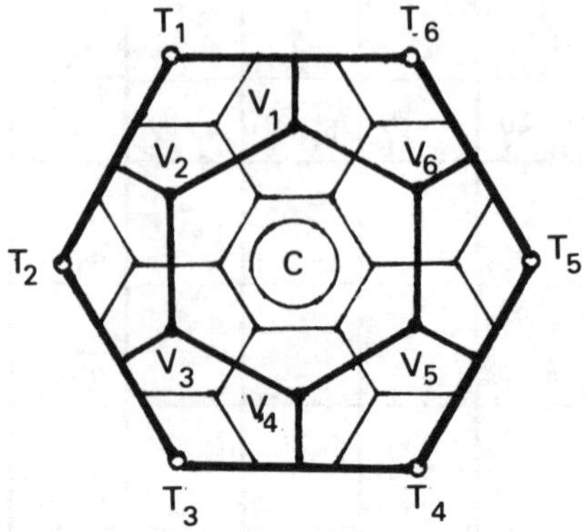

(b)

		Demand											
		T b^1_1	T b^2_1	T b^3_1	T b^4_1	T b^5_1	T b^6_1	V b^1_2	V b^2_2	V b^3_2	V b^4_2	V b^5_2	V b^6_2
Supply	C a_0	c_0	c_0	c_0	c_0	c_0	c_0	c_1	c_1	c_1	c_1	c_1	c_1
	T_1 a_1		c_2				c_3	c_4	c_5				
	T_2 a_1	c_3		c_2					c_4	c_5			
	T_3 a_1		c_3		c_2					c_4	c_5		
	T_4 a_1			c_3		c_2					c_4	c_5	
	T_5 a_1				c_3		c_2					c_4	c_5
	T_6 a_1	c_2				c_3		c_5					c_4

(c)

Figure 4: Construction of the aggregated scheme for the Beckmann-McPherson three-tier model with nesting factors $K_1 = K_2 = 3$, (a) shows the bounded hierarchical model, (b) the computing table, and (c) the aggregated scheme.

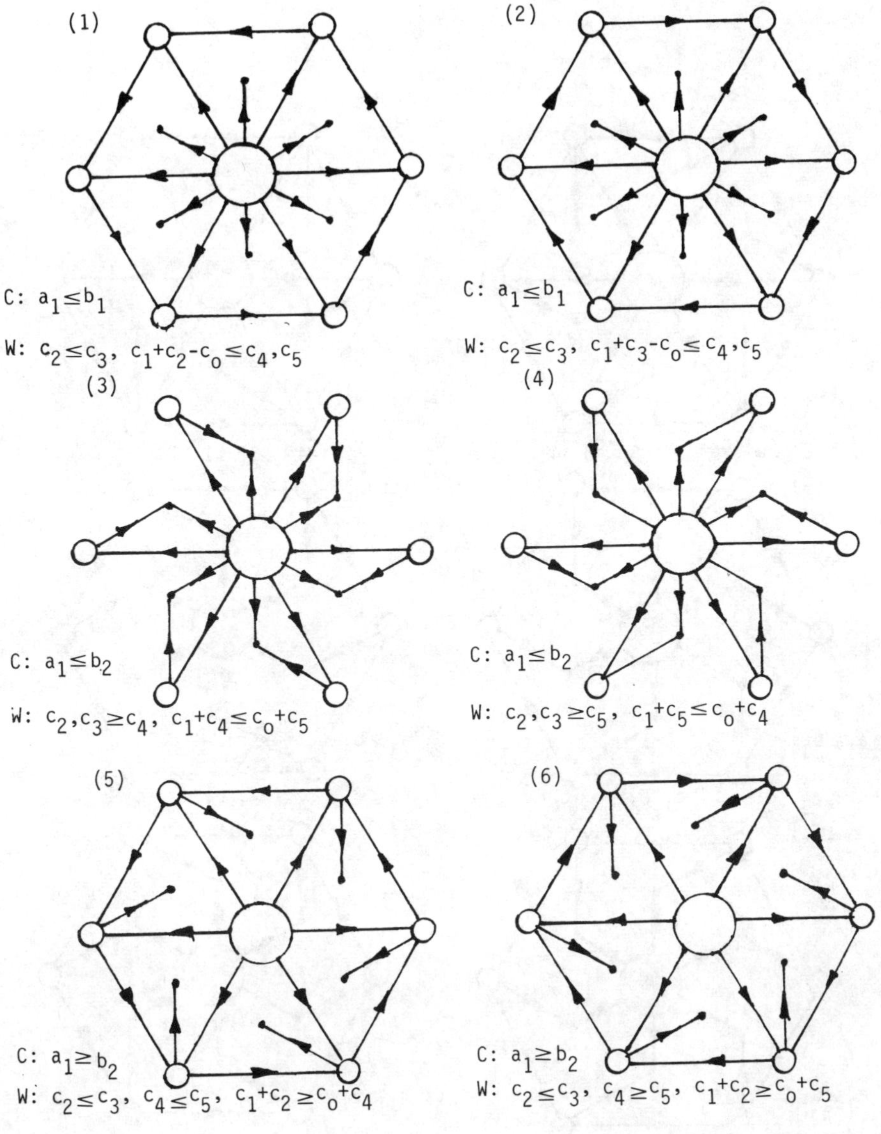

Figure 5: **Complete description of all structurally stable optimal flows and their domains of structural stability for the three-tier hierarchical model with nesting factors $K_1 = K_2 = 3$.**

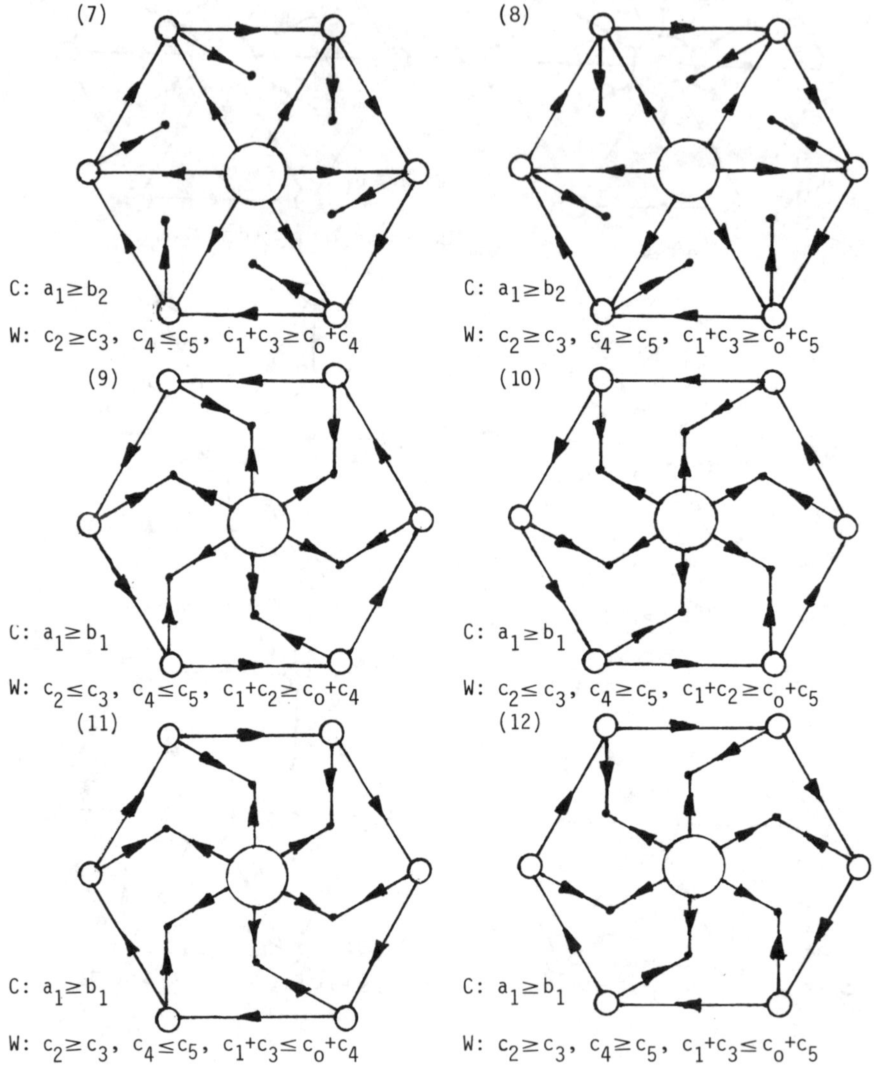

Figure 5: Continued.

(a)

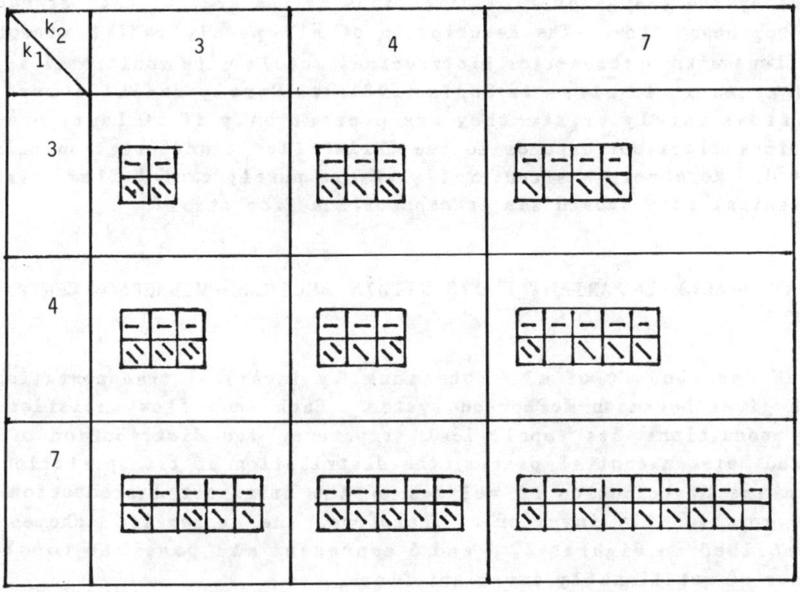

(b)

$k_1 \backslash k_2$	3	4	7
3	12	33	20
4	54	33	52
7	322	204	682

Figure 6: Enumeration of all possible structurally stable flows for two-tier hierarchical models; (a) shows the aggregated schemes, and (b) shows the number of structurally stable flows.

the aggregated scheme includes only four cells with six arcs, and each rotationally invariant tree can be generated by the rotation of only three linkages (M = N = 2, M + N - 1 = 3) corresponding to the choice of six arcs. Thus, there are twenty possible choices of three arcs, but only twelve of these choices correspond to maximal invariant trees (see Figure 5). It is interesting to note that the set of all structurally stable flows for this hierarchical model does not include purely radial flows. A complete description of the aggregated schemes and the numbers of structurally stable 'top-down' flows within three-tier models is given in Figure 6. It is possible to derive the description of optimal structurally stable 'bottom-up' flows by the change of flow directions on the arcs of each structurally stable 'top-down' flow. The description of all purely radial structurally stable flows within three-tier hierarchical models with additional axial and central symmetry is given in Sonis (1982a). Purely radial structurally stable flows rarely exist--they are present only if at least one of the hierarchical tiers corresponds to the Christaller transportation principle, with K = 4. Moreover, a structurally stable purely radial flow exists only if the central city subsidizes transportation from itself.

4. ROTATIONALLY INVARIANT FLOWS WITHIN BECKMANN-MCPHERSON CENTRAL PLACE MODELS

Consider the set of all rotationally invariant transportation flows within a given Beckmann-McPherson system. Each such flow satisfies strong symmetry conditions: its topological structure, the distribution of supply and demand between central places, the distribution of transportation costs, and also the distribution of volumes of the transported production on the arcs are rotationally invariant. Therefore, the aggregated schemes of the type described in Figures 2, 4 and 6 represent all possible topological structures of rotationally invariant flows.

In the preceding section a complete description of rotationally invariant optimal flows was given. These flows are structurally stable and correspond to the vertices of the convex bounded polyhedron, P, of all admissible transportation flows with a fixed supply and demand. This polyhedron also includes the set of all non-optimal rotationally invariant flows. Each rotationally invariant flow is decomposable into rotationally invariant subsets of suppliers and demanders, such that each such subset corresponds to some cell of the aggregated scheme. The essential fact is that for each rotationally invariant subset of suppliers (or demanders) the linear equations, generated by the corresponding rows (or columns) of the computing table, are identical. Therefore, the system of linear equations, determining the rotationally invariant flows, is reduced to a system of M + N linear equations, where M and N are the numbers of rotationally invariant subsets of suppliers and demanders that define the structure (i. e., the number and content of cells). The rank of this reduced system is equal to M + N - 1, because total demand is equal to total supply. The set of all admissible solutions of the reduced system with a fixed supply-demand constitutes the convex bounded polyhedron, $\underline{P}_r$, in an (M + N − 1)-dimensional space. The vertices of this polyhedron are the optimal solutions for the non-reduced transportation problem (i. e., they are equivalent to the vertices of the polyhedron $\underline{P}$, and represent structurally stable optimal transportation flows). Each admissible solution of the reduced system is equivalent to an admissible rotationally invariant solution of the non-reduced system. For example, consider the reduced system of linear equations defining all rotationally invariant transportation flows within a three-tier Beckmann-McPherson system (see Figure 4). Let x_0 and x_1 be the volumes of production supplied by the central city to each town, and to each village. The quantities x_2 and x_3 are the volumes of production that are exchanged between the towns, and x_4 and x_5 are the volumes of production supplied by towns to the nearest villages. The set of suppliers is divided into two rotationally invariant subsets (M = 2). The first of these subsets includes only the central city, whereas the second includes six towns. Similarly the set of demanders also is divided into two rotationally invariant subsets (N = 2). The first includes the towns, and the second includes the villages. Thus, each rotationally invariant flow is associated with a matrix

$$\underline{X} = \begin{pmatrix} x_0 & x_1 \\ x_2 & x_4 \\ x_3 & x_5 \end{pmatrix} ,$$

which satisfies the following reduced system of linear equations:

$$\begin{array}{lllllll} 6x_0 & + 6x_1 & & & & = & a_0 \\ & & x_2 & + x_3 & + x_4 + x_5 & = & a_1 \\ x_0 & & + x_2 & + x_3 & & = & b_1 \\ & x_1 & & & + x_4 + x_5 & = & b_2 \end{array} \quad . \tag{3}$$

The rank of this reduced system is equal to M + N − 1 = 3 because of the equality $a_0 + 6a_1 = 6b_1 + 6b_2$. The matrix for this system, after the last equation has been excluded, will be

$$\underline{A} = \begin{pmatrix} 6 & 6 & 0 & 0 & 0 & 0 \\ 0 & 0 & 1 & 1 & 1 & 1 \\ 1 & 0 & 1 & 1 & 0 & 0 \end{pmatrix} .$$

This matrix includes twelve invertible (basis) matrices associated with twelve structurally stable optimal flows (cf. Figure 5), namely

$$\underline{B}_1 = \underline{B}_2 = \begin{pmatrix} 6 & 6 & 0 \\ 0 & 0 & 1 \\ 1 & 0 & 1 \end{pmatrix} , \quad \underline{B}_3 = \underline{B}_4 = \begin{pmatrix} 6 & 6 & 0 \\ 0 & 0 & 1 \\ 1 & 0 & 0 \end{pmatrix} ,$$

$$\underline{B}_5 = \underline{B}_6 = \underline{B}_7 = \underline{B}_8 = \begin{pmatrix} 6 & 0 & 0 \\ 0 & 1 & 1 \\ 1 & 1 & 0 \end{pmatrix} , \quad \underline{B}_9 = \underline{B}_{10} = \underline{B}_{11} = \underline{B}_{12} = \begin{pmatrix} 6 & 0 & 0 \\ 0 & 1 & 1 \\ 0 & 1 & 0 \end{pmatrix} .$$

which give (with the help of the formula $\underline{X}_0=\underline{B}^{-1}\underline{b}$) the following twelve structurally stable flows:

$$\underline{X}_1 = \begin{pmatrix} b_1-a_1 & b_2 \\ a_1 & 0 \\ 0 & 0 \end{pmatrix}, \quad \underline{X}_2 = \begin{pmatrix} b_1-a_1 & b_2 \\ 0 & 0 \\ a_1 & 0 \end{pmatrix}, \quad \underline{X}_3 = \begin{pmatrix} b_1 & b_2-a_1 \\ 0 & a_1 \\ 0 & 0 \end{pmatrix},$$

$$\underline{X}_4 = \begin{pmatrix} b_1 & b_2-a_1 \\ 0 & 0 \\ 0 & a_1 \end{pmatrix}, \quad \underline{X}_5 = \begin{pmatrix} a_0/6 & 0 \\ a_1-b_2 & b_2 \\ 0 & 0 \end{pmatrix}, \quad \underline{X}_6 = \begin{pmatrix} a_0/6 & 0 \\ a_1-b_2 & 0 \\ 0 & b_2 \end{pmatrix},$$

$$\underline{X}_7 = \begin{pmatrix} a_0/6 & 0 \\ 0 & b_2 \\ a_1-b_2 & 0 \end{pmatrix}, \quad \underline{X}_8 = \begin{pmatrix} a_0/6 & 0 \\ 0 & 0 \\ a_1-b_2 & b_2 \end{pmatrix}, \quad \underline{X}_9 = \begin{pmatrix} 0 & a_0/6 \\ b_1 & a_1-b_1 \\ 0 & 0 \end{pmatrix},$$

$$\underline{X}_{10} = \begin{pmatrix} 0 & a_0/6 \\ b_1 & 0 \\ 0 & a_1-b_1 \end{pmatrix}, \quad \underline{X}_{11} = \begin{pmatrix} 0 & a_0/6 \\ 0 & a_1-b_1 \\ b_1 & 0 \end{pmatrix}, \quad \underline{X}_{12} = \begin{pmatrix} 0 & a_0/6 \\ 0 & 0 \\ b_1 & a_1-b_1 \end{pmatrix}.$$

These twelve flows represent all vertices of the polyhedron $\underline{P}_r$ of rotationally invariant flows within the Beckmann-McPherson three-tier hierarchical system, with nesting factors $K_1 = K_2 = 3$.

The analysis of the structure of each rotationally invariant flow within the general Beckmann-McPherson system is based on the well-known Minkovski-Caratheodory theorem about the center of gravity of the convex bounded polyhedron in a multi-dimensional space (Minkovski, 1910), which may be stated as follows:

> Each point $\underline{Y}_1$ within a convex bounded polyhedron $\underline{P}_r$ in the r-dimensional space can be represented in terms of the center of gravity of the polyhedron $\underline{P}_r$ with the weights $p_1, \ldots, p_k$, $0 \leq p_i \leq 1$, and $p_1 + \ldots + p_k = 1$, hanging on the vertices $\underline{X}_1, \ldots, \underline{X}_k$ $(k \leq r + 1)$. Moreover, the point $\underline{Y}_1$ is the convex combination
>
> $$\underline{Y}_1 = p_1\underline{X}_1 + p_2\underline{X}_2 + \cdots + p_k\underline{X}_k \quad .$$

Thus, each rotationally invariant flow within the convex polyhedron of all rotationally invariant flows can be presented as a convex superposition of a finite number of structurally stable optimal flows. This superposition is not unique, and depends upon the choice of transportation costs (i. e., it depends on the choice of the objective function). The construction of the superposition may be accomplished on the basis of the following theorem (Sonis, 1982b), which is a reformulation of the Minkovski-Caratheodory theorem:

<u>Decomposition theorem</u>: Let $\underline{Y}_1^T = (y_1,\ldots,y_r)$ be an admissible solution of the system of linear equations

$$\left.\begin{array}{lll}\underline{AX} & = & \underline{b} \\ \underline{X}^T & = & (x_1,\ \ldots,\ x_r) \geq \underline{0}^T\end{array}\right\} \quad , \tag{4}$$

[i. e., $\underline{Y}_1$ is the point within the convex r-dimensional polyhedron $\underline{P}_r$ defined by the system of equations (4)]. Let $f(x) = \underline{cX}$ be a linear objective function. Then there is a convex decomposition of $\underline{Y}_1$ such that

$$\underline{Y}_1 = p_1\underline{X}_1 + \ldots + p_s\underline{X}_s + p_{s+1}\underline{Y}_{s+1} \quad , \tag{5}$$

where each vector $\underline{X}_j$ $(j=1,2,\ldots,s)$ is an optimal solution to the optimization problem

$$\begin{array}{ll}\text{MIN:} & \underline{cX} \\ \text{s.t.:} & \underline{AX} = \underline{b} \\ & \underline{X} \geq \underline{0} \quad ,\end{array}$$

with additional zero constraints on coordinates of the vector $\underline{X}$, such that

$$x(i)_1 = x(i)_2 = x(i)_{j-1} = 0 \quad ,$$

and $\underline{Y}_{s+1}$ is an unexplored remainder. Moreover, the number s of terms in the decomposition equation (5) does not exceed the dimension r of the polyhedron $\underline{P}_r$.

The geometrical content of this theorem may be described as follows. Consider the convex polyhedron $\underline{P}_r$, and using the solution to the Linear Programming problem with respect to the objective function $\underline{cX}$ find the vertex $\underline{X}_1$. Further, construct a straight line

$$\underline{Y}(t) = \underline{X}_1 + t(\underline{Y}_1 - \underline{X}_1) \quad .$$

This line passes through the vertex $\underline{X}_1$ and the point $\underline{Y}_1$ [$\underline{Y}(0)=\underline{X}_1$, $\underline{Y}(1)=\underline{Y}_1$], and crosses the opposite face of the polyhedron $\underline{P}_r$ at the point $\underline{Y}_2$. It is known that if some point lies on the face of a bounded polyhedron, defined by a system of linear equations, then one of its coordinates is equal to zero: $x(i)_1 = 0$. It is possible to prove that point $\underline{Y}_2$ is generated by the following value of the parameter t_1:

$$t_1 = \min\,[x_k/(x_k-y_k):\ x_k \leq y_k] \quad ,$$

where x_k and y_k are the coordinates of the points $\underline{X}_1$ and $\underline{Y}_1$, and

$$\underline{Y}(t_1) = \underline{Y}_2 = \underline{X}_1 + t_1(\underline{Y}_1 - \underline{X}_1) \quad .$$

This equation yields

$$\underline{Y}_1 = p_1\underline{X}_1 + (1-p_1)\underline{Y}_2 \quad , \tag{6}$$

where

$$p_1 = 1 - 1/t_1 = \min\,(y_k/x_k\colon\; x_k \leq y_k) \quad .$$

Moreover, the place i_1 of the zero coordinate in $\underline{Y}_2$ is determined by the location of the minimum in the expression for t_1. Furthermore, the point $\underline{Y}_2$ lies on the face of the convex polyhedron $\underline{P}_r$ (i. e., in the polyhedron $\underline{P}_{r-1}$ of lesser dimension, which includes all points from $\underline{P}_r$ with a zero i_1-coordinate).

Again, then, we must find the vertex $\underline{X}_2$ within $\underline{P}_{r-1}$ that is the optimal solution to

$$\begin{array}{ll} \text{MIN:} & \underline{cX} \\ \text{s.t.:} & \underline{AX} = \underline{b} \\ & \underline{X} \geq \underline{0} \;, \quad x(i)_1 = 0 \quad . \end{array}$$

Here we must replace the cost $c(i)_1$ in the objective function $\underline{cX}$ by a very large number, and then solve the Linear Programming problem with an artificial basis (Dantzig, 1963) with respect to the same system of linear constraints. With the help of $\underline{X}_2$ we obtain the decomposition

$$\underline{Y}_2 = q_1\underline{X}_2 + (1-q_1)\underline{Y}_3 \quad ,$$

and after substituting this result into equation (6),

$$\underline{Y}_1 = p_1\underline{X}_1 + p_2\underline{X}_2 + (1-p_1-p_2)\underline{Y}_3 \quad ,$$

where $p_2 = (1-p_1)q_1$, and so on. After a finite number of such steps has been completed, we will obtain the required decomposition equation (5).

Next let us apply the above decomposition procedure to the analysis of the following rotationally invariant flow $\underline{Y}_1$ within the three-tier Beckmann-McPherson model with nesting factors $K_1 = K_2 = 3$:

$$\underline{Y}_1 = \begin{array}{cc} 200 & 100 \\ 150 & 25 \\ 50 & 25 \end{array} \quad ,$$

where the transportation costs are $c_0 = 1$, $c_1 = 3$, $c_2 = 2$, $c_3 = 7$, $c_4 = 10$ and $c_5 = 12$. The reduced system of linear equations (3) defining this flow implies that $a_0 = 1800$, $a_1 = 250$, $b_1 = 400$ and $b_2 = 150$, and the Linear Programming problem becomes

$$\begin{aligned}
\text{MIN:}\quad & \underline{cX} = x_0 + 3x_1 + 2x_2 + 7x_3 + 10x_4 + 12x_5 \\
\text{s.t.:}\quad & 6x_0 + 6x_1 = 1800 \\
& x_2 + x_3 + x_4 + x_5 = 250 \\
& x_0 + x_2 + x_3 = 400 \\
& \underline{X} = (x_0, x_1, x_2, x_3, x_4, x_5) \geq \underline{0} \quad ,
\end{aligned}$$

which has the optimal solution

$$\underline{X}_1 = \begin{pmatrix} 150 & 150 \\ 250 & 0 \\ 0 & 0 \end{pmatrix} \quad .$$

The topological structure of this flow is described in Figure 5(1), and the weight of this optimal flow within $\underline{Y}_1$ is equal to

$$p_1 = \min(100/150,\ 150/250) = 150/250 = 0.6 \quad .$$

Therefore, the actual rotationally invariant flow is the convex superposition (6) of the optimal flow $\underline{X}_1$ and the unexplored remainder $\underline{Y}_2$, namely

$$\underline{Y}_1 = 0.6\underline{X}_1 + 0.4\underline{Y}_2 \quad .$$

From this equation we get

$$\underline{Y}_2 = \begin{pmatrix} 275 & 25 \\ 0 & 62.5 \\ 125 & 62.5 \end{pmatrix} \quad ,$$

which has a zero coordinate $y_2 = 0$.

The Linear Programming problem (7), with the additional requirement that $x_2 = 0$, has the optimal solution

$$\underline{X}_2 = \begin{pmatrix} 150 & 150 \\ 0 & 0 \\ 250 & 0 \end{pmatrix} \quad ,$$

with the topological structure from Figure 5(3). After three steps we obtain the following decomposition of the actual flow $\underline{Y}_1$ into the convex superposition of structurally stable flows:

$$\underline{Y}_1 = \begin{pmatrix} 200 & 100 \\ 150 & 25 \\ 50 & 25 \end{pmatrix} = 0.6\underline{X}_1 + 0.0067\underline{X}_2 + 0.167\underline{X}_7 + 0.0167\underline{X}_8 =$$

$$0.6\begin{pmatrix} 150 & 150 \\ 250 & 0 \\ 0 & 0 \end{pmatrix} + 0.0067\begin{pmatrix} 150 & 150 \\ 0 & 0 \\ 250 & 0 \end{pmatrix} + 0.167\begin{pmatrix} 300 & 0 \\ 0 & 150 \\ 100 & 0 \end{pmatrix} + 0.0167\begin{pmatrix} 300 & 0 \\ 0 & 0 \\ 100 & 150 \end{pmatrix} \quad .$$

The first two optimal flows within this superposition represent the exchange of production between the towns and the supply of production from the central city to towns and villages. These latter two flows represent the exchange of production between towns, and supply from the central city to the towns, and from each town to the nearest village. This superposition is not unique, and can be changed with the help of a suitable choice of transportation costs.

5. SUMMARY AND FINAL COMMENTS

This paper presents a spatial description of all structurally (topologically) stable minimal cost transportation flows within a triangular network associated with the original Christaller hierarchical central place system, and within its generalization to the Beckmann-McPherson bounded central place system. Each such structurally stable flow is invariant with respect to rotations that preserve the transportation network of the central place system. Moreover, the structure of an arbitrary rotationally invariant transportation flow is ascertained. Each such flow is found to be the weighted superposition (convex combination) of structurally stable minimal cost flows.

Although there is no doubt about the conceptual usefulness of the analysis of transportation flows within a Beckmann-McPherson system, its deficiency relates to the possibilities of its application for the analysis of flows within an actual central place system. The decomposition theorem points out one such a possibility: each actual central place system can be presented as a convex superposition of its 'building blocks'--the Beckmann-McPherson bounded hierarchical models (Sonis, 1985). The rationale for this superposition may be summarized as follows. In an actual central place system there are different extreme tendencies pushing toward the actual realization of the marketing, transportation, and administrative optimization principles of market area spatial organization. Each such tendency can be completely expressed in the form of some Beckmann-McPherson hierarchical model. The weight of each Beckmann-McPherson model within the superposition represents a relative realization of the corresponding extreme tendency in the actual central place system. In other words, this weight depends upon the competition and interference between different extreme tendencies in the actual system. Moreover, each weight also gives an evaluation of the degree of development of definite spatial organization within a dynamic real-world central place system.

6. REFERENCES

Beckmann, M. and J. McPherson, 1970, City Size Distribution in a Central Place Hierarchy: An Alternative Approach, Journal of Regional Science, 10: 25-33.

Christaller, W., 1933, Die Zentralen Orte in Suddeutschland, Jena: Gustav Fischer, translated by C. Baskin (1966) as Central Places in Southern Germany, Englewood Cliffs, N. J.: Prentice Hall.

Dantzig, G., 1951, Application of the Simplex Method to a Transportation Problem, in Cowles Commission Monograph 13: Activity Analysis of Production and Allocation, edited by T. Koopmans, New York: Wiley, pp. 359-373.

_____, 1963, Linear Programming and Extensions, Princeton, N. J.: Princeton University Press.

Loesch, A., 1940, Die Raeumliche Ordnung der Wirtschaft, Jena: Gustav Fisher, translated by W. Woglom and W. Stolper (1954) as The Economics of Location, New Haven, Conn.: Yale University Press.

Minkovski, H., 1910, Geometrie der Zahlen, Leptzig-Berlin: Teubner.

Puu, T., 1980, Structural Stability and Change in Geographical Space, Environment and Planning A, 13: 979-989.

Sonis, M., 1982a, Domains of Structural Stability for Minimal Cost Discrete Flows With Reference to Hierarchical Central Place Models, Environment and Planning A, 14: 455-469.

_____, 1982b, The Decomposition Principle Versus Optimization in Regional Analysis--The Inverted Problem of Multiobjective Programming, in The Regions and the Enlargement of the European Economic Community, edited by G. Chiotis, D. Tsoukalos and H. Louri, Athens: Eptalofos, pp. 35-60.

_____, 1985, Hierarchical Structure of Central Place Systems--The Barycentric Calculus and the Decomposition Principle, Sistemi Urbani, 7: 3-28.

STOCHASTIC MIGRATION THEORY AND MIGRATORY PHASE TRANSITIONS

Wolfgang Weidlich and Gunter Haag
University of Stuttgart
Federal Republic of Germany

1. INTRODUCTION

Migration processes are an example of socio-economic dynamics of particular interest for quantitative research. The underlying motivations for the migration of individuals are relatively well defined, and must always result in a clear individual decision to maintain or to change the location in a given interval of time. In contrast, the population numbers and migratory fluxes can be measured and compared with theory. As such it is the objective of a quantitative migration theory to understand the migratory dynamics by connecting quantitatively the micro-level of motivations and decisions with the macro-level of a global migration process.

In this paper I provide a short account only of the general scope of the theory, which describes the migratory dynamics on the stochastic level (master equation), and the quasi-deterministic level (mean-value equations), as well. Then I shall focus on presenting the outlines of a dynamic model for a dramatic migratory process observed in the last century, namely the sudden formation of hugh metropolitan areas all over the world, with special reference to developing countries. This model explains this process by a 'migratory phase transition.' By a phase transition I mean—as in physics—the global change of the macroscopic properties of a system, if some control parameters of that system cross certain 'critical' values. In particular, originally stable states of the system may become unstable and transform into new dynamic modes.

2. THE MIGRATORY DECISION PROCESS AND THE EQUATIONS OF MOTION

The migratory system considered here consists of a homogeneous population of N members migrating between L regions. The 'socio-configuration' $\underline{n} = \{n_1, n_2, \ldots, n_L\}$, with n_i = the number of people in region i, where

$$\sum_{i=1}^{L} n_i = N \quad , \tag{1}$$

describes the 'state' of the system. We seek to understand the dynamics of $\underline{n}(t)$! To achieve this goal we start from the behaviour of individuals, whose decisions to migrate are governed by comparative considerations of the utility of the origin and destination areas of residences.

Therefore, let us introduce utility functions $u_i(n_i)$ for each region i as a measure for the attractivity of that region. In general, this utility can and will depend upon the number n_i of people living there. We assume that

$$u_i(n_i) = \delta_i + \eta_i n_i + \rho_i n_i^2 \quad , \tag{2}$$

where, according to equation (2), the trend parameters δ_i, η_i, ρ_i have the following meanings:

$$\left.\begin{aligned} \delta_i &= \text{the preference parameter,} \\ \eta_i &= \text{the agglomeration parameter, and} \\ \rho_i &= \text{the saturation parameter.} \end{aligned}\right\} \tag{3}$$

But our subsequent procedure deviates from that of many conventional approaches in social science. We do not maximize 'utilities' in order to determine the 'optimal state' of the system, which then is considered as the given (equilibrium) state. Instead, we build the utility functions into a dynamic framework yielding equations of motion.

The first step in this procedure is to introduce 'individual transition probabilities' $p_{ji}(u_j,u_i)$ describing migration from region i to region j. Each of these transition probabilities will depend upon the utilities u_i and u_j of its origin- and destination- area. Since p_{ji} must be a positive definite quantity, by definition, and since one has to expect that $p_{ji} > p_{ij}$, for $u_j > u_i$, the simplest and most plausible form for p_{ji} is

$$p_{ji}(u_j,u_i) = \omega \exp[u_j(n_j+1) - u_i(n_i)] \quad , \tag{4}$$

where ω is a frequency factor scaling the time regime in which the process will take place.

Next, it is easy to construct 'global transition probabilities' $w_{ji}(n_j,n_i)$ for a transition from a socio-configuration $\{n_1,\dots,n_j,\dots,n_i,\dots,n_L\}$ to a neighboring socio-configuration $\{n_1,\dots,(n_j+1),\dots,(n_i-1),\dots,n_L\}$. Assuming that the n_i individuals migrate independently, we obtain

$$\omega_{ji}(n_j,n_i) = n_i\, p_{ji}\,[u_j(n_j+1);u_i(n_i)] \quad . \tag{5}$$

Then we introduce the probability distribution

$$P(n_1,n_1,\dots,n_L;t) = P(\underline{n};t) \geq 0 \quad . \tag{6}$$

By definition, equation (6) is the probability of finding the socio-configuration $\{n_1, n_2, \ldots, n_L\}$ at time t.

The distribution $P(\underline{n};t)$ obeys an equation of motion, which is the following master equation:

$$dP(\underline{n};t)/dt = \sum_{i,j=1}^{L} [\omega_{ij}(\underline{n}^{(ji)})P(\underline{n}^{(ji)};t) - \omega_{ji}(\underline{n})P(\underline{n};t)] \quad , \quad (7)$$

with $\underline{n} = \{n_1, \ldots, n_i, \ldots, n_j, \ldots, n_L\}$,

$\underline{n}^{(ji)} = \{n_1, \ldots, n_i-1, \ldots, n_j+1, \ldots, n_L\}$, and

$\omega_{ji}(\underline{n}) = \omega n_i \exp[u_j(n_j+1) - u_i(n_i)]$.

The master equation (7) has a suggestive interpretation. The change with time of $P(\underline{n};t)$ is caused by

1. the probability flux from $\underline{n}$ to all $\underline{n}^{(ji)}$ (the last term on the right-hand side), and

2. the probability flux from all $\underline{n}^{(ji)}$ to $\underline{n}$ (the first term on the right-hand side).

The solutions of this master equation (7) are discussed by Weidlich and Haag (1983, 1984). Here we use it only to derive equations of motion for mean-values.

The mean-value of a function of $\underline{n}$ is defined by

$$\overline{f(\underline{n})} = \sum_{\underline{n}} f(\underline{n})P(\underline{n};t) \quad , \quad (8)$$

and in particular,

$$\bar{n}_i(t) = \sum_{\underline{n}} n_i P(\underline{n};t) \quad , \quad (9)$$

where the summation extends over all socio-configurations. Making use of equation (7), one obtains the equation of motion

$$d\bar{n}_k/dt = \sum_{\underline{n}} n_k \, dP(\underline{n};t)/dt = \sum_{i=1}^{L} \overline{\omega_{ki}(\underline{n})} - \sum_{j=1}^{L} \overline{\omega_{jk}(\underline{n})} \quad . \quad (10)$$

If the probability distribution $P(\underline{n};t)$ is essentially unimodal, the approximation

$$\overline{\omega_{ki}(\underline{n})} \approx \omega_{ki}(\bar{\underline{n}}) \quad (11)$$

holds, which may be inserted into equation (10) to yield

$$d\bar{n}_k/dt = \sum_{i=1}^{L} \omega_{ki}(\underline{\bar{n}}) - \sum_{j=1}^{L} \omega_{jk}(\underline{\bar{n}}) =$$

$$\sum_{i=1}^{L} \omega\bar{n}_i \exp[u_k(\bar{n}_k)-u_i(\bar{n}_i)] - \sum_{j=1}^{L} \omega\bar{n}_k \exp[u_j(\bar{n}_j)-u_k(\bar{n}_k)] \quad . \tag{12}$$

3. THE DYNAMIC MODEL FOR SUDDEN URBAN GROWTH

The preceding frame of equations, in particular equation (12), is the starting point for a dynamic model for a migratory phase transition describing instabilities of the kind that cause sudden urban growth. For details of the derivations, one should consult Weidlich and Haag (1986). The model assumptions are as simple as possible. The L regions are assumed to be primordially equivalent, which means that no natural preference exists for one region over another, and that all regional agglomeration parameters n_i coincide. This assumption implies that the following trend parameters and utility functions are assumed:

$$\delta_i = 0 \quad , \eta_i = \eta \text{ , and } u_i(n_i) = \eta n_i \quad . \tag{13}$$

Introducing the scaled variables

$$\tau = \omega t \text{ and } \Delta_i = \eta n_i \quad , \tag{14}$$

the mean-value equations (12) then assume the form

$$d\Delta_i/d\tau = F_i(\Delta_1,\ldots\ldots,\Delta_L) \quad , \tag{15}$$

with

$$F_i(\underline{\Delta}) = \exp(\Delta_i) \sum_{j=1}^{L} \Delta_j \exp(-\Delta_j) - \Delta_i \exp(-\Delta_i) \sum_{j=1}^{L} \exp(\Delta_j) \quad , \tag{16}$$

while the constraint equation (1), which is compatible with equation (15), now has the form

$$\sum_{i=1}^{L} \Delta_i = \eta N \quad . \tag{17}$$

3.1. Stationary States of the Model

The stationary states of equation (15), $\underline{\hat{\Delta}} = \{\hat{\Delta}_1,\hat{\Delta}_2,\ldots,\hat{\Delta}_L\}$, are seen to obey the equations

$$S\hat{\Delta}_i = \exp(2\hat{\Delta}_i),\ i = 1,2,\ldots,L \text{ and } S = \text{a constant.} \tag{18}$$

This equation has the following two solutions for $S > 2e$:

$$\hat{\Delta} = \hat{\Delta}_+(S) \quad \text{and} \quad \hat{\Delta} = \hat{\Delta}_-(S) \quad . \tag{19}$$

Let us assume that the first p $\hat{\Delta}_i$ take the values $\hat{\Delta}_+(S)$, and that the $q = (L-p)$ last $\hat{\Delta}_i$ take the values $\hat{\Delta}_-(S)$. This result corresponds to a stationary state of the L-region system, with p densely and q thinly populated regions, whether or not the state is stable. The agglomeration parameter η pertaining to this situation follows from equation (17), such that

$$\eta = \eta_p(S) = [p\hat{\Delta}_+(S) + q\hat{\Delta}(S)]/N \quad . \tag{20}$$

3.2. Computer Solutions of the Model

Due to its simplicity, computer solutions as well as special analytical solutions of the model can be found. In this section we discuss computer solutions. For illustrative purposes we restrict ourselves to the case of $L = 3$ in the first example. Because of the constraint equation (17), the space of variables is two-dimensional only in this case, and can be represented as the interior of a triangle.

In Figures 1, 2, and 3 we show the fluxlines of the three-region-system for a small value η_α, an intermediate value η_β and a large value $\hat{\eta}_\gamma$ of the scaled agglomeration parameter $\eta = \eta/\eta_c$, with $\eta_c = L/2N$. Even for this simple system, it turns out that the three representative values $\hat{\eta}_\alpha < \hat{\eta}_\beta < \hat{\eta}_\gamma$ of the 'control parameter' η lead to three globally different cases of dynamic behaviour. If η crosses certain critical values, a phase transition from one to the other macro-dynamic phase takes place. For small values of $\eta = \eta_\alpha$ (see Figure 1), the system has one stable stationary state, which corresponds to a homogeneous population of the regions. All non-equilibrium states evolve into this equilibrium state. For intermediate values of $\eta = \eta_\beta$ (see Figure 2) the system has 4 stable stationary states, namely the homogeneous state and three other stable states, each with one densely and two thinly populated regions. Evolution of the migratory system now depends upon initial conditions, which determine which of these 4 equilibria emerges. Finally, for large values of $\hat{\eta} = \hat{\eta}_\gamma$ (see Figure 3), still another global situation emerges. The homogeneous state now has become unstable, and only three stable equilibrium states persist, namely those with one dense and two thin regions. Again the system selected evolution trajectories to one of these three equilibria depends upon the initial conditions.

The fluxlines of the first example do not show how the system traverses them with time. Therefore we represent this evolution with time in a second example, for $L = 16$ (see Figure 4). The two systems in Figures 4a and 4b

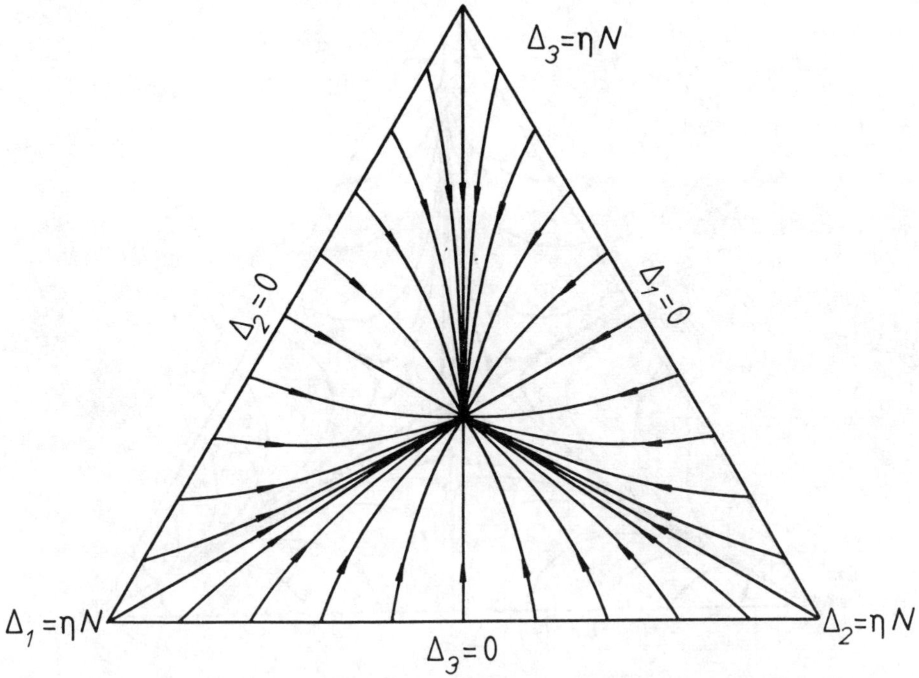

Figure 1: The fluxlines of the three-region-system for a small value η_α.

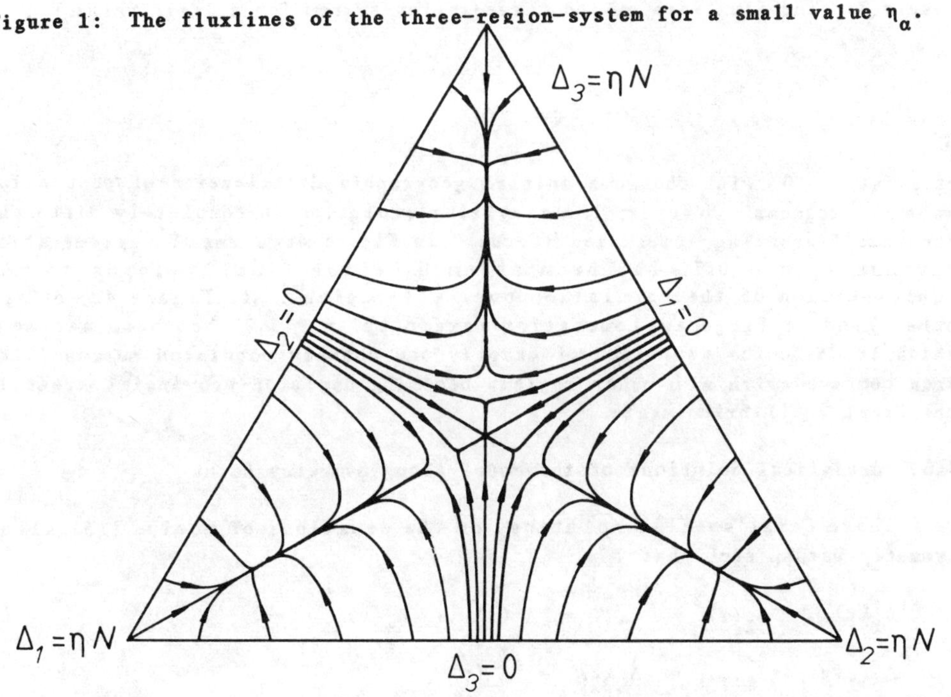

Figure 2: The fluxlines of the three-region-system for an intermediate value η_β.

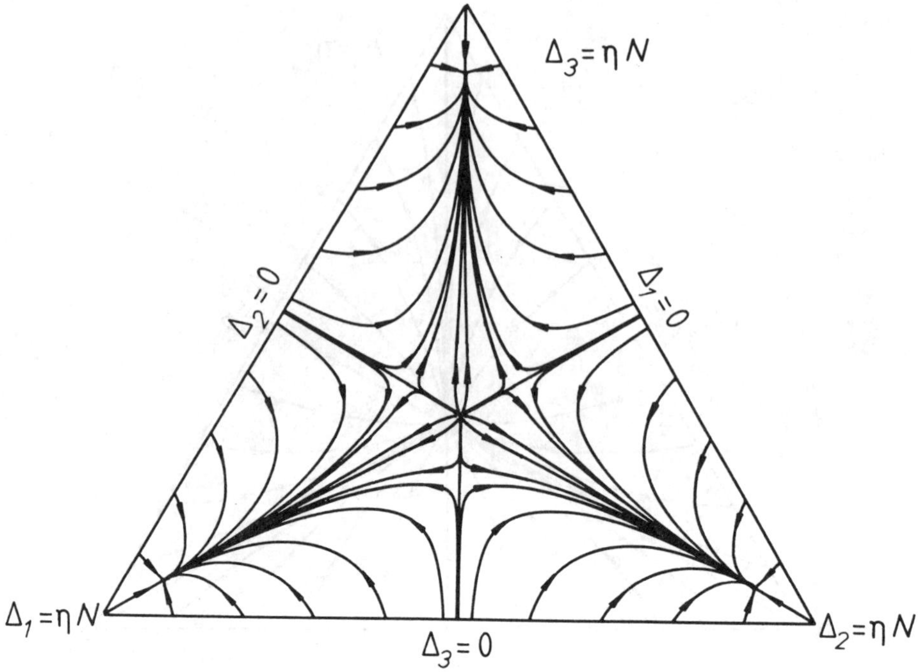

Figure 3: The fluxlines of the three-region-system for a large value η_γ.

start at t = 0, with the same initial geographic distribution of population over the regions. However, the resulting evolution is completely different due to different agglomeration trends. In Figure 4a a small agglomeration parameter, η = 0.7, has been assumed, which finally leads to the equipartition of the population over all regions. In Figure 4b, on the other hand, a large agglomeration parameter, η = 1.2, has been assumed, which leads to the evolution of exactly one densely populated metropolitan area compared with all other regions becoming depleted provincial areas in the final equilibrium state.

3.3. Analytical Solutions of the Model Along Symmetry Paths

There exist special solutions of the equations of motion (15) along symmetry paths, such that

$$\left.\begin{array}{l} \Delta_1(\tau) = \Delta_2(\tau) = \ldots = \Delta_p(\tau) = \Delta_+(\tau) \\ \Delta_{p+1}(\tau) = \ldots = \Delta_L(\tau) = \Delta_-(\tau) \end{array}\right\} , \qquad (21)$$

which can be treated analytically. An exact equation of motion can be

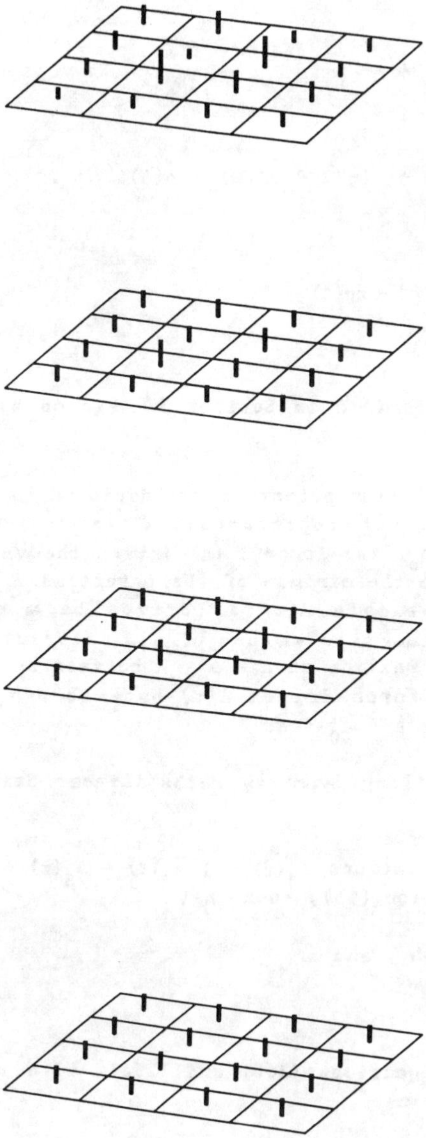

Figure 4a: Evolution over time with L=16 and η=0.7.

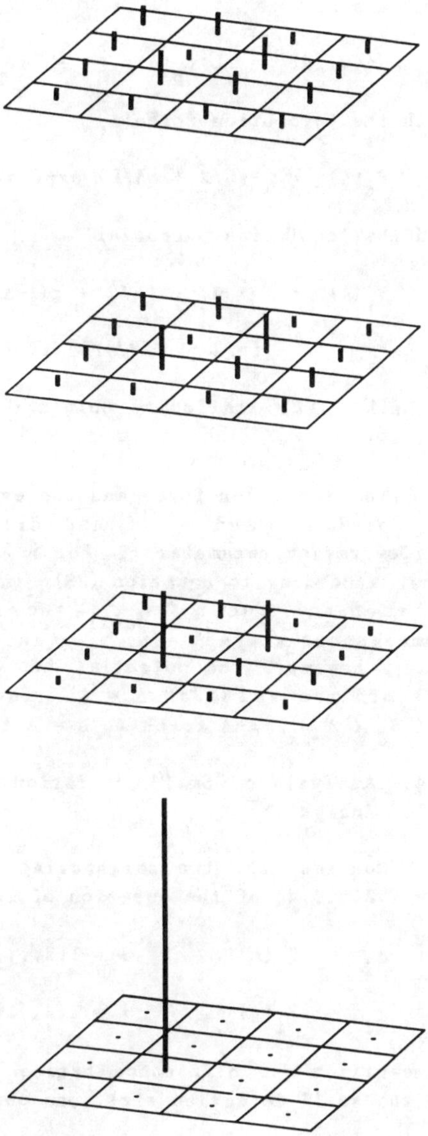

Figure 4b: Evolution over time with L=16 and η=1.2.

derived from equation (15) for the difference variable

$$\Delta(\tau) = \Delta_+(\tau) - \Delta_-(\tau) \quad . \tag{22}$$

It reads as

$$d\Delta(\tau)/d\tau = L\, f_p(\Delta) = -L\, \partial V_p(\Delta)/\partial\Delta \quad , \tag{23}$$

with the 'evolution force'

$$f_p(\Delta) = -(\hat{\eta}/2 + q\Delta/L)\exp(-\Delta) + (\hat{\eta}/2 - p\Delta/L)\exp(\Delta) \quad , \tag{24}$$

and the 'evolution potential'

$$V_p(\Delta) = (\hat{\eta}+1) - [\hat{\eta}/2 + p(1-\Delta)/L]\exp(\Delta) - [\hat{\eta}/2 + q(1+\Delta)/L]\exp(-\Delta) \quad . \tag{25}$$

[Remark: All stationary points determined in Section 3.1 lie on symmetry paths.]

The evolution force and the evolution potential are depicted in Figure 5, for $L = 6$ and $p = 1$ and different representative values of the agglomeration parameter $\hat{\eta}$. For $\hat{\eta} > \hat{\eta}_c$, the force $f_p(\Delta)$ drives the variable $\Delta(\tau)$, according to equation (23), into the minimum of the potential. On the other hand, for $\hat{\eta} < \hat{\eta}_{c(1)}$ the variable $\Delta(\tau)$ is driven back to the homogeneous state $\Delta = 0$. For intermediate values of $\hat{\eta}_i$, between $\hat{\eta}_c$ and $\hat{\eta}_{c(1)}$, however, the potential has a maximum at $\Delta = \Delta_{max}$, between $\Delta = 0$ and its minimum value at $\Delta = \hat{\Delta}$. The force drives $\Delta(\tau)$ back to $\Delta = 0$ for $0 < \Delta_0 < \Delta_{max}$, and forth to $\Delta = \hat{\Delta}$ for $\Delta < \Delta_0 < \hat{\Delta}$.

3.4. Analysis of Small Deviations Along Symmetry Paths (Linear Stability Analysis)

Consider the two neighboring solutions $\Delta_i(\tau)$,and $\tilde{\Delta}_i(\tau) = \Delta_i(\tau) + \varepsilon_i(\tau)$, $i = 1,2,\ldots,L$, of the equation of motion (15), such that

$$\dot{\Delta}_i = F_i(\underline{\Delta}) \quad , \quad i = 1,2,\ldots,L \text{ , and} \tag{26a}$$

$$\dot{\tilde{\Delta}}_i = F_i(\underline{\tilde{\Delta}}) \quad , \quad i = 1,2,\ldots,L \text{ .} \tag{26b}$$

Linearizing the difference between equations (26b) and (26a), with respect to the small deviation $\varepsilon(\tau)$, one obtains

$$\dot{\varepsilon}_i = \sum_{k=1}^{L} F_{i|k}[\underline{\Delta}(\tau)]\, \varepsilon_k \quad , \tag{27}$$

or, in concise Dirac notation,

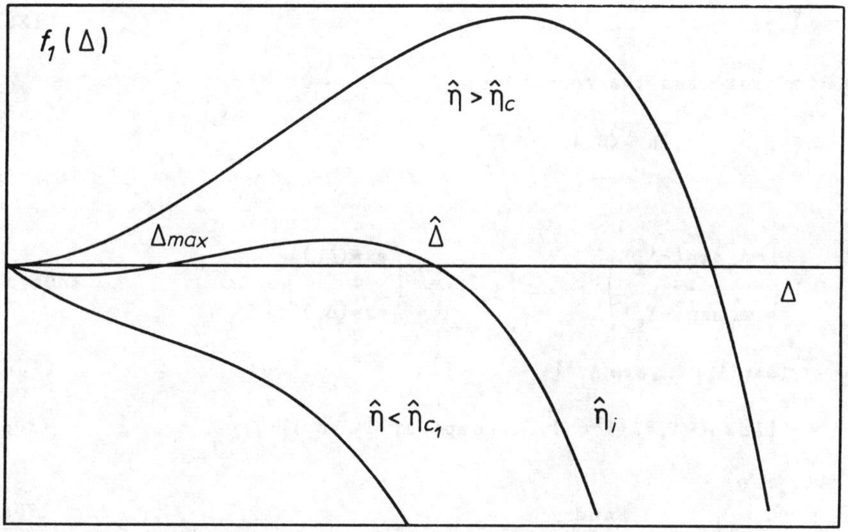

Figure 5a: The evolutionary force for L=6 and p=1.

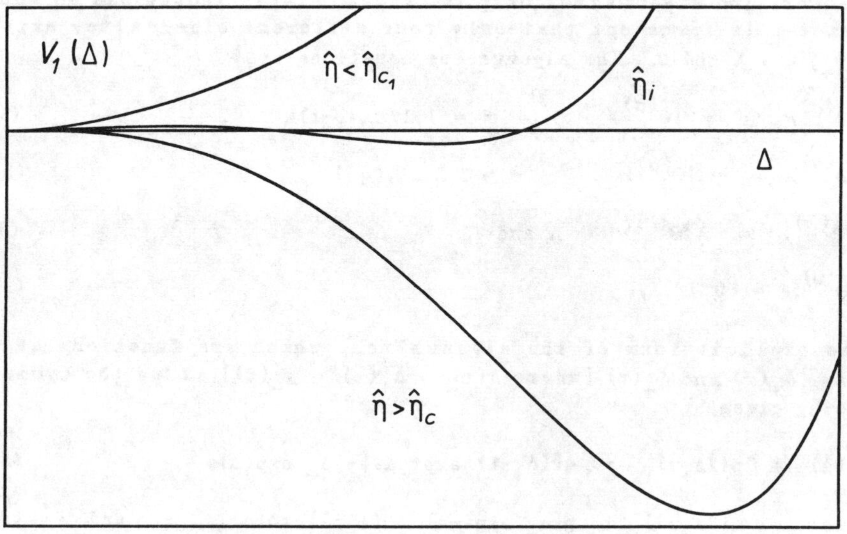

Figure 5b: The evolution potential for L=6 and p=1.

$$|\mathring{\varepsilon}\rangle = \underline{\Psi}\,|\varepsilon\rangle \quad . \tag{28}$$

The 'dynamic matrix' has the form

$$\underline{\Psi} = |\alpha_1\rangle\langle\beta_1| + |\alpha_2\rangle\langle\beta_2| + \underline{\Gamma} \quad , \tag{29}$$

with

$$|\alpha_1\rangle = \begin{pmatrix} -\Delta_1\exp(-\Delta_1) \\ \vdots \\ -\Delta_L\exp(-\Delta_L) \end{pmatrix} \quad , \quad |\alpha_2\rangle = \begin{pmatrix} \exp(\Delta_1) \\ \vdots \\ \exp(\Delta_L) \end{pmatrix} \quad , \tag{30a,b}$$

$$\langle\beta_1| = \{\exp(\Delta_1),\ldots,\exp(\Delta_L)\} \quad , \tag{30c}$$

$$\langle\beta_2| = \{[\exp(-\Delta_1)](1-\Delta_1),\ldots,[\exp(-\Delta_L)](1-\Delta_L)\} \quad , \tag{30d}$$

$$\underline{\Gamma} = \begin{pmatrix} \gamma_1 & & 0 \\ & \ddots & \\ 0 & & \gamma_L \end{pmatrix} \quad , \text{ and} \tag{30e}$$

$$\gamma_i = (\Delta_i-1)\exp(-\Delta_i) \sum_{j=1}^{L} \exp(\Delta_j) + \exp(\Delta_i) \sum_{j=1}^{L} \Delta_j\exp(-\Delta_j) \quad . \tag{30f}$$

In the following discussion assume that $\underline{\Delta}(\tau)$ is a symmetry path solution described by equation (21).

Calculation of the solution to equation (28) is facilitated if the eigenvectors and eigenvalues of $\underline{\Psi}$ are known. These indeed can be found. Furthermore, it turns out that only four different eigenvalues exists, namely γ_+, γ_-, λ and 0. The eigenvector equations are

$$\underline{\Psi}|\theta_+^{(\sigma)}\rangle = \gamma_+|\theta_+^{(\sigma)}\rangle \quad , \quad \sigma = 1,2,\ldots,(p-1) \quad , \tag{31a}$$

$$\underline{\Psi}|\theta_-^{(\sigma)}\rangle = \gamma_-|\theta_-^{(\sigma)}\rangle \quad , \quad \sigma = 1,2,\ldots,(q-1) \quad , \tag{31b}$$

$$\underline{\Psi}|\theta^{(\lambda)}\rangle = \lambda|\theta^{(\lambda)}\rangle \quad , \text{ and} \tag{31c}$$

$$\underline{\Psi}|\theta^{(0)}\rangle = 0 \quad , \tag{31d}$$

with the explicit form of the eigenvalues, which are functions of the variables $\Delta_+(\tau)$ and $\Delta_-(\tau)$ [where $\Delta(\tau) = \Delta_+(\tau) - \Delta_-(\tau)$] along the symmetry path, being given by

$$\gamma_+(\Delta) = p(2\Delta_+-1) + q[(\Delta_+-1)\exp(-\Delta) + \Delta_-\exp(\Delta)] \quad , \tag{32a}$$

$$\gamma_-(\Delta) = q(2\Delta_--1) + p[\Delta_+\exp(-\Delta) + (\Delta_--1)\exp(\Delta)] \quad , \text{ and} \tag{32b}$$

$$\lambda(\Delta) = [L\Delta_--p]\exp(\Delta) + [L\Delta_+-q]\exp(-\Delta) = -L\partial^2 V_p(\Delta)/\partial\Delta^2 \quad . \tag{32c}$$

The explicit form of the eigenvectors is not reported here. Rather, we note the following illustrative meaning of these modes. First, the eigenvectors $|\theta_+^{(\sigma)}\rangle$ describe the deviation of two dense regions (in opposite directions) from the symmetry path. There exist (p−1) linearly independent deviation modes of this kind. Second, the eigenvectors $|\theta_-^{(\sigma)}\rangle$ describe, correspondingly, deviations of two thin regions from the symmetry path. There exist (q−1) = (L−p−1) linearly independent modes of this kind. Third, the eigenvector $|\theta^{(\lambda)}\rangle$ describes a collective deviation mode of all dense and thin regions. This implies an acceleration/retardation on the symmetry path. Finally, $|\theta^{(0)}\rangle$ is a mode that cannot appear in the following expansion of $|\varepsilon\rangle$ of equation (33), since this would lead to a forbidden deviation $|\varepsilon\rangle$, with $\sum_{j=1}^{L} \varepsilon_j \neq 0$.

Let us now expand the deviation vector $|\varepsilon\rangle$ as a linear combination of the eigenvectors of the dynamic matrix $\underline{\Gamma}$:

$$|\varepsilon(\tau)\rangle = s_\lambda(\tau)|\theta^{(\lambda)}\rangle + \sum_{\sigma=1}^{p-1} s_\sigma^+(\tau)|\theta_+^{(\sigma)}\rangle + \sum_{\sigma=1}^{q-1} s_\sigma^-(\tau)|\theta_-^{(\sigma)}\rangle \,. \quad (33)$$

The omission of the mode $|\theta^{(0)}\rangle$ insures that equation (33) satisfies the constraint

$$\sum_{j=1}^{L} \varepsilon_j(\tau) = 0 \quad , \quad (34)$$

which must hold, since $\underline{\Delta}(\tau)$ and $\underline{\Delta}(\tau)$ both fulfill equation (17).

The substitution of equation (33) into (29) yields the decoupled equations of motion for the deviation amplitudes $s_\sigma^+(\tau)$, $s_\sigma^-(\tau)$ and $s_\lambda(\tau)$, and their solutions

$$\dot{s}_\sigma^\pm = \gamma_\pm(\tau)s_\sigma^\pm \quad \text{, with solution} \quad (35a)$$

$$s_\sigma^\pm(\tau) = s_\sigma^\pm(0)\exp[\Gamma_\pm(\tau)] \quad \text{, where } \Gamma_\pm(\tau) = \int_0^\tau \gamma_\pm(\tau')d\tau' \quad , \quad (35b)$$

and $\dot{s}_\lambda = \lambda(\tau)s_\lambda$, with solution (35c)

$$s_\lambda(\tau) = s_\lambda(0)\exp[\Omega(\tau)] \quad \text{, where } \Omega(\tau) = \int_0^\tau \lambda(\tau')\,d\tau' \quad . \quad (35d)$$

The stability of a symmetry path solution now depends upon the increase or decrease of the perturbation modes in $|\varepsilon(\tau)\rangle$. This behaviour, in turn, depends upon the sign of the eigenvalues $\gamma_{\pm}(\tau)$ and $\lambda(\tau)$ belonging to $s_{\pm}(\tau)$ and $s_{\lambda}(\tau)$, respectively.

In particular we are interested in the stability of the stationary points lying on the symmetry paths. Let us first consider the stationary point

$$\Delta_{+} = \Delta_{-} = \bar{\Delta} = \eta N/L \quad , \text{ where } \hat{\Delta} = \Delta_{+} - \Delta_{-} = 0 \quad , \tag{36}$$

describing the homogeneous distribution of population geographically distributed over the L regions. At this point the eigenvalues $\gamma_{\pm}(\Delta)$ and $\lambda(\Delta)$ degenerate into

$$\lambda = \gamma_{+} = \gamma_{-} = L(\hat{\eta}-1) \tag{37}$$

with $\hat{\eta} = \eta/\eta_c$, where $\eta_c = L/2N$.

Since the degenerate eigenvalue given by equation (37) changes sign if $\hat{\eta}$ crosses the value 1, we conclude that the homogeneous population of regions is stable for $\hat{\eta} < 1$ (low agglomeration trend), but unstable for $\hat{\eta} > 1$ (high agglomeration trend). In other words, a phase transition of the migratory system takes place, if $\hat{\eta}$ crosses 1 from below, rendering the originally stable homogeneous state unstable.

In order to see which of the other stationary states on symmetry paths are stable, we make use of the relation

$$\exp(\Delta_{+}-\Delta_{-}) = (\Delta_{+}/\Delta_{-})^{1/2} \quad , \tag{38}$$

which is valid for stationary states because of equation (18). The eigenvalues given by equation (32) then can be rewritten in the form

$$\gamma_{+}(\hat{\Delta}) = (p\rho_{+} + q\rho_{-})(\rho_{+} - 1/\rho_{+}) > 0 \quad , \tag{39a}$$

$$\gamma_{-}(\hat{\Delta}) = (p\rho_{+} + q\rho_{-})(\rho_{-} - 1/\rho_{-}) < 0 \quad , \text{ and} \tag{39b}$$

$$\lambda(\hat{\Delta}) = q\rho_{-}(\rho_{+} - 1/\rho_{+}) + p\rho_{+}(\rho_{-} - 1/\rho_{-}) = -L\partial^2 V_p(\hat{\Delta})/\partial\hat{\Delta}^2 \lessgtr 0 \ , \tag{39c}$$

$$\text{with } \rho_{+} = (2\hat{\Delta}_{+})^{1/2} > 1 \text{ and } \rho_{-} = (2\hat{\Delta}_{-})^{1/2} < 1 \quad , \tag{40}$$

Because $\gamma_{+}(\hat{\Delta}) > 0$, we conclude that only stationary points with $p = 1$ [one dense region, (L-1) thin regions] can be stable, since the corresponding expansion given by equation (33) does not contain the modes $s_{\sigma}^{+}(\tau)|\Theta_{+}^{\sigma}\rangle$. Furthermore, only those stationary points on the symmetry path $p = 1$ are stable that correspond to a minimum of the evolution potential [i. e., $\partial^2 V_p(\Delta)/\partial\Delta^2 > 0$, or to $\lambda < 0$ (see Figure 5)].

4. CONCLUSIONS

Summarizing the main results, our (clearly over-simplified) model leads to three main conclusions. First, starting from model equations containing one universal agglomeration parameter η only, the stationary solutions can be determined. They consist of geographical population distributions with p likewise dense and q = (L-p) likewise thin regions, where p = 0,1,2,... . One stationary state of particular interest is the homogeneous state consisting of L equally populated regions. Second, further analysis of the dynamic evolution, including a stability analysis, yields several interesting results. The homogeneous state is stable for $\eta < \eta_c$, but unstable for $\eta > \eta_c$ (in other words, if the agglomeration trend η is larger than a critical value η_c). In the latter case there arises a 'migratory phase transition,' that is the evolution towards a new global settlement pattern. Finally, this evolution can end only in one type of final stable stationary population pattern, namely that with exactly one densely populated region (to be interpreted as the metropolitan area) and the remaining (L-1) thinly populated regions (to be interpreted as provincial areas).

5. REFERENCES

Haag, G. and W. Weidlich, 1984, A Stochastic Theory of Interregional Migration, Geographical Analysis 16: 331-357.

Weidlich, W. and G. Haag, 1983, Concepts and Models of a Quantitative Sociology, Berlin: Springer-Verlag.

_____, and G. Haag, 1986, A Dynamic Phase Transition Model for Spatial Agglomeration Processes, Journal of Regional Science, in press.

SECTION 2

TRANSFORMATIONS OF URBAN SYSTEMS

Urban systems have been a major focal point of much geographic research over the decades. Urban places constitute the elements of an inter-urban system, whereas neighborhoods, distance zones, or individual parcels of land constitute the elements of an intra-urban system. Flows amongst these elements are both abstract, such as spatial competition or agglomeration forces, and concrete, such as movements of information, goods and people. And, system links include physical channels, such as roads, railroads and waterways, telecommunication networks, and relative space itself (for spillover effects). Mechanisms governing the transformation over space and through time of these urban system components still need to be distilled. Nevertheless, many of the rudimentary ingredients already have been uncovered. Latent configurations molding prevailing autoregressive influences are defined by system links. The autoregressive processes themselves arise from those influences pulsating over these links. Non-linearities are to be found in the laws of motion describing flows that travel over system links. Clearly dynamics are implicit here. A mixture of these three ingredients provides the necessary, but not sufficient, conditions for bifurcation points to materialize. Obviously studies of these ingredients will illuminate the nature of the elusive transformation mechanisms being sought, as well as eventually lead to the formulation of bona fide evolutionary geographical models.

In the first paper Huff et. al. deal with inter-urban systems, and discuss features of these systems that relate to dynamics. The primary source of autocorrelation in this case is spatial competition between neighboring urban places. Because physical space is assumed to be homogeneous and isotropic, distance decay governs autoregressive mechanisms. The purpose of these authors is to highlight shortcomings of central place theory, in an attempt to elevate it to the status of a dynamic model. Attention is called to the more restrictive assumptions underlying the theory, and possible modifications are suggested that should open the way for addressing issues concerned with structural change within a central place hierarchy. The notion of describing observed hierarchical structures as mixtures of ideal Christaller structures, together with its implications for a dynamic central place theory, is discussed first. Next the geometry of market areas is reviewed and extended. The focus here is on the fluctuations of market boundaries implied by less restrictive assumptions concerning consumer and firm behavior. The third theme of this paper has to do with the chronic balance of trade problem implied by top-down flows of goods in a central place system, and with the failure of the classical central place model to capture interdependencies between transport infrastructure and the spatial organization of an urban hierarchy. One salient conclusion reached is that the transformation of a hierarchy within the mathematical space of its possible states, and changes in the dimension of this mathematical space represent non-catastrophic as well as catastrophic discontinuities. These two types of discontinuities reflect both rapid change and locational inertia within an urban system. Clearly, then, bifurcation points are latent in this contribution to central place dynamics. Non-linearities emerge in the modified formulation of market boundaries, as well as hierarchical trade flow patterns. These transformational properties are articulated in the last section of this paper, which critically evaluates the Allen and Sanglier urban system model.

Bennett, in the second paper, developes a non-linear representation of the profit impacts brought about by local government fiscal policies. His paper addresses relationships between the spatial structure of local municipal government partitioning of economic space, the geographical pattern of local tax expenditure decisions, and the spatial process of welfare and profitability for individual or company businesses. The main goal is to formulate a non-linear model to describe the spatial process that is built upon gradients of spatial variation in the financial stimulus of business profitability. This model uncovers feedbacks of the spatial process, which constitute one of the prevailing spatial autoregressive mechanisms. One prominent finding in this paper is that high tax and low tax areas occur in close proximity to one another in the U. K. Clearly the necessary conditions are present here for bifurcation points to exist.

The third paper of this section, by Birkin and Clarke, is concerned with intra-urban systems. The primary objective of this paper is to raise some of the issues concerning the applicability and relevance of dynamic models to an understanding of and as a tool for helping to solve contemporary urban problems, and then to outline some rational responses to these issues. The authors argue that most dynamic spatial models formulated to date need to be couched in a framework that incorporates both macro- and micro-approaches to dynamics, and allows for a more detailed treatment of policy issues. They point out that the combination of macro- and micro-models within a single framework poses some difficulties, especially in relation to aggregation, while macro-micro interdependencies present complications for equilibrium theory. Notwithstanding these problematic features, a comprehensive framework is proposed that combines macro- and micro-scales. The model put forth consists of four subsystems that are captured in the supply-side structure of variables, namely the geographic distribution of industrial activities, the housing stock, the retail and service activities, and the transport network coupled with transport costs. The demand-side is represented by a demographically complemented geographic distribution of population. Both supply- and demand-side structure variables are assumed to have a slow dynamics, and form the inputs into a series of spatial interaction models (i. e., journey-to-work, journey-to-shop and movement of goods) constituting fast dynamics. In this context the micro-simulation framework furnishes an efficient means for storage and accounting. Again the non-linear and autoregressive structures are conspicuous, implying the presence of bifurcation points. The fundamental conclusion reached in this paper is that an integrated framework is feasible, even though at present many non-trivial technical problems remain unsolved.

Lombardo's paper discusses a dynamic model of urban retailing in which attention is paid not only to the choice behavior of consumers, but also the costs of developers. Time enters into the model as consumer choice changes with prevailing supply conditions as comparisons are made between centers. The actions of developers change as they respond to changes in such things as expectations and costs. Lombardo compares the model with the Harris and Wilson model through simulation experiments. The novelty of spatial forms and the important differences in transitory behavior suggest that this model represents an important generalization of earlier urban retailing models.

Consequently, these four papers illustrate to some degree the presences and roles of non-linear structures, bifurcation points, and autoregressive structures in the characterization of space-time transformations of urban systems. Their respective authors try to articulate these three concepts in an attempt to specify the nature of the transformation mechanisms in question. Considerable insight should be gleaned from these papers.

DYNAMIC CENTRAL PLACE THEORY: AN APPRAISAL AND FUTURE PROSPECTS

James Huff
University of Illinois
United States

Daniel A. Griffith
State University of New York at Buffalo
United States

Michael Sonis
Bar-Ilan University
Israel

Lloyd Leifer
State University of New York at Buffalo
United States

Debra Straussfogel
Pennsylvania State University
United States

1.0. INTRODUCTION

While central place theory suffers from a number of shortcomings, one of the most limiting is that this theory is static. It tells us nothing of the generative or the redistributive processes underlying the emergence, transformation, or stabilization of an urban hierarchy. And yet the theory stands as one of the few attempts to explicitly characterize and explain the spatial regularities in the structure of an urban system. In fact the ability to summarize the spatial structure of an urban system with a limited number of parameters promises a means of exploring relationships between structure and process implicit in dynamic, but largely aspatial, models of urban and regional growth and change.

In this paper we call attention to the more restrictive assumptions underlying classical central place theory, and suggest a series of possible modifications to the classical theory that open the way for subsequent work on a theory capable of addressing issues pertaining to structural change within the central place hierarchy. We begin with a discussion of the restrictive nature of the strict nesting relationship posited by Christaller. Sonis (1982, 1985) has been working on the problem of describing observed hierarchical structures as mixtures of ideal Christaller structures. Some of the results from his work, together with their implications for dynamic central place theory, are summarized in the next section. The discussion then turns to an analysis of the geometry of market areas. Griffith (1985) and Leifer (1985) have been studying the restrictive assumptions giving rise to the rigid hexagonal market areas characteristic of classical central place theory. Their research focuses on the fluctuation of market boundaries implied by less restrictive assumptions concerning consumer and firm behavior. Huff (1985) is concerned with the pattern of trade flows implied by the assumptions of the classical model.

The first issue is the chronic balance of trade problem implied by the top-down flow of goods in the system; and, the second issue is the failure to recognize interdependences between the structure of the transportation network and the spatial structure of the central place hierarchy. Finally, Straussfogel (1985) takes on the task of critically evaluating the Allen and Sanglier model, which is selected as a provocative example of recent advances in the modelling of urban system dynamics.

A surprising amount of effort has been expended on classical central place theory since its initial formulation, but only a small portion of this work is of relevance to the current focus on flows and structural change within the central place hierarchy. In addition, the need to summarize this literature already has been met, to a large extent, by Mulligan's (1984) excellent review article on agglomeration and central place theory. Consequently, our treatment of the central place literature is decidedly eclectic, and is narrowly focused on the work directly pertaining to the particular issue of dynamics.

2.0. SUPERPOSITIONAL CENTRAL PLACE HIERARCHIES

Central place theory, as derived from the original work of Christaller (1966) and Loesch (1954), posits a hierarchically organized system of centers, or central places, whose primary function is the distribution of goods and services to the population in the surrounding market areas, or hinterlands, dominated by these centers. Each center possesses two main sets of attributes, namely the number and type of functions provided to the surrounding hinterland, and the center's location *vis-a-vis* other central places in the system. The structure of a central place system is defined by a set of rules mapping functions $\{b_k\}$ onto locations, coupled with a set of rules defining the hinterland dominated by a central place located at (i,j), say h_{ijk}, providing function b_k. In the classical model, each function b_k defines a sub-lattice L(k) of the triangular lattice L, and each hinterland h_{ijk} is a Dirichlet region centered on a member of this sub-lattice. Christallerian structures, say C[k,m], are restricted to the set of nested sub-lattices

$$L(k^n) \text{ where } k = x^2 + xy + y^2,$$

$$x, y, n \text{ are integers, and}$$

$$1 \leq n \leq m.$$

The rigid structure implied by the classical theory has hampered efforts to come to grips with the issue of structural change within a central place hierarchy, as reflected in the transformation of the underlying geometry of the system. Sonis (1982, 1985), in particular, has been working on a solution to this problem involving the description of observed hierarchical structures as mixtures of ideal Christallerian structures. Sonis' superpositional model represents a conclusive link in a long line of research beginning with Christaller and Loesch, passing through Woldenberg (1968), Beckmann-McPherson (1970), Parr (1973, 1978, 1981), and culminating in a model of a hierarchical central place structure that allows the empirical evaluation and the incorporation of central place hierarchical dynamics. Two main principles constitute the basis of this superpositional hierarchical model: (1) the optimization principle, and (2) the superposition principle.

The optimization principle interprets Christaller's marketing, transportation and administration principles as principles of optimal organization of central place market areas; moreover, the unbounded Loesch hexagonal landscape is interpreted as a set of all possible optimal spatial organizations of market areas on one hierarchical level. Furthermore, the Beckmann-McPherson hierarchical models are interpreted as the realizations of the optimal arrangement of market areas on the final number of different hierarchical levels.

The superposition principle means that in an actual central place hierarchy the different optimal arrangements of market areas on different hierarchical levels are superimposed in the form of the simultaneous action of different extreme tendencies; each extreme tendency finds its optimal expression in the form of some Beckmann-McPherson model, and the simultaneous action of different optimal tendencies is the weighted sum (i. e., convex combination) of the corresponding Beckmann-McPherson models. From a viewpoint of the underlying scientific principles, this scheme of hierarchical analysis is an adoption and elaboration of the decomposition principle of analysis of actual states of a regional system, and, therefore, is a solution of the inverted problem of multi-objective programming [Sonis, 1982].

The proposed superpositional model consists of the representation of an actual central place hierarchy as a 'weighted sum' of the Beckmann-McPherson models, which are the best approximations of the actual hierarchy, and its deviations from the sequential cumulative sums of the best-fitting models. The main idea of the proposed hierarchical analysis is the immersion of the actual central place hierarchy into the space of all admissible central place hierarchies whose average nesting factors lie between the boundaries defined by theoretical nesting factors. Geometrically, this space is a convex bounded polyhedron--a multidimensional rectangular parallelepiped--whose vertices correspond to the Beckmann-McPherson models, and the actual central place hierarchy is a center of gravity of these models. The weight of each Beckmann-McPherson model represents the relative realization of the corresponding extreme tendency in the actual central place hierarchy.

2.1. Construction of the Superpositional Hierarchical Model

The numerical description of the actual central place hierarchy can be given by the sequence of market area frequencies

$$(m_1^o, m_2^o, \dots, m_{n-1}^o, 1) \quad ,$$

where m_j^o is the frequency of market areas from the j-th hierarchical level, and n is the number of different levels. Meanwhile, the ratios

$$k_j^o = m_j^o / m_{j+1}^o$$

are the average nesting factors. Obviously,

$$m_j^o = k_j^o k_{j+1}^o \dots k_{n-1}^o \quad .$$

Thus, the actual central place hierarchy can be equally represented by the sequence of its average nesting factors

$$(k_1^o, k_2^o, \dots, k_{n-1}^o). \tag{1}$$

When the above sequence (1) represents a Beckmann-McPherson theoretical model, the nesting factors k_j^o will be the Loesch-Dacey numbers $x^2 + y^2 + xy$, where x and y are integers.

For an evaluation of the hierarchical structure of an actual central place system, this system will be mapped into the space of all admissible central place hierarchies. In order to achieve this goal we will choose, for each hierarchical level j, the Loesch-Dacey theoretical nesting factors K_j and K_j' in such a way that the segment $[K_j, K_j']$ will include the average nesting factors k_j^o:

$$K_j \leq k_j^o \leq K_j'$$

This choice of theoretical nesting factors defines the space (convex polyhedron) of all admissible central place hierarchies. In other words, it includes all sequences of average nesting factors

$$(k_1, k_2, \dots, k_{n-1})$$

such that

$$K_j \leq k_j \leq K_j' \ ; \ j = 1,2,\dots,n-1. \tag{2}$$

The set of all positive solutions of the system of inequalities given in (2) forms (n-1)-dimensional rectangular parallelepiped whose vertices have the integer coordinates K_j and K_j'. Hence, these vertices correspond to the Beckmann-McPherson central place models. The actual central place hierarchy defined by (1) corresponds to an inner point of this polyhedron.

Next consider the slack variables

$$y_j = k_j - K_j; \ z_j = K_j' - k_j; \ j = 1,2,\dots,n-1 \quad .$$

Then each admissible central place hierarchy may be represented by means of a three-row non-negative matrix

$$X = \begin{pmatrix} k_1 & k_2 & \dots & k_{n-1} \\ y_1 & y_2 & \dots & y_{n-1} \\ z_1 & z_2 & \dots & z_{n-1} \end{pmatrix},$$

and the actual central place system $(k_1^o, k_2^o, \dots, k_{n-1}^o)$ may be represented by the matrix

$$X_o = \begin{pmatrix} k_1^o & k_2^o & \dots & k_{n-1}^o \\ k_1^o - K_1 & k_2^o - K_2 & \dots & k_{n-1}^o - K_{n-1} \\ K_1' - k_1^o & K_2' - k_2^o & \dots & K_{n-1}' - k_{n-1}^o \end{pmatrix}.$$

Accordingly, an hierarchical analysis of the actual central place system, represented by the matrix X_o, is reduced, (i. e., <u>decomposition</u> of the matrix X_o), to the weighted sum (convex combination or center of gravity) of matrices $X_1, X_2, \dots, X_{r+1}$, $r \leq n$, such that

$$X_o = p_1 X_1 + p_2 X_2 + \dots + p_{r+1} X_{r+1}, \tag{3}$$

where each matrix X_i corresponds to some Beckmann-McPherson model, and the weights p_i have the property

$$p_1 + p_2 + \dots + p_{r+1} = 1;\ 0 < p_{r+1} \leq \dots \leq p_2 \leq p_1 < 1 \quad .$$

We interpret the decomposition (3) as a superposition of substantially significant tendencies toward an optimal arrangement of market areas in the form of Beckmann-McPherson structures. These tendencies define the set of matrices X_i, and the weights p_i are the probabilities of the entry of the corresponding Beckmann-McPherson structures into the actual central place hierarchy. The most noteworthy feature is the <u>non-uniqueness</u> of the decomposition shown in equation (3), which ensues from the fundamental statement that the description of an actual state of a regional system depends upon the point of view of the investigator (Sonis, 1982). Our viewpoint is based on the ideas of the 'best approximation' of an actual central place hierarchy by the 'closest' Beckmann-McPherson model. The proposed numerical algorithm of decomposition proposed here generalizes Parr's (1978) best fitting procedure, and gives a unique decomposition (3) with the largest first weight p_1.

The best-fitting procedure may be summarized as follows:

for each hierarchical level j, the inequality

$K_j \leq k_j^o \leq K'_j$ generates the corresponding nesting factor k_j^1 from the best fitting BeckmannMcPherson model $(k_1^1, k_2^1, \ldots, k_{n-1}^1)$ such that

$$k_j^1 = \begin{cases} K_j & \text{if } k_j^o \leq (K_j+K'_j)/2 \\ K'_j & \text{if } k_j^o > (K_j+K'_j)/2 \end{cases}$$

with the corresponding three-row matrix X_1.

In this procedure the values $(K_j+K'_j)/2$ define the <u>domains of structural stability</u> of the decomposition (3). The weight p_1 of the Beckmann-McPherson model X_1 is the largest positive p_1 $(0<p_1<1)$ satisfying the property

$$X_o - p_1X_1 \geq 0 \quad ,$$

or, in coordinate form,

$$\begin{aligned} p_1 &= \max_j \{ p \mid 0<p<1,\ k_j^o-pk_j^1\geq 0,\ (k_j^o-K_j)-p(k_j^1-K_j)\geq 0, \\ &\qquad (K'_j-k_j^o) - p(K'_j-k_j^1)\geq 0\} \\ &= \max_j \{ p \mid 0<p<1,\ p\leq k_j^o/k_j^1,\ p\leq (k_j^o-K_j)/k_j^1-K_j), \\ &\qquad p\leq (K'_j-k_j^o)/(K'_j-k_j^1)\} \\ &= \min_j \{ 1,\ k_j^o/k_j^1,\ (k_j^o-K_j)/(k_j^1-K_j),\ (K'_j-k_j^o)/K'_j-k_j^1)\} \end{aligned} \tag{4}$$

The place of the components of the matrices X_o and X_1, giving the minimum in (4), represents the hierarchical level on which there exists the strongest <u>interdiction</u> to the extreme tendency corresponding to the chosen Beckmann-McPherson model, with regard to other tendencies operating in the actual central place system.

2.2. Structural Stability and Structural Change in Central Place Hierarchical Dynamics

The construction of the decomposition of an actual hierarchy, arising from the consideration of the convex polyhedron of all admissible central place hierarchies, gives the possibility of geometrically describing evolution of a central place hierarchy. The basis for dynamic hierarchical changes is the transfer of a few central places from one hierarchical level to another. These changes imply a movement of the point X_o of the actual hierarchy within the space of admissible central place hierarchies. The geometrical realization of the proposed algorithm of decomposition is given by Figure 1, which has been applied to the example of South Bothnia, Finland (see Palomaki, 1964). This diagram illustrates that the movement of the point X_o within the shaded triangle X_1FD will give a decomposition of the

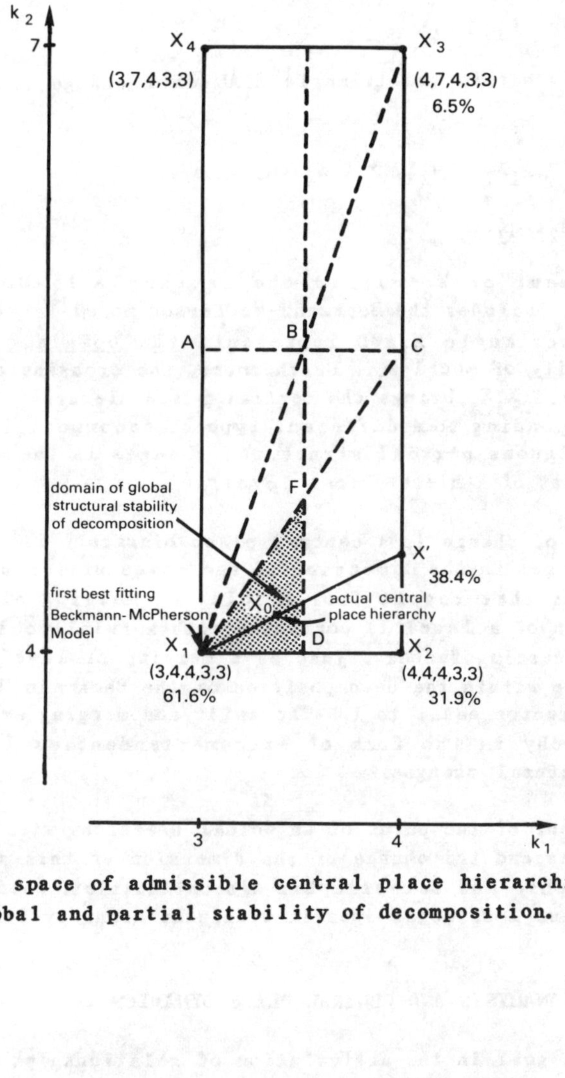

Figure 1: The space of admissible central place hierarchies and domains of global and partial stability of decomposition.

type

$$X_o = p_1X_1 + p_2X_2 + p_3X_3 \quad , \tag{5}$$

where $p_1 \geq p_2 \geq p_3$.

Therefore, this triangle represents <u>the domain of global structural stability</u> of the decomposition (5).

The movement of the actual hierarchy X_o within the triangle X_1BF will give the same decomposition (5), but it will correspond to the condition

$$p_1 \geq p_3 \geq p_2.$$

The movement of X_o within the triangle X_1AB will change the decomposition (5) to

$$X_o = p_1X_1 + q_2X_3 + q_3X_4 \quad ,$$

where $p_1 \geq q_2, q_3$.

Thus, the movement of X_o within the rectangle X_1ABD will give a decomposition that includes the Beckmann-McPherson model X_1 with the largest weight p_1. The rectangle X_1ABD represents <u>the domain of the partial structural stability</u> of model X_1. Furthermore, the crossing of the boundary of the rectangle $X_1X_2X_3X_4$ brings the central place hierarchy X_o into another rectangle corresponding to a different type of decomposition. This wide range of discontinuous partial structural changes is the consequence of modifying the extent of a hierarchical level.

Another type of change in a central place hierarchy is affiliated with <u>catastrophic</u> changes in the dimension of the space of admissible central place hierarchies that result from a split or merging of hierarchical levels. The split of a level is connected with a relative increase in the suitable average nesting factors, just as a merging of levels is connected with the appearance within the decomposition of the Beckmann-McPherson model with the nesting factor equal to 1. The split and merging are developed in an actual hierarchy in the form of extreme tendencies leading to the catastrophic structural changes.

The movement of the point of an actual hierarchy within the space of its possible states and the change of the dimension of this space represent the non-catastrophic and catastrophic discontinuities, which reflect both the rapid change and locational inertia within an urban system.

3.0. MARKET AREA ANALYSIS AND CENTRAL PLACE DYNAMICS

The prominent goal in the articulation of relationships between market boundary changes and central place dynamics is to gain insight into both latent and conspicuous patterns of price changes through space and time. The first step in achieving this goal is to understand salient implications about spatial price differentials drawn from the market area geometry literature.

3.1. Market Area Shapes

One of the first papers to deal with market area shapes under a Euclidean metric was authored by Fetter (1924), who, utilizing two markets, deduced that the indifference boundary, the locus of points where consumer costs to the two markets are equal, must trace a hyperbolic curve.

Accordingly, total consumer costs have two components, namely price of a good, and transportation expenditures calculated by straight line distance. If a price differential exists between two markets, then the indifference boundary maps out the locus of points where the cost differentials for transportation to the two markets offset the difference in price. In other words, points on the market area boundary represent all locations for which

$$rd_b - rd_a = k, \tag{6}$$

where r = unit transportation costs;

d_b = distance to market B;

d_a = distance to market A; and,

k = price differential (price at market A - price at market B).

After factoring, equation (6) becomes:

$$d_b - d_a = k/r. \tag{7}$$

Thus, locations on the indifference boundary represent points whose difference in distance to the markets is a constant (k/r). By definition, these locations must form a hyperbolic curve.

In the case of equal prices the resultant curve is the perpendicular bisector of that line connecting the two markets. As the price differential increases, the respective hyperbolic curve becomes flatter, as it bends around that market having the higher price, say A, delineating a smaller market area. If the increased price differential is exactly equal to the transportation cost between the centers, the indifference boundary becomes a straight line emanating from center A in the direction away from B, that market having the lower price. At this point, any further increase in the price differential will force A out of the geographic market.

It should be noted that only one hyperbolic curve appears as the indifference boundary, whereas equation (7) dictates that two separate curves exist. This is because equation (7) states that the difference in distance is equal to plus and minus a constant. However, in the case of market boundaries, the difference is either plus or minus a constant, but not both. This constant is the price differential that exists between the two centers. The single hyperbolic curve is bent around the center with the higher price.

In 1950, Hyson and Hyson generalized Fetter's theory by relaxing a basic assumption. Instead of positing equal transportation costs for two centers, the authors allowed for a differential in market accessibility. Therefore, they wrote equation (6) as

$$rd_b - sd_a = k, \tag{8}$$

where r and s are the unit transportation costs for travel to and from markets B and A, respectively. Dividing by r and setting h=s/r and K=k/r, equation (8) becomes

$$d_b - hd_a = K, \tag{9}$$

the equation of a DesCartes Oval. As previously described, Fetter showed that when h=1 (i. e., s=r), the DesCartes oval becomes a hyperbolic curve. If, in addition, K=0 because no price differential exists, the indifference boundary degenerates into a straight line. If, however, K=0 but the unit transportation costs vary, then the indifference boundary is a circle that surrounds that center with the greater transportation cost.

When both prices and transportation rates are allowed to vary, the resultant DesCartes oval does not degenerate to a more simple mathematical function. Instead, the indifference boundaries seem to be a combination of both hyperbolic and circular arcs. The hyperbolic shapes are more prominent for indifference points that are relatively close to the two market centers where the price differential is of greater significance than the difference in transportation cost. Conversely, indifference points form more circular arcs when they are located at significant distances away from the market centers, rendering the transportation cost differential of greater importance. It should be noted that as long as a transportation rate differential exists, no matter how insignificant, the indifference boundary will delineate a closed surface around the less accessible market center.

Shape analysis was extended to a multicenter setting by Boots, among others, with his treatment of weighted Thiessen polygons (Boots, 1980). Boots expands the traditional definition of Thiessen polygons, where all points are weighted equally, by allowing each point to have an additive (e. g., price) and a multiplicative (e. g., transportation rate) (dis)advantage. Thus, Boots' tessellation of weighted Thiessen polygons, as shown in Figure 2, can be interpreted as market areas for centers having price and transportation rate differentials.

3.2. Sources of Fluctuations in Market Boundaries

One of the basic assumptions of classical central place theory is that a given good or service is offered at a single price (the same fob price for all retailers). But this assumption is untenable given the cluster of writings describing Weberian industrial location theory. A constant fob price from retailers would hold only in the unlikely event that all inputs are ubiquitous and production is not centralized. As soon as production becomes centralized, and certainly when inputs become non-ubiquitous, the delivered price to the retailer includes assembly, production, and variable distribution costs incurred by the supplying manufacturer. If there is a single producer of the good, then price differentials may arise across the

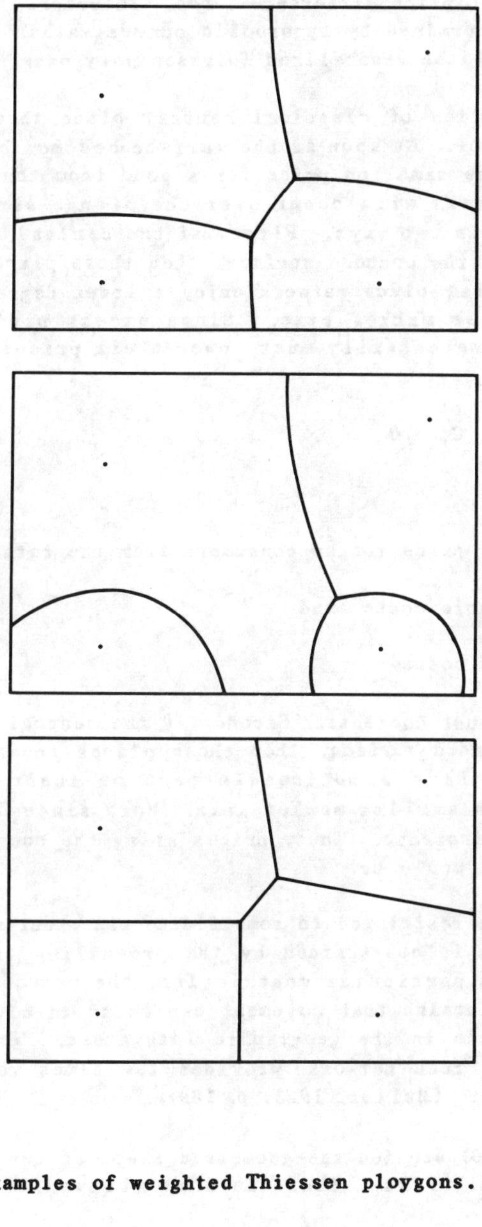

Figure 2: Examples of weighted Thiessen ploygons.

geographic landscape, and are attributable to distribution costs for the supplier. These price differentials emerge as long as the producer does not sell to retailers at a uniform delivered price. In this latter case, although the fob price to consumers from retailers is constant, non-zero location rents will materialize in much the same way that they do in the linear programming transportation problem, or Thunian agricultural location theory. If multiple producers exist, then their respective assembly costs will contribute to spatial price differences, too. Therefore, the resulting market shapes will be determined by hyperbolic curves rather than straight lines, and accordingly will be generalized Thiessen polygons.

Another basic assumption of classical central place theory posits an infinite isotropic surface. As soon as the surface becomes bounded, then even if retailers face the same fob price for a good from their suppliers, spatial price differentials will occur over the planar surface. These differentials may arise in two ways. First, if the central place system does not completely cover the bounded surface, then those places located on the periphery of the central place network enjoy a lower degree of spatial competition, and a larger market size. Since excess profits are not permitted, these places necessarily must lower their prices. Moreover, since

$$D(p_{fob} - c_v) - C_f = 0 \quad , \tag{10}$$

where D = demand,

p_{fob} = the fob price to the consumers from the retailer,

c_v = the variable costs, and

C_f = the fixed costs,

if D increases then p_{fob} must decrease. Second, if the central place system completely covers the bounded surface, then those places located along the regional boundary will have a noticeable part of their demand cone truncated, resulting in a smaller market size. Here since D in equation (10) decreases, p_{fob} must increase. Thus, prices along the boundary will be higher than they otherwise would be.

Not only is a surface restricted to some finite size, but also consumer movement over the surface is constrained by the prevailing transportation network. Because of this particular restriction, the use of a Euclidean distance metric, which maintains that movement can occur in all directions, has received much criticism in the geographic literature. Thus, physical distance that ignores the road network 'provides few clues regarding such factors as...time and cost' (Muller, 1982, p. 189).

Eaton and Lipsey (1980) studied the geometric shape of two market areas under a Manhattan metric. First, these authors allowed only prices to differ. This situation, which gives rise to hyperbolic shaped boundaries under a Euclidean metric, creates a piecewise linear indifference boundary with respect to a rectangular metric. The shape of the boundary is dictated by the minimum bounding rectangle (MBR). When the indifference boundary intersects this MBR, the boundary continues in either a strictly horizontal or vertical direction. If the intersection occurs at one of the corners of this MBR, then a two-dimensional indifference boundary, or a 'zone of indifference', is created. In the Appendix to their paper, the authors allow transportation rates to differ, too. The resultant boundary is a closed polygon, around the market center with the higher transportation rate, having either four, six, or eight sides, depending upon the significance of the price differential.

Many studies have attempted to find an appropriate distance metric to describe a road system. Frequently, these analyses have calculated a 'best fit' function for a set of nodal distances that is based on a road network. Two studies by Love and Morris (1972, 1979) fall into this category. In their first study, the authors looked at seven metrics and identified the one that was optimal for modelling intercity road distances. The results from this study showed that the three-parameter Minkowskian metric

$$k\,(|X_1 - X_2|^p + |Y_1 - Y_2|^p)^{1/s} \quad , \tag{11}$$

with parameters k for scale, p, and s, provided the best approximation to road distances. However, the more traditional Minkowskian metric,

$$k\,(|X_1 - X_2|^p + |Y_1 - Y_2|^p)^{1/p} \quad , \tag{12}$$

with only two parameters, provided a similar approximation. Therefore, the authors concluded that the two-parameter metric may be the best choice because the function is less 'expensive' to fit (Love and Morris, 1972, p. 69).

In their second study, the authors found the best distance metric in the same manner for travelling within rather than between cities. The major finding of importance of their studies to this discussion concerns the rejection of the assumption of rectangular distance with respect to intraurban travel. In fact, with the exception of one city, Euclidean distance provided a more accurate approximation than did rectangular distance. In addition, the authors concluded once again that the two-parameter Minkowskian metric (12) 'was not decidedly inferior' (Love and Morris, 1979, p. 135) to either the more general metric (11), or to the elliptical distance function that was found to be optimal.

Two studies that have incorporated an alternative distance metric into a market area analysis compared market area shapes under the Manhattan and Euclidean metrics (von Hohenbalken and West, 1984; Gambini et. al., 1967). The first of these two studies is of particular significance to this section. These authors computed market areas for a set of supermarkets under an Euclidean and a Manhattan metric. Prices at the supermarkets were allowed to vary, but accessibility remained constant over the study region. Thus, the indifference boundaries were hyperbolic curves under the Euclidean metric and piecewise linear curves under the Manhattan metric. Comparing the two sets of polygons, the authors found that although some shape and topological differences did exist, 'the market areas calculated using the Euclidean metric provide good approximations to those using the Manhattan metric' (Von Hohenbalken and West, 1984, p. 27). These findings have been substantiated in another recent study by Leifer (1985).

3.3. Selected Components of Central Place Dynamics

The preceding sections allude to some of the laws of motion that govern change in a central place system. The most obvious of these stems from alterations in transport rates, prices of raw materials, production costs, or the spatial distribution of demand. These kinds of changes may well create a reallignment of retailers and industrial producers, and in many cases will result in structural shifts that manifest themselves in changes in market shapes.

In the case of a bounded but not completely covered surface, there will be a greater propensity for expansion of the central place system to take place along the periphery than at interstitial locations, due to the lower degree of spatial competition. Once a bounded surface is covered, however, the central place system may well be in a state of disequilibrium, since those central places along the regional boundary often will be relocating in an attempt to capture a larger market. Incentives for such movements are motivated by the multiple objectives a firm tries to optimize. For instance, a retail outlet will attempt to maximize its market in order to have security for its long-run survival, as well as to maximize its aggregate profit; or the outlet will seek a market size whose affiliated demand [D in equation (10)], will yield an fob price that permits it to coexist with its spatial competitors. This sort of jumping around on the planar surface does not occur spontaneously, though, because the fixed cost term, C_f, of equation (6) is positive. This fixed cost represents an inertia in the system, and requires a central place to be committed to its location for a minimum amount of time. In fact, once $\int_0^T F(v,t)\,dt$ is expended, then a retail outlet has some incentive to operate as long as $p_{fob} - c_v \geq 0$ [see equation (6)]. In this situation, the outlet would have as one of its objectives pure cost minimization.

The presence of a spatially differentiated price surface furnishes another source of central place dynamics. Because a retailer is a profit maximizer, and hence a market share maximizer, he seeks to change his cost structure in such a way that c_v decreases. This, in turn, will allow p_{fob} to decrease if equation (10) is to hold. One way to achieve this goal is through the adoption of technological innovations. Such innovations will decrease C_f, though, and so

$$D(c_{v,old} - c_{v,new}) \geq C_{f,old} - C_{f,new} \quad ,$$

in order for the innovation to be adopted. In keeping with Scott (1976), retailers having the greatest incentive to adopt will be those enjoying the largest location rents under a uniform pricing scheme. In the more general case, those retailers having the highest p_{fob} will have the most to gain from an innovation. Again, the result of adoption will involve alterations in the prevailing geometric pattern of market boundaries.

A further source of central place dynamics comes from the mapping of bundles of goods and services onto the levels in the hierarchy. Given the cost structure prevailing in the central place landscape, equation (10) can be solved for the price term, p_{fob}. If the cost structure is infeasible, then p_{fob} lies in the complex numbers domain (see Griffith, 1984). But the corresponding goods and services still may be provided in the central place system because of competition! Such goods and services may serve, for example, as 'loss leaders.' Similarly, retailers may offer goods and services that normally map onto a higher hierarchical level, for the same sorts of reasons. This new mapping causes bundles of goods and services to compete for levels in the central place hierarchy, and will alter the allignment of market areas to central places. Meanwhile, the loss experienced by supplying these goods becomes incorporated into the c_v for other goods provided by the retailer. Increased c_vs in turn cause the corresponding p_{fob}s to increase, ultimately changing the geometric pattern of market areas.

Although studies to date focusing on central place geometry have been anything but enlightening, relations between central place geometry and dynamics suggest several issues that would be worthwhile to pursue. Market area tessellations and their geometric properties appear to be a manifestation of certain features of central place dynamics, and as such should furnish a tool for monitoring changes in the central place landscape, especially those having to do with the underlying transport network and its associated metric space. These geometric patterns also should reflect temporal variations in production costs of retailers and suppliers (due to the diffusion of innovations), the impact of regional boundaries, and competitive strategies employed by retailers that affect the hierarchical structure of central places. Similarly, an understanding of central place geometry should facilitate the execution of simulation or comparative static studies by allowing initial conditions to be set. Therefore, rather than exploring static geometric properties of central places, which apparently is both a barren and a futile endeavor, geometric properties of market shapes need to be studied whose exploitation would lead to illuminations of central place dynamics.

4.0. TRADE FLOWS WITHIN A CENTRAL PLACE HIERARCHY

Although trade is the life blood of any central place system, trade and the associated transport network are the sources of two thorny problems for anyone working with classical central place theory. In the case of trade, a chronic balance of payments problem is implied by the one-way flow of goods down the hierarchy. The transport network poses a problem because it most certainly should be a structural component of any urban system. And yet its role in Christaller's version of the theory is ambiguous at best, whereas Loesch nullifies the influence of a discrete transport network by assuming a uniform (per unit distance) transport cost surface. Only the balance of trade problem is addressed in this section, though.

4.1. A Model of Balanced Trade Within a Central Place Hierarchy

The trade model outlined below is patterned after Tinbergen's (1967) hierarchical trade model in that it is specifically designed to generate a balance of trade between exports and imports for each level of the central place hierarchy. The income and consumption accounts are balanced as well in the model just as was the case for the Tinbergen model. Finally, the population distribution implied by the distribution of income across levels in the hierarchy is shown to be a special case of the Beckmann-McPherson (1970) central place population distribution discussed earlier in this paper.

The proposed trade model differs from and is a decided improvement over Tinbergen's original effort and Mulligan's (1984) elaboration of Tinbergen's model. Tinbergen's model is quite restrictive from a trade standpoint, since a place is assumed to export only those goods that are not produced by smaller places in the hierarchy. A much more reasonable assumption forming the basis of the trade model outlined in this section is that a place can and will export some portion of all goods produced in that place. This less restrictive condition on the pattern of trade flows is consistent with the assumptions of classical central place theory.

Two sets of assumptions serve as the basis for the model. One set consists of assumptions creating the Christaller central place production and distribution system. The other set is comprised of assumptions pertaining to income and consumption patterns. These assumptions may be stated as follows.

(1) The settlement system is described by a Christaller structure $C(m,z)$ where m is the number of levels in the hierarchy and z is the nesting factor. Fundamental rural areas around each central place produce only primary goods called zero order goods. An n-th order place provides n distinct bundles of goods and it provides all goods and services (except primary goods) supplied by any lower order place, $1 \leq n \leq m$. Every central place (or rural area) of the same order or level in the hierarchy has the same

population size.

(2) There is a fixed total income, Y, in the system. Per capita income is the same everywhere. And, every individual has the same propensity to consume goods in bundle i. At each time period all income is spent, while a_i is the proportion of the income spent on goods in bundle i such that

$$\sum_{i=0}^{m} a_i = 1.$$

Given the above assumptions we can determine

(i) the income, $Y^{(n)}$, earned through the sale of bundle n goods,

(ii) the income, Y(n), earned by each level n in the hierarchy, and

(iii) the income, Y_n, earned by each central place of order n (i.e., a member of the n-th level of the hierarchy).

As indicated in Assumption 2, the income earned through the sale of bundle n goods is

$$Y^{(n)} = a_n Y.$$

The income earned by each level in the hierarchy is calculated recursively. Since the rural areas are the sole producers of order zero goods,

$$Y(0) = Y^{(0)} = a_0 Y.$$

First-order goods are supplied by all central places in the system, and $z^{m-2}(z-1)$ of the z^{m-1} central places in a C(m,z) hierarchy are first-order places. Therefore,

$$\begin{aligned} Y(1) &= W(1) + E(1) \\ &= a_1\, Y(1) + a_1\, Y(0)\, z^{m-2}(z-1)/z^{m-1} \\ &= (1-a_1)^{-1}\, a_1\, Y(0)\, (z-1)/z \quad , \end{aligned}$$

where W(1) is the value of Bundle 1 goods consumed in first-order places, and E(1) is the value of exports from first-order places to the fundamental rural areas surrounding each first-order place. The total income generated by the second level in the hierarchy is Y(2) = W(2) + E(2), such that

$$\begin{aligned} Y(2) &= (1-a_1-a_2)^{-1}\, \{a_2[Y(0) + Y(1)]\, (z-1)/z + a_1\, Y(0)\, (z-1)/z^2\} \\ &= (1-a_1-a_2)^{-1}\, \{a_2[Y(0) + Y(1)]\, (z-1) + (1-a_1)\, Y(1)\}/z \quad . \end{aligned}$$

For the n-th level in the hierarchy,

$$Y(n) = A_n[a_n \hat{Y}(n-1)\ (z-1) + Y(n-1)/A_{n-1}]/z, \quad 1<n<m \ ,$$

and for the m-th level

$$Y(m) = A_m[a_m \hat{Y}(m-1) + Y(m-1)/A_{m-1}\ (z-1)] \quad , \tag{13}$$

$$\text{where} \quad A_n = [1 - \sum_{i=1}^{n} a_1]^{-1} \quad , \text{ and}$$

$$\hat{Y}(n-1) = \sum_{j=0}^{n-1} Y(j) \quad .$$

Since per capita income is constant throughout the system, the size of the population, P(n), living in places constituting the n-th level of the hierarchy necessarily varies directly with Y(n). If P is the total population in the system and $\hat{P}(n) = \sum_{i=0}^{n} P(i)$, then it follows from (13) that

$$P(0) = a_0 P \quad ,$$

$$P(1) = A_1 a_1 P(0)\ (z-1)/z \quad , \text{ and}$$

$$P(n) = A_n[a_n \hat{P}(n-1)\ (z-1) + P(n-1)/A_{n-1}]/z, \quad n>1 \quad . \tag{14}$$

In the original Tinbergen model, each place exports only the highest order good provided by that place. Under Tinbergen's conditions,

$$\left.\begin{aligned} Y(n) &= A_n\, a_n\, \hat{Y}(n-1) \quad , \text{ and} \\ P(n) &= A_n\, a_n\, \hat{P}(n-1) \quad . \end{aligned}\right\} \tag{15}$$

Table 1 provides a comparison of the population distributions generated by Tinbergen's model as given in equation (15), and the proposed modification as given in equation (14). The parameters $\{a_i\}$ are estimated by Tinbergen (1967) from French employment data and are common to both models. The central place structure used is a Christaller C (7,4) structure. As might be expected, the two models generate very similar population distributions since the current model is a modification of the earlier model. The central place model certainly does no worse than the original Tinbergen model when predicted and observed values are compared. Both models, however, tend to underestimate the population in lower levels of the hierarchy, and to over estimate population in the higher levels (Paris being a notable exception). This pattern suggests the presence of serial correlation in the rank-size rule distribution of population.

TABLE 1

TINBERGEN AND CENTRAL PLACE POPULATION DISTRIBUTIONS AS COMPARED TO FRANCE'S ACTUAL POPULATION DISTRIBUTION[a]

Rank	a_i	Tinbergen populations ('0000)	Central Place populations ('0000)	Actual populations ('0000)
7	.027	5100.8	6748.2	7813.9
6	.023	3575.9	5040.6	2626.8
5	.096	10040.9	9559.6	4725.3
4	.178	9335.8	8140.2	5317.0
3	.216	5507.8	4556.	4615.3
2	.125	1959.7	1720.4	3671.7
1	.109	1314.9	986.2	5174.9
0	.225	10722.6	10722.6	13613.5

[a]Source: The values in the table are either taken directly from Tinbergen [1967] or are estimates derived from Tinbergen's data.

The modified Tinbergen model avoids the rather restrictive assumption that a place only exports the highest order good supplied by that place, and the predicted population distribution fits an actual population distribution as well as Tinbergen's original model. But the modification also must generate a balance of trade at each level in the hierarchy if it is to be considered as a viable alternative to Tinbergen's model. A balance of trade is achieved when the value of exports is equal to the value of imports at each level. The value of imports, Q(n), into the n-th level of the hierarchy is given by

$$\left.\begin{aligned} Q(0) &= \sum_{i=1}^{m} a_i Y(0), \\ Q(n) &= \sum_{i=n+1}^{m} a_i Y(n) + a_0 Y(n), \quad 1 \leq n < m, \text{ and} \\ Q(m) &= a_0 Y(m) \quad . \end{aligned}\right\} \tag{16}$$

The value of exports, E(n), from the n-th level of the hierarchy is given by

$$\left.\begin{aligned} E(0) &= a_0\,[\hat{Y}-Y(0)] \\ E(n) &= [a_n\, Y(n-1)\,(z-1) + Y(n-1)/A_{n-1}]/z,\ 1 \le n \le m \quad . \end{aligned}\right\} \tag{17}$$

A comparison of equations (13) and (17) indicates that

$$E(n) = Y(n)/A_n \tag{18}$$

and equation (16) implies that $Y(n) = Q(n)/(\sum_{i=n+1}^{m} a_i + a_0)$. Since

$\sum_{i=0}^{m} a_i = 1$, it follows that $A_n = [\sum_{i=n+1}^{m} a_i + a_o]^{-1}$ and $Y(n) = Q(n)\, A_n$.

Returning to equation (18), we have

$$E(n) = Q(n)A_n/A_n = Q(n),$$

thus completing the proof that a balance of trade is preserved at all levels in the hierarchy.

The discussion thus far has focused upon the properties of the levels in the hierarchy. Similar results may be derived for specific places in the system, since all places of the same order are assumed to have the same population size, and places of the same order export goods to identical hinterlands. The implication here is that the value of internally produced and consumed goods, W_i, and the value of exports, E_i, are identical for all order i central places.

In a C(m,z) hierarchy there are z^{m-i-1} (z-1) order i central places, $1 \le i < m$, one order m place, and z^{m-1} fundamental rural areas. The income generated by an n-th order place is, therefore,

$$Y_n = \begin{cases} Y(0)/z^{m-1}, & n=0 \\ Y(i)/z^{m-i-1}(z-1), & n = i,\ 1 \le i < m \\ Y(m), & n=m \quad . \end{cases}$$

A similar expression exists for P_n, the population size of an n-th order place. A balance of trade also exists for all central places since each i-th order place has the same value of exports, E_i, and the same value of imports, Q_i. Furthermore, the model generates a trade pattern that minimizes transportation costs in the sense that imports come from the nearest possible producer.

It should be noted that the exports from a fundamental rural area do not necessarily go exclusively to the nearest central place. The distribution system is such that a rural area receiving Bundle i goods from a central place will export $a_i E_0 = a_i (1-a_0) Y_0$ primary goods to that place.

Finally, the central place populations generated by the present model are shown to be directly related to the general population distribution derived by Beckmann and McPherson (1970) and the central place population described by Alao _et al_. (1977).

The income of an n-th order place has been stated as a function of the total income generated by its corresponding level in the hierarchy. It also is possible to state Y_n as a function of the value of the exports to lower order places:

$$Y_n = A_n \sum_{i=1}^{n} a_i T_i \tag{19}$$

where

$$T_i = \sum_{j=1}^{i-1} (z-1)\, z^{i-j-1} Y_j + z^{i-1} Y_0,\ 1<i<m \quad \text{and}$$

$$T_1 = Y_0 \ .$$

Since population is proportional to income, it follows from equation (19) that

$$P_n = A_n \sum_{i=1}^{n} a_i S_i \tag{20}$$

where

$$S_i = \sum_{j=1}^{i-1} (z-1)\, z^{i-j-1} P_j + z^{i-1} P_o,\ 1<i<m \quad \text{and}$$

$$S_1 = P_o \quad \text{, and}$$

$$P_0 = a_0 P/z^{m-1}$$

The proof of example 10 and example 12 in Alao _et al_. (1977, p. 212) demonstrates that the population distribution described by equation (20) is a special case of the general Beckmann-McPherson distribution as well as the population distribution described in Theorem 4 by Alao _et al_. (1977, p. 207). The above result implies that several existing central place population models are consistent with a modified Tinbergen model of the urban hierarchy under certain restrictive conditions. The conditions are that the 'multiplier' associated with each bundle i of goods in the economic base models is set equal to a_i, the propensity to consume goods in bundle i, such that

$$\sum_{i=1}^{m} a_i = 1-a_0 \quad , \text{ and } \quad a_0 = z^{m-1}P_0/P \quad .$$

Now the city size distribution implied by equation (20) may be determined, given estimates of $\{a_i; i=1,2,\dots,m\}$, the total population, P, in the system, and the structure of the urban hierarchy as expressed by the number of levels in the hierarchy, m, and the appropriate nesting factor, z. For example, if France's city system is approximated by a C(7,4) hierarchy and Tinbergen's estimates of $\{a_i; i=1,2,\dots,m\}$ are used, then the predicted city size distribution for France is as shown in Table 2.

The apparent disparity between Tinbergen's approach to the population distribution within an urban hierarchy and the approach taken by central place theorists has been resolved hopefully to the mutual benefit of both lines of enquiry. The solution is particularly attractive since it opens the way for the application of Keynesian type trade multipliers within a well defined, albeit highly stylized, spatial context.

5.0. SIMULATING CENTRAL PLACE DYNAMICS

The earlier sections of this paper have dealt with specific limitations of current theory that hamper progress toward a dynamic model, but we have only touched upon efforts to actually model the processes giving rise to changes in the distribution of goods, jobs and people within an urban hierarchy. The strategy taken in this section is first to discuss in some detail the Allen and Sanglier (1979, 1981a, 1981b) model.

5.1. The Allen and Sanglier Model

A model was developed during the late 1970s and early 1980s, by Allen, Sanglier and their colleagues at the Free University of Brussels, which seeks to go beyond the traditional models of central place systems that, for the most part, have assumed a static regional system with readily classifiable economic sectors and a population homogeneous with respect to income, demand for goods and services, mobility, and the motivation to optimize economic returns (Allen and Sanglier, 1979, 1981a, 1981b). This model approaches some of these problems in a way unique in its theoretical underpinnings. The structuring of a central place region is described as a dynamic relationship between the availability of employment, services, and amenities, and the consequent effect on the migratory pattern of the regional population, such that a hierarchy of places forms. Originally posed for physico-chemical systems by Prigogine, in the field of non-equilibrium thermodynamics (Nicolis and Prigogine, 1977; Prigogine, 1980; Prigogine and Stengers, 1984), Allen and Sanglier's model simulates regional change in the following way. Overall, the regional system experiences shifts in population as people respond to the spatial distribution of employment opportunities. This dynamic is the 'average' behavior of

TABLE 2

PREDICTED CITY SIZE DISTRIBUTION FOR FRANCE
BASED UPON THE MODIFIED TINBERGEN MODEL FOR A C(7,4) HIERARCHY[a]

Rank	a_i	# of places at each rank	predicted population - P_n
7	.027	1	6,748,200
6	.023	3	1,680,200
5	.096	12	796,633
4	.178	48	169,588
3	.216	192	23,729
2	.125	768	2,240
1	.109	3072	321
0	.225	4096	2,618

[a]Source: The values in the table are estimates derived from Tinbergen's data [1967].

consumers and entrepreneurs, which is deterministic and predictable from theories of economic behavior and spatial interaction. However, fluctuations, from either endogenous or exogenous sources with respect to the system, will affect the current stability of the system's space-time trajectory. These fluctuations bring the system to a bifurcation point, where the system's response can cause a shift to a new trajectory. At each bifurcation point, if indeed a new trajectory results, the system is said to have undergone structural change. An important point is that the current structure of a system is contingent upon the previous structure, and it in turn affects the form of a future structure. In this way, the system is said to have a history, and to be evolving toward future structures through a process of self-organizing behavior labelled by Prigogine 'order through fluctuation'. Each successive structure of the system is called a dissipative structure because its maintenance necessitates interaction with the rest of the system's environment. These dissipative structures are characteristic of an open system existing far from thermodynamic equilibrium. 'The symbol of such a situation is a tree, for which there is neither a single deterministic path nor total freedom of action for the future, but instead a branching of possibilities for evolution from wherever one starts.' [Allen, 1982,p. 98]

Two differential equations characterize the model, and are based on the following Verhulst equation, used in describing the growth of a biological population in a given locality:

$$dx/dt = bx(N-x) - mx. \tag{21}$$

The Verhulst equation describes a logistic curve for the growth of a population, x, toward a ceiling value, N, called the carrying capacity, which is the maximum population that can be sustained at a given locality. The parameters b and m are the birth and death rates of this population, respectively. Following by analogy from equation (21), Allen and Sanglier assume that the rate of change in the population at a center i is

$$dx_i/dt = bx_i[\sum_i (J_i^o + \sum_k J_i^k)(R_{ij}/\sum_j R_{ij}) - x_i] - mx_i \tag{22}$$

The carrying capacity is denoted here as total jobs, with J_i^o denoting primary and secondary sector jobs, and J_i^o denoting tertiary sector jobs that are divided into k levels or bundles. The term J_i^o does not change. Only jobs generated by the tertiary sector are assumed to change and affect the system's evolution.

Basically, equation (22) states that if there are more jobs than people at center i, then people will move to i in order to fill the vacant jobs. Total jobs are apportioned to each of the i centers according to a residential attractivity term, $R_{ij}/\sum_j R_{ij}$, which is a function of the density of a center and the ease of moving from it to other places in the system, as perceived by the resident population. The second equation of the model of regional evolution describes the way in which numbers of jobs in the tertiary sector change with respect to time, such that

$$dJ_i^k/dt = \alpha J_i^k (M_i^k - J_i^k) \quad . \tag{23}$$

This equation also describes a logistic curve and states that the number of jobs required to support each tertiary activity, k, will change as a function of the rate of entrepreneurial reaction to changes in the market demand for their goods, and M_i^k, the number of jobs that <u>could</u> be generated from production of bundle k goods in place i. Finally, a critical feedback loop is introduced into the model by assuming that M_i^k is proportional to the potential demand for bundle k goods from place i, D_i^k, such that

$$M_i^k = \eta_i^k D_i^k \tag{24}$$

where η_i^k is a constant.

The potential demand, D_i^k, is directly related to the total number of people and how much of the good they require, and is indirectly related to the price of the good in question and its transport costs. This potential demand is weighted by a consumer attractivity term, which indicates how attractive a place is perceived to be as a shopping center. In turn, this attractivity term is a function of a center's population size with respect to the threshold for a good, and the prices and transport costs of goods. Potential demand then is translated into potential jobs by the term η_i^k, and jobs are added to or subtracted from a center according to whether the current number is greater or lesser than the potential demand for these jobs.

The points to be emphasized about the model are, first, it is not strictly a deterministic model. Random fluctuations play a large role in the simulated development, and there is no assumption of a known or expected outcome. Second, the posited relationships are dynamic ones in that there are feedback mechanisms that are nonlinear in character. A third, and related point to the model's dynamic nature, is that the regional system attains no equilibrium state, only a constant evolution of structures characterized, not by quantitative increments over time, but by successive stages in a process of self-organization and self-maintenance propelled by both chance and deterministic events. The fourth important point is that the changes that emerge are qualitative and structural, not just quantitative. It is not an accounting-type model, but really a simulation model in the most literal sense. Finally, and of critical importance, is that the region modelled has a history, and the conditions of this history impinge upon and direct the future structure of the region.

Allen and Sanglier's model, and the underlying theory of dissipative structures, provide one of the first frameworks for the analysis of central places as explicitly dynamic systems. It is attractive because it appears 'to offer an improved form of representation and explanation of urban structure over time' (Couclelis, 1981, p. 15). However, it is a model in the early stages of its development, and as such, skepticism about it is warranted. As Couclelis cautions, the model's ability to offer a 'plausible form of explanation should not mislead us to think that we are about to penetrate the secrets of urban evolution.' Many claims and implicit assumptions are made about the definition and dynamics of central place systems, and these need to be critically examined.

The first claim has to do with one of the key elements of the evolutionary theory--the nature and source of the fluctuations so important to the mechanism of structural change. In the working algorithms of the model, fluctuations are instigated either by fluctuating the population values of points in a random way to simulate fluctuation from a source exogenous to the system, or by 'stopping' the simulation and changing parameter values (which is essentially re-setting the initial conditions) to simulate endogenous system fluctuations. In other words, there is no explicit mechanism for injecting fluctuation or invoking parameter (and, therefore, structural) change. This shortcoming presents an interpretive problem in identifying the underlying processes of change. Futhermore, the exogenously generated fluctuations are the only aspect of the model that allow it to be considered an 'open' model, which leads directly to the question of whether this form of <u>open model</u> is a model of an <u>open system</u>, and ultimately raises questions about whether this model adequately represents, or is an appropriate analogy to, Prigogine's original theoretical formulations.

A second, and related issue, is whether the model can in fact be considered dynamic. As it is written, the two main feedback loops are the only specifically defined constructs that give the model a dynamic character. Without direct intervention, the structural parameters remain constant, and it is possible that the system will equilibrate, a situation which is in violation of the original theory of dissipative structures.

A third set of issues is related to generating results from the model. The results to date are, for the most part, simulations on triangular lattices. The maps produced do in fact represent plausible configurations of central place-like structures. Preliminary work with non-regular lattices also produced hierarchies of points (Straussfogel, 1985). But a caution must be injected here. The model is extremely difficult to calibrate, even to small data sets. Furthermore, results are extremely sensitive to initial conditions. Hence, there is some question as to the representativeness of these results. Allen and Sanglier (1981a, 1981b) have had some success in applying the model to replicate the development of a central place system in the Bastogne region of southern Belgium. However, it is difficult to evaluate this success from the published report.

A fourth and final set of issues relates to the assumptions about the nature of the structure and dynamics of a central place system, and to the more behavioral assumptions implicit in the model. Much attention has been given in traditional central place studies to the geometry of central place systems. Although this may be the proverbial beating of a dead horse, it is perhaps not out of place to discuss briefly how Allen and Sanglier's model treats this underlying geometry. In fact, the locations of the settlements are assumed as an initial condition. Points may grow or decline, but they do not appear or disappear. That is to say, the geometry of the system is predefined, and does not change. That means, therefore, that geometric change is precluded as a form of structural change. Accordingly, structural change is defined and limited to economic and, to some extent, demographic change. The place where geometry does play a role is in the location of centers with respect to each other in the system, in terms of the economic and demographic dynamics. Yet, an explicit connection between centers, namely a transport network, also is missing.

In essence, then, structural change is defined as the redistribution of functions, jobs and people over a set of fixed centers within a bounded region. People go to jobs that are generated by central place functions, which in turn are generated by population demand. Although on the surface this dynamics mechanism is intuitively appealing, there are several factors that are either missing from the model, or are implicitly assumed. First, the supply of labor is assumed, essentially, to equal the population.

Second, the population is assumed to migrate to an indicated excess of jobs like iron filings to a magnet; and there is no time-lag between the availability of jobs and the redistribution of the population. Finally, the demographics of the model are sketchy, at best. Immigration and births, and emigration and deaths are lumped together, and are outside the main feedback loops that drive the model. Hence, although the Allen and Sanglier model may be a plausible representation of system evolution, in general, it does not in its current form incorporate any aspects of demographic or migration theory in order to make it a more robust representation of the specific type of human system of which it is meant to be a model.

Regardless of these drawbacks, the Allen and Sanglier model embraces several strong points, too. It is an attempt to place the dynamic analysis of a complex and inherently spatial system into a coherent and unified structure. The theory as propounded is a general and flexible statement about how a system of central places may be expected to evolve. For the most part, location theories have tended to exclude time, while theories of economic growth have tended to exclude a distinctly spatial element (Beaumont, 1982). The Allen and Sanglier model includes explicit spatio-temporal assumptions about central place systems. In addition, it attempts to link mico-level behavior of the system with macro-structural outcomes. The mathematics of the current version of the model are simple, thereby exemplifying rather simple assumptions about the behaviors of the actors in the system. However, the theory is sufficiently adaptable to allow alterations or additions to its mathematical formulation. Moreover, the generality and intuitive soundness of a theory of system evolution based on the processes of self-organization and order through fluctuation, and the, at least, provocativeness of the results of modelling systems in various experimental contexts constitute motivation for continued investigation.

6.0. CONCLUDING REMARKS

As this paper is organized around the on-going research interests of the authors, it is necessarily eclectic in its perspective on central place dynamics. However, there are several common themes running through our work and that of others in the general area. First, it is very difficult to deal simultaneously with changes in the structure of the urban system, and changes in the flows within that system. As Straussfogel points out, if we wish to work on problems involving changes in both structure and process, we appear to be relegated to the realm of speculative simulation models of the sort proposed by White and, Allen and Sanglier.

A less ambitious tack is one in which we continue to systematically strip away the restrictive assumptions underlying classical central place theory, thereby increasing the generality of the theory as well as its potential application as a representation of actual urban systems. Sonis' work on mixed hierarchies, and the work on generalized market areas by Griffith and Leifer, are good examples of attempts to break away from the restrictive geometric structures implied by classical central place theory. Similarly, Huff's trade model provides a means of solving the chronic balance of trade problem endemic to the urban system implied by the classical theory.

In view of the above developments, it is evident that the prospect of a dynamic central place theory is coming more clearly into focus. The pieces are beginning to fall into place. The geometry of the system at any point in time will reflect the cumulative effects of locational decisions made over a potentially long prior history. The influence of the transportation and communication networks on the evolving settlement pattern must be addressed directly. Central place geometry, as it has been defined to date, is a set of dominance relations between points on a plane (and the implied size and shape of the corresponding Thiessen polygons). Unfortunately this geometry does not identify the structure of the network or networks linking these points into a system. Christaller argued that the K=4 hierarchy as the system characterized by an efficient transportation network, but Loesch, in assuming a uniform transportation surface, effectively derailed any further work on central place transportation networks and their associated geometry.

The necessity and the importance of explicitly treating the transportation/communication network as an integral part of the central place system is particularly evident when one turns to a discussion of structural change within the hierarchy since such changes are often bound up with a transformation of the transportation system (Borchert, 1967; Janelle, 1968). Parr (1979, 1981, 1985) in particular has attempted to come to grips with the issue of structural change within a central place hierarchy--change that encompasses the transformation of the transport network. Arguing along lines similar to those of Parr, Huff (1985) shows (in a heuristic fashion) that a nested hierarchy of central places and the transport network linking places in the hierarchy can be 'grown' in tandem via a 'budding and branching' process. The evolving pattern of functionally differentiated nodes (central places) and links is shown to be a fractal having the following properties:

(a) market areas defined by the hierarchy are closed in that no market areas are truncated by the regional boundary and all central places are served by one and only one central place providing function k goods;

(b) the transport network is such that every central place is directly connected to the nearest larger central place while maximizing the number of central places per unit length of the network; and,

(c) as the system 'grows', new lower order central places emerge at a distance s/n along the transport routes emanating from each higher order central place, where s is the unit distance between central places prior to the emergence of the new interstitial nodes, and n is a measure of the trade-off between centrifugal (n small) and centripetal (n large) forces operating on the settlement process.

In addition, the central place hierarchy is evolving within a bounded region and the size and shape of the region is likely to have a profound effect upon the emerging settlement pattern. Finally, when dealing with the economic and demographic portions of the model, it is evident that spatial agglomeration and congestion must be considered in greater detail, since these phenomena appear to be the primary positive and negative feedback processes in the system.

7.0. REFERENCES

Alao, N., et al., 1977, Christaller Central Place Structures: An Introductory Statement, Evanston, Ill: Northwestern University.

Allen, P., 1982, Evolution, Modelling and Design in a Complex World, Environment and Planning B, 9: 95-111.

_____, and M. Sanglier, 1979, A Dynamic Model of Growth in a Central Place System, Geographical Analysis, 11: 256-72.

_____, 1981a, A Dynamic Model of a Central Place System II, Geographical Analysis, 13: 149-64.

_____, 1981b, Urban Evolution, Self-organization and Decision-making, Environment and Planning, A, 13: 167-183.

Beckmann, M., and J. McPherson, 1970, City Size Distribution in a Central Place Hierarchy: An Alternative Approach. Journal of Regional Science, 10: 25-33.

Beaumont, J., 1982, Towards a Conceptualization of Evolution in Environmental Systems, International Journal of Man-Machine Studies, 16: 113-145.

Boots, B., 1980, Weighted Thiessen Polygons, Economic Geography, 56: 248-257.

Borchert, J., 1967, American Metropolitan Evolution, Geographical Review, 57: 301-32.

Christaller, W., 1966, Central Places in Southern Germany, Englewood Cliffs: Prentice Hall, (translated by C. Baskin from Die Zentralen Orte in Suddeutschland, Jena: Gustav Fisher, 1933).

Couclelis, H., 1981, The Notion of Prior Structuring in Urban Modelling. paper for the Department of Civil Engineering, University of Waterloo.

Eaton, R., and R. Lipsey, 1980, The Block Metric and the Laws of Markets, Journal of Urban Economics, 7: 337-347.

Fetter, F., 1924, The Economic Law of Market a Areas, Quarterly Journal of Economics, 38: 520-529.

Gambini, R., D. Huff and G. Jenks, 1967, Geometric Properties of Market Areas, Papers of the Regional Science Association, 20: 85-92.

Griffith, D., 1984, The Construction of Central Place Structures Using Demand Cones, unpublished paper, Department of Geography, SUNY/Buffalo.

Huff, J., 1985, Dynamic Central Place Theory, unpublished manuscript, Department of Geography, University of Illinois.

Hyson, C., and W. Hyson, 1950, The Economic Law of Market Areas, Quarterly Journal of Economics, 64: 319-327.

Janelle, D., 1968, Central Place Development in a Time-space Framework, Professional Geographer, 20: 5-10.

Leifer, L., 1985, A Two-step Model for Market Area Analysis, unpublished manuscript, Department of Geography, SUNY/Buffalo.

Loesch, A., 1954, The Economics of Location, New Haven: Yale University Press (translated by W. Woglom and W. Stopler from Die Raumliche Ordnung der Wirtschaft, Jena: Gustav Fischer, 1944).

Love, R., and H. Morris, 1972, Modelling Inter-city Road Distances by Mathematical Functions, Operations Research Quarterly, 23: 61-71.

______, 1979, Mathematical Models of Road Travel Distances, Management Science, 23: 130-139.

Muller, J., 1982, Non-Euclidean Geographic Space: Mapping Functional Distances, Geographical Analysis, 14: 189-203.

Mulligan, G., 1982, Tinbergen Type Central Place Systems, International Regional Science Review, 7: 83-91.

______, 1984, Agglomeration and Central Place Theory: A Review of the Literature, International Regional Science Review, 9: 1-42.

Nicolis, G., and I. Prigogine, 1977, Self-organization in Non-equilibrium Systems, New York: Wiley.

Palomaki, M., 1964, The Functional Centers and Areas of South Bothnia, Finland, Fennia, 88: 1-235.

Parr, J., 1973, Size and Structure in the Urban System of Loesch, Economic Geography, 49: 185-212.

_____, 1978, An Alternative Model of the Central Place System, London Papers in Regional Science, 8: 31-45.

_____, 1979, Regional Economic Change and Regional Spatial Structure: Some Interrelationships, Environment and Planning, A, 11: 825-37.

_____, 1981, Temporal Change in a Central Place System, Environment and Planning, A, 13: 97-118.

_____, 1985, Square Market Areas, Loeschian Numbers and Spatial Organization, Geographical Analysis, 17: 284-301.

Prigogine, I., 1980, From Being to Becoming: Time and Complexity in the Physical Sciences, San Francisco: Freeman.

Prigogine, I., and I. Stengers, 1984, Order Out of Chaos, New York: Bantom Books.

Scott, A., 1976, Land Use and Commodity Production, Regional Science and Urban Economics, 6: 147-160.

Sonis, M., 1982, The Decomposition Principle Versus Optimization in the Regional Analysis--The Inverted Problem of Multiobjective Programming, in The Regions and the Enlargement of the European Community, edited by G. Chiotis, D. Tsoukalas and H. Louri, Athens: Eptalofos, pp. 35-60.

_____, 1985, Hierarchical Structure of Central Place System--Barycentric Calculus and the Decomposition principle, Sistemi Urbani, 1: 3-28.

Straussfogel, D., 1985, An Assessment and Critique of Allen and Sanglier's Model of the Evolution of Dynamic Central Place Systems, npublished manuscript, Department of Geography, The Pennsylvania State University.

Tinbergen, J., 1967, The Hierarchy Model of the Size Distribution of Cities, Papers of the Regional Science Association, 20: 65-68.

von Hohenbalken, B., and D. West, 1984, Manhattan Versus Euclid: Market Areas Computed and Compared, Regional Science and Urban Economics, 14: 19-35.

Woldenberg, M., 1968, Energy Flow and Spatial Order--Mixed Hexagonal Hierarchies of Central Places, Geographical Review, 58: 552-574.

NON-LINEAR REPRESENTATION OF THE PROFIT IMPACTS OF LOCAL GOVERNMENT TAX AND EXPENDITURE DECISIONS[1]

R. J. Bennett
London School of Economics
England

1. INTRODUCTION

This paper addresses the relationship between the spatial structure of local government partitioning of economic space, the spatial pattern of local tax expenditure decisions, and the spatial process of welfare and profitability for the individual or company business. Attention is restricted here to analysis of effects on business profitability. A non-linear model is developed that permits the measurement of spatial gradients of profit between locations. Then this model is estimated for the UK in 1980.

2. STRUCTURE, PATTERN AND PROCESS

This paper considers the interrelationship of local government structure, spatial patterns of spending, and processes of change of individual and business welfare. Spatial structure in this case can be viewed as the outcome of a political and administrative system that partitions economic space into a set of autonomous of quasi-autonomous jurisdictions with local decision-making councils. This definition of spatial structure is thus a constitutional one with defined geographical boundaries. However, it is also a political and legal structure involving the administrative location of responsibilities and independence of taxing, expenditure and other powers. Administrative location also may be an aspect of spatial structure; however, it, too, normally is framed in constitutional terms as a division of powers between central and local government, or federal, state and local government in federal systems. Spatial structure in this sense is a combination of geographical factors (such as those from the Christaller approach to political administrative systems), and non-spatial factors that determine the territorial division of power within the state.

Spatial pattern involves the effect of the distribution of people and economic activity on the decisions of local government. Primary attention is concentrated here on issues of public finance where decisions on taxing

[1]This research has been supported, in part, by the Anglo-German Foundation.

and expenditures are of primary concern. In a previous paper to this series of NATO Institutes (Bennett, 1984) the models that can be used to characterize local tax and expenditure decisions have been fully reviewed. They fall into six broad classes:

(i) demand models,
(ii) political models,
(iii) demand and supply models,
(iv) bureaucratic models,
(v) models based of explicit utility functions, and
(vi) dynamic elasticity models.

The models that can be developed for tax-expenditure decisions are not independent of spatial or constitutional structure, since the division of tax and expenditure responsibilities depends upon the role of central viz a viz local government (e. g., in tax raising powers, role of grants, administrative discretion). The definition of local government boundaries is also normally the legal responsibility of higher level government that can modify boundaries, taking into account spatial patterns of taxing and spending, among other factors. Spatial structure and spatial patterns are thus as strongly interrelated in studies of local government finance as they are in other fields.

Spatial process in local finance concerns (i) _in situ_ changes to welfare of individuals and economic activity, and (ii) spatial interactions and relocations as a result of financial stimuli, as well as other factors. The financial counterparts of spatial processes of trade flows, income diffusion, regional growth and decline, or spatial interaction are effects on welfare, such as benefit and tax incidence, influences on profitability, employment and wages, and stimuli to migration or industrial movement. Such processes involve aspects that are defined both within and between local government jurisdictions. Their importance will depend upon the financial _gradient_ between locations, which itself may be spatially ordered in a variety of ways. The most important forms of spatial ordering for subsequent discussion are

(i) spatial autocorrelation and neighborhood effects,
(ii) hierarchical structures between ordered centres,
(iii) urban structures, such as those between urban cores and hinterlands (e. g., urban-rural shift), and
(iv) large scale regional change (e. g. regional shift).

The spatial ordering requires a coding of neighbourhoods and potential spatial interrelations of alternative inter-site dependence.

The main purpose of this paper is to develop a model to describe the spatial process underlain by the gradient of spatial variation in financial stimuli, in this case applied to business profitability. This model then is used as a basis to begin to explore the feedback affects of spatial process on spatial pattern by modification in marginal tax base and hence tax rate settings.

3. MEASURING THE SPATIAL GRADIENT

The measurement of the gradient of financial stimuli for either business relocation or profit differences is developed below in terms of the fiscal wedge between pre-tax and post-tax rates of return on investments judged at the margin. The approach adopted here follows a similar 'hypothetical project' method to that developed by King and Fullerton (1984). As far as possible results are kept comparable with King and Fullerton, but considerable development is required for assessment of local taxation systems.

The fiscal wedge measures the effect of the tax system in modifying the real rate of return to capital investment. It is defined by

$$w = p - s \quad , \tag{1}$$

where the wedge w is the difference between the pre-tax real rate of return (net of book depreciation) p, and the post-tax real rate of returns. In both cases the returns are considered to relate to a marginal increment of one unit of investment. The size of this wedge is likely to vary considerably between industries, different types of assets, different sources of finance, different categories of investors, and different localities. To determine macro-economic effects King and Fullerton aggregate a series of 'hypothetical projects', where the marginal investment is considered to be one possible combination of assets, industry, financial sources, and investors. The following three categories of each possibility are examined:

(1) <u>assets</u> - machinery (including plant, equipment and vehicles),
buildings,
inventories;

(2) <u>industry</u> - manufacturing,
commerce (non-financial services, distribution),
other (excluding extractive industry);

(3) <u>financial sources</u> - debt (banks and bank borrowing),
new share issues,
retentions; and,

(4) <u>investors</u> - households (including mutual funds and banks),
tax exempt institutions (mainly pension funds),
non-exempt institutions (mainly life insurance companies).

This taxonomy allows 81 possible combinations to be estimated for a given project. It also allows aggregate estimates by combining the 81 combinations, taking a weighted average, with weights related to the existing distribution of capital between assets, industries and investors. Thus the macro-economic average fiscal wedge is defined as

$$w = \sum_{k=1}^{81} (p_k - s_k)\alpha_k \quad , \tag{2}$$

where k denotes the project, and α_k is a stock weight relating to asset-industry-investor distribution (subject to $\Sigma\,\alpha_k = 1$). It also is possible to develop sums over subsets of industry, assets, financial sources, and investors. This analysis is restricted to non-extractive and non-financial business. No international tax effects are analyzed; no public sector expenditures or investments are included; all tax are assumed known and to be constant in the future from the base year of data chosen for analysis, businesses are assumed to respond with certainty, inflation is assumed uniform across the economy, and all relevant tax allowances are assumed to be claimed.

The hypothetical project approach outlined above allows the effective tax rate to be computed from the tax wedge in equation (1). Two measures of tax rates are normally employed. The first is the 'gross tax rate', or 'tax exclusive' measure:

$$t_g = (p - s)/p \quad , \tag{3}$$

as developed by the Meade Committee (1978) and Hall (1981). The second effective tax rate is the 'net tax rate' or 'tax inclusive' measure:

$$t_n = (p - s)/s \quad . \tag{4}$$

These two measures are identical in that they both include the fiscal wedge as the numerator, but differ as to whether the denominator includes or excludes tax effects.

It should be noted that both measures are non-linear functions of p or s, which vary simultaneously. Hence the behavior of these equations is complex. Bradford and Fullerton (1981) note various problems that may arise, with two of these being relevant to the present analysis. First these measures are sensitive to the interest rate r employed (i. e., they are sensitive to the time preference for income). Generally different interest rates and hence effective tax rates will be faced by different projects. For an individual investment project a unique solution can be found yielding a single p and s (or r). But in the aggregated case [equation (2)] we are concerned with finding a solution over a wide range of schedules for p and s. One way to solve this problem is to assume a fixed value of either p or s (i. e., either allow r to vary between projects and hold p constant, or fix r and evaluate the effect of variable p). This perspective results in the following two cases:

(i) fixed-p case--p is held constant and the effect of variable r calculated through its impact on s; and,

(ii) fixed-r case--a common value of r is set for all projects (i. e., assume constant post-tax rate of return), and hence all projects have the same s and all sources of finance are treated similarly.

The real-world interpretation of this latter case is that 'arbitrage' between investment projects and sources of finance reduce all returns to the same point. This interpretation can only be upheld in specific circumstances.

The second problem encountered here is that it is possible in the calculation of an effective tax rate that p becomes zero or negative. In such cases these measures of the effective tax rate becomes unstable. The situation in which p has these properties arises when the tax system allows credits on accelerated depreciation that equal or outweigh the corporate tax effects. In such cases the rates of return required from the investment will be zero or negative, and all returns to sources are received from tax subsidies. In this case the gross tax rate, t_g, is undefined, and the net tax is -100%.

It also is possible for s to approach zero. This situation can arise where investments produce very low rates of return that are cancelled completely by the tax system. In the case where either p or s becomes zero or negative, it is only possible to report the fiscal wedge (p - s). Bradford and Fullerton (1981, p. 269) interpret this circumstance as a 'property tax' rate; the percentage of asset value paid in tax each year on a unit investment. The fiscal wedge provides a comparable measure to the effective tax rate, which depends upon the interest rate, the inflation rate and the tax system. Thus it is normal to quote w as well as t_g and t_n in empirical analysis.

3.1. Post-tax Real Rate of Return

The evaluation of post-tax returns involves an assumption about who is the business decision maker, and hence who is affected by the tax system. The assumption made is that the returns to debt-holders are the base for assessment. Accepting this position, the post-tax real rates of return depend upon personal income tax and personal wealth tax. This dependency leads to defining s in equation (1) as

$$s = (1 - m)(r + \pi) - \pi \quad , \tag{5}$$

where m is the marginal income tax rate on interest income, and r is the real rate of interest in the market after allowing for inflation rate π and the nominal market rate of interest i; that is, $r = i - \pi$. This expression permits evaluation of the returns to the saver after allowance for tax effects.

3.2. Pre-tax Real Rate of Return

The concern is to evaluate the minimum level of return necessary to support an investment project under a given regime of non-domestic rates expressed for the marginal unit increase in investment as

$$p = MRR - \delta \quad , \tag{6}$$

where MRR is the gross marginal rate of return, and δ is the book value of depreciation. This depreciation figure will be assumed to be constant over industries and over time. It differs from depreciation allowed for tax purposes (which will be discussed later), and is assumed to be $\delta = 2/L$, where L is the average asset life of industry derived from national incomes accounts.

The approach to be followed here estimates investment worth as the value of future cash flows derived from a project discounted to present values (NPV) calculated by discounted cash flow analysis. A calculation is made of the minimum level of cash flows necessary to support a project in the pre-tax and post-tax situations. This cash flow is dependent upon the firm's time preference for income (the rate at which it discounts the value of future earnings), defined as ρ, which will depend on the market rate of interest, the rate of inflation π, and other factors. It is assumed that ρ is exponentially declining with time, (i. e., present earnings are valued more favorably than future earnings, in a geometrical series). This is the standard approach of macro-economic investment appraisal (see, inter alia, Lund, 1976; Hall and Jorgenson, 1967; Jorgenson, 1963, 1967; Nickell, 1978) and is useful for simplification purposes. If depreciation also is assumed to be taken as exponential (according to the declining-balance method), the discounted presented value V is the sum of future benefits given by

$$\begin{aligned} V &= \int_0^{\infty} (1-\tau)\ MRR\ \exp\{[-(\rho+\delta-\pi)]u\}du \qquad (7) \\ &= (1-\tau)MRR/(\rho + \delta - \pi) \quad , \end{aligned}$$

where τ is the rate of corporation tax and the integral equation has been solved subject to $(\rho + \delta - \pi) > 0$ (which is not restrictive, provided that the inflation rate is not high). MRR, as defined in equation (6), can be substituted into equation (7) to give

$$V = (1 - \tau)(p + \delta)/(\rho + \delta - \pi) \quad .$$

Thus the pre-tax rate of return necessary is:

$$p = V(\rho + \delta - \pi)/(1 - \tau) - \delta \quad , \quad (8)$$

This last equation can be solved for p by setting V to unity, (i. e., by assuming that marginal rates of return are evaluated on a unit investment).

The general method of evaluating the pre-tax rate of return p given in equation (8) now needs to be expanded to take into account the full range of complexities of the taxes on capital. Specifically, the following seven features are discussed below: (i) standard depreciation of business assets for tax purposes, (ii) accelerated depreciation and other forms of capital allowances, (iii) wealth and capital gains taxes, (iv) valuation of inventories, (v) the non-domestic rates, (vi) forms of investment finance (e. g., equity shares, retentions), and (vii) ownership of returns (households, tax exempt institutions and non-exempt institutions).

The tax base for non-domestic rates is an assessment of the net annual value of the property based upon an historical valuation v_t^k at time t. This valuation is not usually kept fully up to date. There are differential levies on land, property and equipment components, and each locality levies a different rate poundage. Lack of sufficient data on distribution of assets prevents a full assessment of these effects, but tax rate and valuation data can be included directly in the calculation of the NPV of necessary pre-tax rates of returns. The corporation tax rate is defined by τ. Let A be the level of capital allowances against corporation tax, v the proportion of inventories taxed on historical cost principles, d_2 the proportion of inventories in total assets, d_1 a dummy variable equalling one where the rates are deductible from corporation tax and zero otherwise, τ_{pi}^k the local rate poundage in locality i on assets of type k, and v_t^k the proportion of taxable property assets of type k in a given investment. Rates of return on business are assumed to be separate from any tax on wealth. Consequently, the resulting model for the UK tax system inclusive of local non-domestic rates is given by

$$\rho_i = (1-A)[(\rho+\delta-\pi) + (1-d_1\tau)\tau_{pi}^k v_t^k + d_2\tau v\pi]/(1-\tau) - \delta \quad , \quad (9)$$

This result is a measure of the variation of rates of return by locality that result from non-domestic rates. The expression includes deductibility of property tax from corporation tax bills.

The expession for pre-tax rates of return assumes that the rates act as a tax on improvements or increases in wealth (i. e., they raise the user costs of capital). As a result, changes in non-domestic rate poundage are equivalent to a change in the level of interest rates (Jorgenson, 1968; Bischoff, 1969). In a wider analysis the effect of property taxes will depend on the elasticity of demand for capital and for particular types of property that normally will vary very considerably between different spatial markets.

3.3. The Spatial Gradient and Process Feedback

The profit gradient now can be specified either in terms of the effective tax rates given by the tax inclusive measure t_g or the tax exclusive measure t_n. Alternatively the fiscal wedge w = p − s can be used as the gradient measure. Without loss of generality one can restrict attention to the fiscal wedge, which may be defined as w_i for local government i. The spatial gradient now can be defined in terms of w_i and w_j as

$$D = w_i - w_j = f(t_i^1, t_j^1, t_i^2, t_j^2, \ldots) \quad , \qquad (10)$$

where t_i^k are the k tax rates for each tax base (in a multi-tax system) at location i. It is assumed that once the fiscal gradient D exceeds a threshold of renewal and upheaval costs α that potentially a movement of investment will take place between i and j, i. e.,

$$B_i^k - B_j^k = f'(D) \quad , \text{ s. t.: } w_i - w_j \geq \alpha \quad , \qquad (11)$$

where B_i^k is the total base for tax k in locality i. Thus it is possible to obtain an explicit expression for tax base modification as a result of relocated investment. This is an expression for the influence of spatial process or spatial pattern. In turn tax base change will influence tax rate setting, i. e.,

$$t_i = f''(B_i^1, B_i^2, \ldots) \quad , \qquad (12)$$

thus completing the full feedback loop to further modify tax wedges. This full feedback process is explored in subsequent research. For the present, attention is restricted to analysis of the spatial gradient of the fiscal wedge and effective tax rates.

4. EMPIRICAL ASSESSMENT OF THE SPATIAL GRADIENT

The example of the spatial gradient assessed here is for the UK local authorities in 1980. Actual inflation rates of 13.6% are assumed and the only local tax to be considered is the property tax or 'rates.' Only the maximum spatial gradient is quantified in the first instance. This is derived from measuring the national range of rate poundages, which in 1980/1 was from 80.0 to 176.52 pence to the pound, a ratio on 1:2.21. The average for England and Wales as a whole was 119.71. Property taxation is treated as a wealth tax, in the first stage of analysis. The average rate poundage then translates into a corporate wealth tax at a rate equal to the ratio of the total yield of rates to the value of private net capital stock of buildings other than dwellings. This gives an average rate of 2.46% in 1980. The range of rate poundages then translates into a range of corporate wealth tax rates of 1.64% to 3.63%.

TABLE 1

RANGE OF EFFECTIVE TAX RATES AND POST-TAX RATE OF RETURN NECESSARY TO EARN A 10% PRE-TAX REAL RATE OF RETURN NET OF CORPORATION TAX AND RATES (FIXED-p CASE, 1980 TAX SYSTEM)

Category	p	s		p − s		100(p−s)/p		Difference in (p−s)/p
Asset								
Machinery	10.0	13.6	—	− 3.6	—	− 36.2	—	0.0
Buildings	10.0	6.8	5.4	3.2	4.6	32.2	46.4	14.2
Inventories	10.0	6.2	—	3.8	—	38.5	—	0.0
Industry								
Manufacturing	10.0	11.2	10.7	− 1.2	−0.7	− 11.6	− 6.7	4.9
Other	10.0	10.7	10.4	− 0.7	−0.4	− 6.8	− 3.8	3.0
Commerce	10.0	6.7	6.2	3.3	3.8	33.2	38.4	5.2
Source of Finance								
Debt	10.0	20.4	19.7	−10.4	−9.7	−103.8	−97.0	6.8
Shares	10.0	10.6	10.1	− 0.6	−0.1	− 6.3	− 1.5	4.8
Retentions	10.0	7.1	6.7	2.9	3.3	28.6	32.8	4.2
Owner								
Households	10.0	6.0	5.6	4.0	4.4	40.2	44.1	3.9
Tax exempt	10.0	14.7	14.2	− 4.7	−4.2	− 47.4	−41.6	3.8
Insurance Cos.	10.0	10.9	10.4	− 0.9	−0.4	− 9.0	− 4.1	4.9
Total	10.0	9.8	9.4	0.1	0.6	1.5	6.2	4.7

Note: The figures in this table represent the range resulting from the maximum and minimum non-domestic rate poundages.

The results of applying this range of corporate wealth tax rates to corporate investments under the UK tax system in 1980 are shown in Tables 1 and 2 for the fixed p and fixed r cases, respectively. In each case the full set of results for pre-tax rates of return p, post tax rates of return r, fixed wedge (p−s), effective tax rates (p−s)/p, and spatial gradients are reported. For the fixed p case a constant pre-tax rate of 10% is assumed.

TABLE 2

RANGE OF EFFECTIVE TAX RATES AND PRE-TAX RATE OF RETURN NECESSARY TO EARN A 5% POST-TAX REAL RATE OF RETURN NET OF CORPORATION TAX AND RATES (FIXED-r CASE, 1980 TAX SYSTEM)

Category	p		s	p − s		100(p−s)/p		Difference in (p−s)/p
Asset								
Machinery	2.4	---	3.9	-1.4	---	- 57.5	---	0.0
Buildings	8.0	10.0	3.9	4.2	6.2	52.0	61.5	9.5
Inventories	7.1	---	3.9	3.3	---	45.9	---	0.0
Industry								
Manufacturing	4.1	4.6	3.9	0.2	0.8	5.7	17.1	11.4
Other	4.2	4.6	3.9	0.4	0.7	8.8	16.2	7.4
Commerce	8.1	9.1	3.9	4.3	5.3	52.7	57.8	5.1
Source of Finance								
Debt	-3.0	- 2.3	3.3	-6.3	-5.7	---	---	---
Shares	3.6	4.3	4.0	-0.3	0.3	- 9.4	7.4	16.8
Retentions	7.4	8.1	4.0	3.4	4.1	46.3	50.7	4.4
Owner								
Households	4.3	5.0	-0.2	4.5	5.2	104.9	104.2	0.7
Tax exempt	6.4	7.1	9.0	-2.6	-1.9	- 40.2	- 27.2	13.0
Insurance Cos.	5.4	6.0	4.8	0.5	1.2	10.1	20.0	9.9
Total	5.2	5.9	3.9	1.4	2.0	26.4	34.6	8.2

Note: The figures in this table represent the range resulting from the maximum and minimum non-domestic rate poundages. Further, in the case of debt finance, p is negative and the effective tax rate is undefined.

Hence the results of Table 1 report the values of post-tax rate of returns necessary to cover the 10% pre-tax real rate of return net of corporation tax and non-domestic rates. The dominant features of Table 1 are the major subsidy to machinery as a result of the influence of the first year allowance against corporation tax that were available in 1980. This very considerably raises the rate of return required in commercial investment. A further general feature is the influence of tax concessions in reducing the rates of return necessary for investment from debt and tax exempt institutions. The effect of the spatial gradient of rates payments is assumed to be borne by building assets (although in reality some effect is borne by rated plant and machinery). The spatial gradient of local rates modifies the required post-tax rates of return by 1.4% for pure building investments. This translates into a gradient in effective tax rates of 14.2%. The effect of this gradient in required rates of return then becomes translated into other forms of investment, depending upon the ratio of buildings to other assets. Because of reduced benefits from corporation tax deductibility of the rates, the largest gradients of effective tax rates are for debt finance (differences of 6.8%), and tax exempt institutions (differences of 5.8%). A considerable gradient of effective tax rates also exists for commerce (5.2%) and manufacturing (4.9%). Overall the range of rate poundage produces a gradient of 0.5% in the level of post-tax rate of return required across the whole range of industry, assets, sources of finance, and their ownership. This is equivalent to a gradient of total effective tax rates of 4.7%.

Turning to the results for the fixed-r case in Table 2, these report the pre-tax rates of return necessary to earn a specified 5% post-tax real rate of return net of corporation tax and rates. The impact of rates is restricted to building assets and induces a 2.0% gradient in the pre-tax rate of return required, or a 9.5% gradient in effective tax rate, between the localities with the highest and lowest rate poundages. This translates into a 16.8% gradient of effective tax rates for equity finance, a 13.0% gradient for tax-exempt institutions, and a 11.4% gradient for manufacturing industry. Overall the gradients in pre-tax rates of return required is 0.6%. This translates into a gradient of 8.2% in effective tax rates. Evaluated from the point of view of the post-tax rates of return required (fixed-r case), therefore, non-domestic rates produce impacts that have a larger gradient than in the fixed-p case, but produce very comparable spatial gradients in the overall required rates of 0.5% (fixed-p) and 0.6% (fixed-r).

An important special case in these results arises with the fixed-r case for debt finance. In this case the pre-tax real rate of return is negative for all local authorities (varying between -2.3% and -3.0%). This finding illustrates the real non-linearities in the effective tax rate assessment. It can be interpreted as follows: with low interest rates i, or high tax rates, or both, the effect of accelerated depreciation and other tax allowances increases. As interest rates increase or tax rates decrease the effective tax rate increases, as shown in the left-hand side of Figure 1. However, beyond the point at which the pre-tax real rate of return $p \leq 0$, then the effective tax rate becomes negative and essentially is undefined. Hence, in this case, which applies to the UK for debt finance (fixed-r), tax subsidies outway the costs necessary to produce the specified 5% post-tax

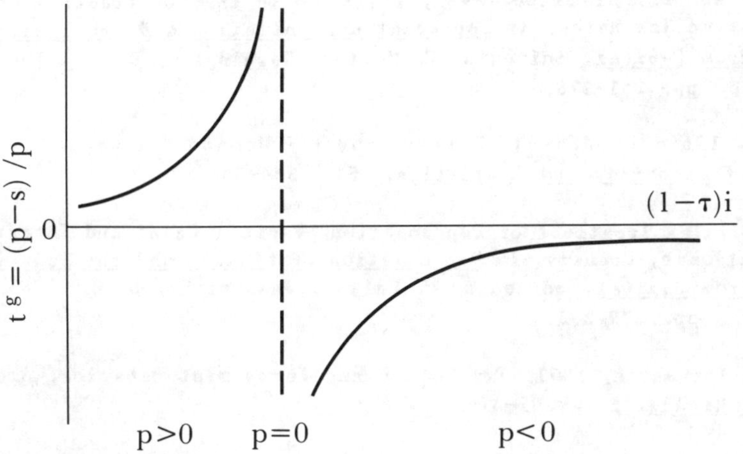

Figure 1: Alternative values for the effective tax rate t_g, depending upon the corporate tax rate τ and the interest rate i, with no inflation.

rate of return. Thus negative real rates of return will still yield positive returns to savers in the debt market.

5. CONCLUSION

Clearly the results reported in this paper are only preliminary measures of the spatial gradient, and require considerable development of spatial detail. However, high tax and low tax areas occur in close proximity in the UK, mainly in the metropolitan areas. Hence the estimates given do provide a realistic assessment of the actual profitability gradients that do arise in the metropolitan areas. Subsequent research will seek to relate these results to the actual patterns of local economic growth and relocation of businesses by expansion of the non-linear model that has been presented.

6. REFERENCES

Bennett, R., 1984, Dynamic Models of Spatial Variation in Local Government Expenditures, in Evolving Geographical Structures, edited by D. Griffith and A. Lea, The Hague: Martinus Nijhoff, pp. 400-412.

Bradford, D., and D. Fullerton, 1981, Pitfalls in the Construction and Use of Effective Tax Rates, in Depreciation, Inflation and the Taxation of Income from Capital, edited by C. Hulten, Washington, D. C.: The Urban Institute, pp. 251-278.

Bischoff, C., 1969, Hypothesis Testing and the Demand for Capital Goods, Review of Economics and Statistics, 51: 354-368.

Hall, R., 1981, Tax Treatment of Depreciation, Capital Gains and Interest in an Inflationary Economy, in Depreciation, Inflation and the Taxation of Income from Capital, edited by C. Hulten, Washington, D. C.: The Urban Institute, pp. 279-293.

_____, and D. Jorgenson, 1967, Tax Policy and Investment Behavior, American Economic Review, 57: 391-414.

Jorgenson, D., 1963, Capital Theory and Investment Behaviour, American Economic Review, 53: 247-259.

_____, 1967, The Theory of Investment Behaviour, in Determinants of Investment Behaviour, edited by R. Ferber, New York: National Bureau of Economic Research, pp. 129-145.

King, M., and D. Fullerton, 1984, The Taxation of Income From Capital: A Comparative Study of the United States, United Kingdom, Sweden and West Germany, Chicago: Chicago University Press.

Lund, P., 1976, The Econometric Assessment of the Impact of Investment Incentive, in The Economics of Industrial Subsidies, edited by T. Whiting, London: Dept. of Industry, HMSO, pp. 41-57.

Meade Committee, 1978, The Structure and Reform of Direct Taxation: Report of a Committee Chaired by Professor J. E. Meade, London: Allen and Unwin and the Institute for Fiscal Studies.

Nickell, S., 1978, The Investment Decisions of Firms, London: Nisbet and Cambridge University Press.

COMPREHENSIVE DYNAMIC URBAN MODELS: INTEGRATING MACRO- AND MICRO-APPROACHES

Mark Birkin and Martin Clarke
University of Leeds
England

1. INTRODUCTION

1.1. Context

Over the last ten years the most important single development in urban modelling has been the construction and operationalisation of dynamic models. Progress, on many fronts, has been swift, often involving elegant mathematics and novel insights into the dynamic properties of models that have been used for some time, as well as new models that have been proposed. The field also has become fashionable, attracting researchers from other disciplines to try their hand at this new and growing subject area. As a consequence, there now exists an impressive body of literature concerning the development of dynamic urban models (for example, Wilson, 1981; Griffith and Lea, 1983; Dendrinos and Mullally, 1985).

Yet despite all the progress made there do remain some unanswered questions, which largely concern the applicability and relevance of dynamic models to understanding and helping solve problems of contemporary urban society. In this paper we aim to raise some of these issues and attempt to respond to them in a number of ways. What we argue is that the dynamic models that have been developed so far need to be extended into a framework that both incorporates two different, but complementary, approaches to dynamics, and allows for a more detailed assessment of policy issues. We do not pretend that this extension is an easy one or that we have solved all the problems associated with it. Instead, we present some early results as well as some topics for further research.

1.2. Models of Spatial Structure

The majority of dynamic urban models have been constructed at the aggregate (or meso-) scale, and largely have focussed on spatial structure, examining how changes in stock variables by location take place over time. A summary of the development in this area can be found in Clarke and Wilson (1983a, 1985), Clarke, Wilson and Birkin (1985) and Wilson (1981). Two main developments can be singled out for limited attention. The first is the work built upon the approach outlined in Harris and Wilson (1978), who have demonstrated how a conventional retail trade model could be reformulated into an equilibrium model of retail size and location, and how dynamics could be introduced through changes in exogenous variables and parameters. It was shown that for small and smooth change in any of these variables, discrete change in the structural variables could occur. Early numerical experiments demonstrated not only the subtlety of the mechanisms of dynamic change, but also the analytical intractability of these models (Wilson and Clarke, 1979). The role of numerical experiments in non-linear analysis proved particularly important. As Campbell, et. al. (1985, p. 374) point out 'in the past NON-LINEAR was nearly synonymous with NON-SOLVABLE'.

The behaviour of both comparative static and genuine disequilibrium or dynamic versions of the Harris- Wilson model were demonstrated by Clarke and Wilson (1983a). Important research tasks still are associated with these models, however, and four strands of development may be briefly outlined. The first of these is the need to calibrate the model parameters, at an appropriate level of disaggregation. Such work has begun for Leeds (Birkin and Clarke, 1985) and for Rome (Lombardo and Rabino, 1984). Secondly, by formulating appropriate mathematical programming versions of the models (e.g. Clarke, 1981) it is possible to specify alternative model structures with different kinds of properties (Wilson, 1985). This argument has an analogue for disequilibrium modelling in the sense that alternative dynamic mechanisms also may be developed (Birkin and Wilson, 1985a). Finally, it is possible to extend the ideas from the retailing subsystem to a wide variety of other sectors of the urban economy, including housing and residential location (Clarke and Wilson, 1983b), industrial location (Birkin and Wilson, 1986b,c), and agricultural location (Wilson and Birkin, 1984). These extensions are particularly important for the development of comprehensive models, which we discuss below.

A related and complementary approach to the dynamic modelling of socio-economic processes is found in the work of Allen and his colleagues in Brussels. Here numerical simulation again is used to model the evolution of sets of deterministic difference equations with small random terms. In view of non-linearities in the difference equations, it is possible that varying the random number strings can lead to the generation of different dynamic trajectories, according to the paths adopted at bifurcation points in the system. Applications of these principles have ranged across intra- and inter- urban evolution, population dynamics and energy modelling.

1.3. Modelling of the Demand-Activity System Using Micro-Simulation

One of the main problems in using the extended spatial interaction models discussed above is the adequate representation of the heterogeneity and variability that is found in many socio-economic systems. A conventional aggregate occupancy matrix approach tends to become unwieldy and inefficient, as the number of attributes under consideration increases. The size of the occupancy matrix will be dependent upon the number of attributes and the number of classes assigned to each attribute. Formally this number, N^1, is given by:

$$N^1 = \sum_{u=1}^{m} a^u \tag{1.1}$$

where a^u is the number of categories associated with the u-th attribute. Typically N^1 will be very large for even a moderate number of attributes and categories. Moreover, the occupancy matrix will be very sparse--a large proportion of states will not be occupied by ANY individual or household. The micro-level representation simply will consist of the number of individuals and households in the population or sample, multiplied by the number of attributes considered, or

$$N^2 = (S^i X_I) + (S^h X_H) \quad , \tag{1.2}$$

where S^i and S^h refer to the number of individual and household attributes, respectively, and X_I and X_H are the number of individuals and households in the sample. Furthermore, for many purposes the sample size needed for a satisfactory representation of the whole population is relatively small, and therefore typically

$$N^1 >> N^2 \quad . \tag{1.3}$$

One possibility for improving the efficiency of the occupancy matrix representation is to decompose the matrix into smaller sub-matrices. It is commonly remarked that considerable information loss is the major result of this process.

The storage efficiency and information loss arguments have been used for a strong justification of the micro-level approach. There are, however, a number of crucial considerations that need to be examined before this argument can be fully accepted. These relate to the nature of the joint distribution of characteristics over the population, in particular the amount of interdependence that exists (for a full discussion of these issues, see Clarke, 1985, Chapter 5). However, let us assume for present purposes that it is deemed appropriate to use a micro-level representation. The first step in the modelling process is to obtain an initial population or sample of individuals and households specified by an appropriate set of attributes. Ideally, this should take the form of a real sample of micro-units, but for a variety of reasons this may not be available or not available with the required attribute set. In this case the production of a synthetic sample from contingency tables may prove a suitable solution to the problem. The theoretical basis for generating entries to a full contingency table consistent with available conditional and marginal probability distributions is long established (see, for example, McFadden, et. al., 1977).

The method in principle is straightforward, but in practice, due to the lack of detailed conditional probabilities, leads to some problems of inconsistency. The basic idea is to build up household and individual attribute lists on the basis of conditional probabilities obtained from published information, such as the 1981 Census of Population (OPCS, 1980), Family Expenditure Survey (Department of Employment, 1982), General Household Survey (OPCS, 1983), and the such. The basic procedure takes the form of producing a joint probability distribution $p(\underline{x})$ as a product of conditionals, such as

$$p(\underline{x}) = p(x_1)\ p(x_2|x_1)\ p(x_3|x_2,x_1)\ \ldots\ldots\ p(x_M|x_{m-1},\ \ldots,\ x_1) \quad , \quad (1.4)$$

and adopting approximations that impose a simplified structure on the conditional dependencies, because of missing information. The estimation of missing information is of course the province of information theory, and several well-known techniques are available to assist us in this task, notably entropy maximising and information minimising (Shannon, 1948). These techniques can be used for matrix infilling where row and column totals are known, and where a matrix configuration may have to satisfy certain criteria. The scheme we have adopted is based on Feinberg (1970), and is fully described in Clarke (1985).

Once an initial population has been generated, it is possible to incorporate household dynamics by LIST-PROCESSING. This simply involves examining each household and individual on the list and determining eligibility for certain events or transitions (such as birth, death, migration, redundancy). If eligibility is established, the appropriate conditional probability of the event occurring is determined, and Monte Carlo sampling once again invoked. Further, the effect of a transition from one state to another may have secondary effects. For example, if a woman is deemed to have given birth in a time period not only do we have to 'create' a new individual, but in addition the female will be taken out of the labour force for a period of time, and this itself may change the total household income that will instead be eligible for certain welfare payments. Household size also will change, which in turn may increase the likelihood of a residential move taking place. In modelling both the housing and employment sub-systems. for instance, we have to face up to the difficult problem of 'chaining.' This phenomenon occurs when an individual or household is both a 'supplier' and a 'demander' of a particular good. An individual who moves job occupies a vacancy and, usually, also creates a vacancy on his or her departure. To model these types of systems in a consistent manner we have to aggregate linked supply and demand pools, and usually solve an appropriate mathematical program. A full discussion of this family of models can be found in Williams, Keys and Clarke (1986).

Although the most recent developments in micro-simulation have come from economics (Orcutt, et. al., 1976; Havemann and Hollenbeck, 1980), there is growing evidence that its application in spatial analysis is being pursued by a number of workers (e. g., Bonsall, 1979; Clarke, et. al., 1981; Kain and Apgar, 1977; Wegener, 1983). We shall suggest in later sections of this paper that for addressing many contemporary issues of urban policy and planning, micro-simulation has much to offer. We also are convinced that an integration of the two main approaches described in this section is a potentially rewarding step forward. We now outline a framework within which this goal can be achieved.

2. A FRAMEWORK FOR THE INTEGRATION OF MACRO- AND MICRO-APPROACHES

The combination of macro- and micro-models within the same overall framework poses some difficult problems, particularly in relation to aggregation. The question of aggregation has, of course, been a central theme of economic theory for a long time (e. g., Fisher, 1922) and continues to attract the attention of urban modellers (e. g., Couclelis, 1985). In general terms the aggregation problem relates, on the one hand, to how micro-relations or behaviour can be transformed into aggregate relations, and on the other hand, to how, say, allocations of resources to aggregate groups are allocated to individuals and households. The problem is particularly acute when strong interdependencies exist between micro- and macro-processes. For example, the price of a given product is determined by the interaction of demand and supply, both of which are themselves a function of price. Thus the demand for a good at a given price cannot be simply taken as an aggregate of individual decisions since the resultant aggregation may yield a change in the price that the producer sets.

This micro-macro interdependence presents crucial problems for equilibrium theory. A classic example is Alonso's (1964) model of the housing market, which is essentially driven by the need to solve simultaneously and unambiguously the micro- and macro-models. However, the problems are equally pervasive in aspatial economic theory (e. g., the Walras-Wald model of general economic equilibrium is an approach to the same kind of problem). One of the key features of our own style of research is the argument that reality does not operate in this way. Thus systems typically are not in equilibrium, and if they are, the equilibrium only results from some combination of dynamic processes reaching a steady state. The explicit recognition of the dynamic nature of spatial-economic processes provides an alternative route of attack to these difficult problems, while also providing new kinds of insight into system behaviour (e. g., Wilson, 1981).

The framework we propose is set out in simplified form in Figure 2.1, and we now describe each part in turn. Starting with a micro-level specified population of individual and household attributes, it is possible to generate a measure of the 'demand' for certain goods and services by aggregating over appropriate individual and household attributes, and applying probabilistic models to determine whether, in a given time period, an individual will make use of a certain service or purchase a particular good. To take a health care example, an individual with a given set of age, sex and socio-economic attributes will have a certain probability of becoming a hospital in-patient in a year. This process then can be repeated to produce an estimate of demand for any of the other sub-systems being

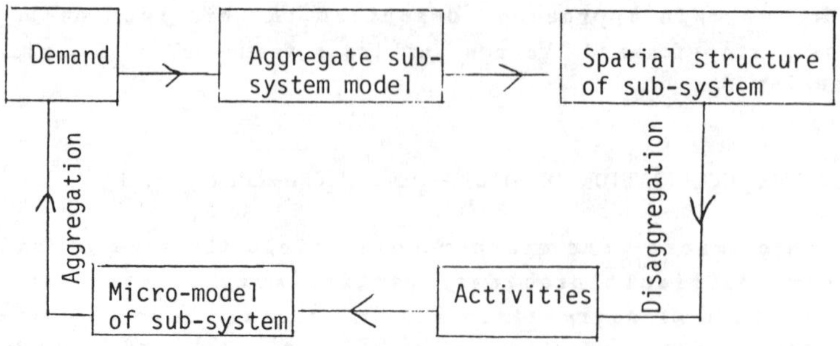

Figure 2.1: Structure of the integrated framework.

studied. The technical aspects of this approach were summarised in Section 1.3. The aggregate demand generated in this way then forms the input into the supply-side models, which are specified at the meso-scale. For example, the Harris and Wilson retail location model requires the zonal distribution of demand for different types of goods as a model input (and of course the model solution is sensitive to changes in the distribution of demand). The appropriate supply-side models then can be solved to produce the new spatial distribution of (say) facilities as well as associated prices and rents. We explain in detail how this is done in the next section.

The next step in the argument is to determine how these changes in spatial structure, prices, rents, and so on, influence the activities of individuals and households in the population. This step can be achieved in a number of ways. First the resultant interaction matrices from the supply model can be converted into probabilities. This conversion would allow, say, the shopping location of each household for particular types of goods to be determined, again using Monte Carlo sampling. The effects of changing prices may be examined by making changes in the household budget, along the lines presented in Wilson and Pownall (1976). It may be possible to account for constraints on activities--for example, to ensure no non-car owning households are allocated to out of town shopping centres--but this is not always straightforward and is a topic we return to later. Some example applications are discussed further in Section 3. The household and individual attributes then can be updated through list processing to account for demographic and other change. This completes the cycle and we can once more aggregate over the new set of attributes to provide the next estimate of demand.

This is the framework in its simplest form and we hope to elaborate the argument in subsequent sections of the paper, building up to the specification of a comprehensive or multi-subsystem model in Section 4. The approach is exploratory as well as flexible. For example it might be thought that the 'activity' side of the model has a faster dynamic than the spatial structure component, and this could be reflected by suitable alteration to the above scheme.

3. AN EXAMPLE OF A COMBINED MACRO-MICRO FRAMEWORK

3.1. Introduction

In this section we describe an example of the use of the integrated framework we proposed in Section 2 for the City of Leeds. Essentially it combines the aggregate structural models of Section 1.2 with the micro-model described in Section 1.3, once again using retailing as our example. An important point to emphasise at the outset is that, typically, the use of micro-models in this context is justified for reasons of efficiency and practicality, and not on 'philosophical' grounds. In most cases micro-models will exist as an ALTERNATIVE SOLUTION METHOD for a set of dynamic equations, and the merits or otherwise of the equivalent macro-model also should be examined. Only in cases where a model is SPECIFIED and ESTIMATED at the micro-level can we realistically speak of a separate class of model, and in some situations such classes may be equivalent to a related aggregate model (see Clarke and Williams, 1985 and Anas, 1983 for further details). Nevertheless, as we have pointed out earlier, we believe that there are potentially a wide-ranging number of applications where the integrated framework will have much to offer.

3.2. Specification of the Initial Population

As we observed in Section 1.3, the first step in the micro-simulation procedure is to generate an initial population, which involves the following attributes: for the household--household size, location, age of head, total household income, tenure type, number of children in the household and a household identification label; and, for each individual--age, sex, race, marital status, occupation, earned income, education status, weeks worked in present year and an identification label. The conditional probabilities that enable the joint distribution of these attributes to be generated are obtained from a number of sources: the 1981 Census of Population (OPCS, 1980), the Family Expenditure Survey (Department of Employment, 1982), the General Household Survey (OPCS, 1983), and so forth. To produce our synthetic population we use Monte Carlo sampling and create, with the use of the computer, an appropriate number of linked households and individuals.

Perhaps the most important attributes to generate are those relating to income. This goal is achieved in the following way. For all individuals in an occupation we need to derive a weekly wage, this being calculated on the basis of age, sex and occupation. Unfortunately, a complete age, occupation and sex-wage distribution is not available for the 18 occupation classes at the regional or national level in the U.K. What is available consists of age and sex distributions for manual and non-manual workers. The approach we adopted, therefore, consisted of fitting a non-linear curve to these distributions and then modifying these curves by an appropriate occupation multiplier. That is, we assume (in the absence of further information) that the curve takes the same general shape for all the occupations in the manual or non-manual class, for a particular sex. What we then do is to adjust the y-intercept and the maximum value of the curve on the y-axis by an appropriate occupation multiplier that is easily calculated for each occupation/sex category. We then can sample from the resultant adjusted distribution to obtain an age-sex-occupation specific wage.

This, however, is only the first stage of the procedure. We also have to take account of the dispersion that is clearly evident around an age-sex-occupation specific wage. This is done in the following way. The <u>New Earnings Survey</u> (Department of Employment, 1981) publishes occupation, sex and region specific wage distributions in the form of deciles. What we do when we have obtained an age-sex-occupation specific wage, say W_i^{so}, is to assume that there is an equal probability of the wage dispersion lying between the lowest decile and W_i^{so}, and between W_i^{so} and the highest decile. Thus we calculate an appropriate cumulative probability distribution between the lowest and highest decile, and sample from this interval in the usual way. So, suppose W_i^{so} was found to be 65 and the highest and lowest decile for that occupation were 40 and 150, respectively; there would be a 0.5 chance that the wage lay in the interval 65- 150. Total household income can now be obtained by summing over individual members' wages and adding any state benefits they are entitled to. Because we have details of individual attributes, most of these sums are relatively easy to calculate.

Zonal incomes then may be estimated simply by aggregating households across the relevant spatial units. We have assumed here a system of 30 demand zones (census wards for Leeds in 1971). To calculate the zonal demand for different types of retail goods we use expenditure multipliers from the <u>Family Expenditure Survey</u>. These multipliers give the proportion of household income spent on different goods by household income and type. Again each household is assigned an annual expenditure total based on these multipliers, and the relevant aggregation produces an estimate of the zonal demand for retail goods. The zonal income and expenditure distributions for our base year of 1975 are given in Table 3.1.

3.3. A Dynamic Integrated Model

The structure of the demographic model and associated interdependencies is given in Figure 3.1. The model has been run for a ten-year period (1975-84), and can be used to generate new expenditure and income distributions, as are reported in Table 3.2. Unfortunately, corresponding distributions for the supply-side structure variables (i. e., retail floorspace by type of good) initially were unavailable, although some data now are becoming available (G. Clarke, 1985). For our first set of results we use an intermediate income and expenditure distribution for the year 1979, and solve the structural model for the three orders of good. This model is

TABLE 3.1

1975 INCOME AND EXPENDITURE DISTRIBUTION (IN MILLIONS OF POUNDS)

Zone	Income	Good 1	Good 2	Good 3	Zone	Income	Good 1	Good 2	Good 3
1	34.39	5.45	1.83	1.82	16	40.62	6.38	2.20	2.15
2	35.61	5.34	1.87	1.75	17	43.12	6.68	2.28	2.22
3	40.89	6.52	2.15	2.03	18	35.31	5.35	2.11	1.95
4	34.05	5.80	1.95	1.88	19	34.67	5.56	1.89	1.73
5	24.41	4.02	1.23	1.13	20	31.02	5.14	1.75	1.65
6	23.62	4.16	1.33	1.25	21	46.45	7.08	2.54	2.48
7	30.61	5.08	1.62	1.54	22	47.64	7.31	2.54	2.42
8	20.10	3.54	1.14	1.10	23	32.32	5.05	1.74	1.64
9	37.56	5.48	1.93	1.93	24	40.68	6.12	2.12	2.01
10	28.27	4.75	1.60	1.52	25	45.88	6.42	2.40	2.22
11	36.82	5.68	1.95	1.88	26	35.19	5.86	2.10	1.98
12	40.32	5.88	2.29	2.13	27	26.43	4.31	1.47	1.36
13	29.70	4.71	1.57	1.46	28	35.09	5.28	1.78	1.79
14	47.06	7.17	2.39	2.23	29	28.53	4.34	1.43	1.48
15	2.10	3.67	1.21	1.13	30	40.90	8.71	4.82	4.28

specified as follows:

$$S_{ij}^{mg} = A_i^g \, e_i^{mg} \, P_i^m \, W_j^g \exp(- \beta^g \, c_{ij}^{mg}) \quad , \tag{3.1}$$

where S_{ij}^{mg} is a cash flow involving goods type g purchased at location j by population type m at location i (P_i^m), having expenditure characteristics e_i^{mg}. The term c_{ij}^{mg} is a measure of the transport cost associated with this interaction, and β^g is a set of distance deterrence parameters. A_i^g are balancing factors specified such that the following demand constraints are always met:

$$\sum_j S_{ij}^{mg} = e_i^{mg} \, P_i^m \quad . \tag{3.2}$$

W_j^g is a 'combined attractiveness factor' with a dependence on the output of g (W_j^g) and on the availability of other goods at j, such that

$$W_j^g = (W_j^g)^{\alpha g} \, (\sum_g W_j^g)^{\alpha^* g} \quad . \tag{3.3}$$

TABLE 3.2

1984 INCOME AND EXPENDITURE DISTRIBUTION (IN MILLIONS OF POUNDS)

Zone	Income	Good 1	Good 2	Good 3	Zone	Income	Good 1	Good 2	Good 3
1	52.86	8.69	2.85	2.78	16	62.95	9.42	3.40	3.19
2	54.22	8.52	2.85	2.76	17	63.94	9.21	3.14	2.92
3	66.84	10.75	3.58	3.45	18	66.38	9.65	3.51	3.36
4	53.82	7.95	2.72	2.46	19	56.44	8.59	2.86	2.79
5	45.11	7.72	2.55	2.38	20	46.05	7.35	2.53	2.31
6	31.63	5.67	1.69	1.63	21	69.82	10.12	3.78	3.66
7	52.89	7.94	2.72	2.64	22	70.17	9.83	3.41	3.67
8	36.39	6.18	2.03	1.95	23	49.21	7.93	2.80	2.53
9	56.93	8.57	3.13	2.96	24	52.12	8.86	2.99	2.78
10	45.00	7.79	2.66	2.37	25	66.49	10.60	3.67	3.48
11	57.72	9.73	3.35	3.07	26	54.59	8.30	3.02	2.82
12	67.11	9.24	3.12	3.14	27	42.53	6.84	2.22	2.05
13	47.36	7.37	2.52	2.24	28	49.02	8.30	2.85	2.87
14	71.20	12.07	4.19	3.84	29	44.20	6.95	2.38	2.34
15	35.05	5.73	1.88	1.81	30	59.45	9.67	4.15	3.96

The problem then is to solve the revenue-cost equations

$$\sum_{im} S_{ij}^{mg} = k_j W_j^g \qquad (3.4)$$

in the facility distributions $\{W_j^g\}$, where k_j is an average cost at j.

In the absence of either structural or interaction data, it is necessary to estimate a plausible set of parameters based upon previous experience. The model was solved under three sets of β^g values $\{\beta^1=0.5, \beta^2=0.4, \beta^3=0.3\}$, $\{\beta^1=0.4, \beta^2=0.3, \beta^3=0.2\}$ and $\{\beta^1=0.3, \beta^2=0.2, \beta^3=0.1\}$. In each case the values of α^g were $\alpha^1=1.05$, $\alpha^2=1.1$ and $\alpha^3=1.15$ and the value of the combined attractiveness function parameters α^{*g} were $\alpha^{*1}=0.1$, $\alpha^{*2}=0.05$ and $\alpha^{*3}=0.02$. The results from the equilibrium model are shown in Figures 3.2, 3.3 and 3.4. As we would anticipate, lower β^g values give a more concentrated pattern of facility distribution (Clarke and Wilson, 1983a). The spatial system consists of the 30 demand zones and, potentially, 900 supply zones that consist of a 30-by-30 regular lattice of zone centroids

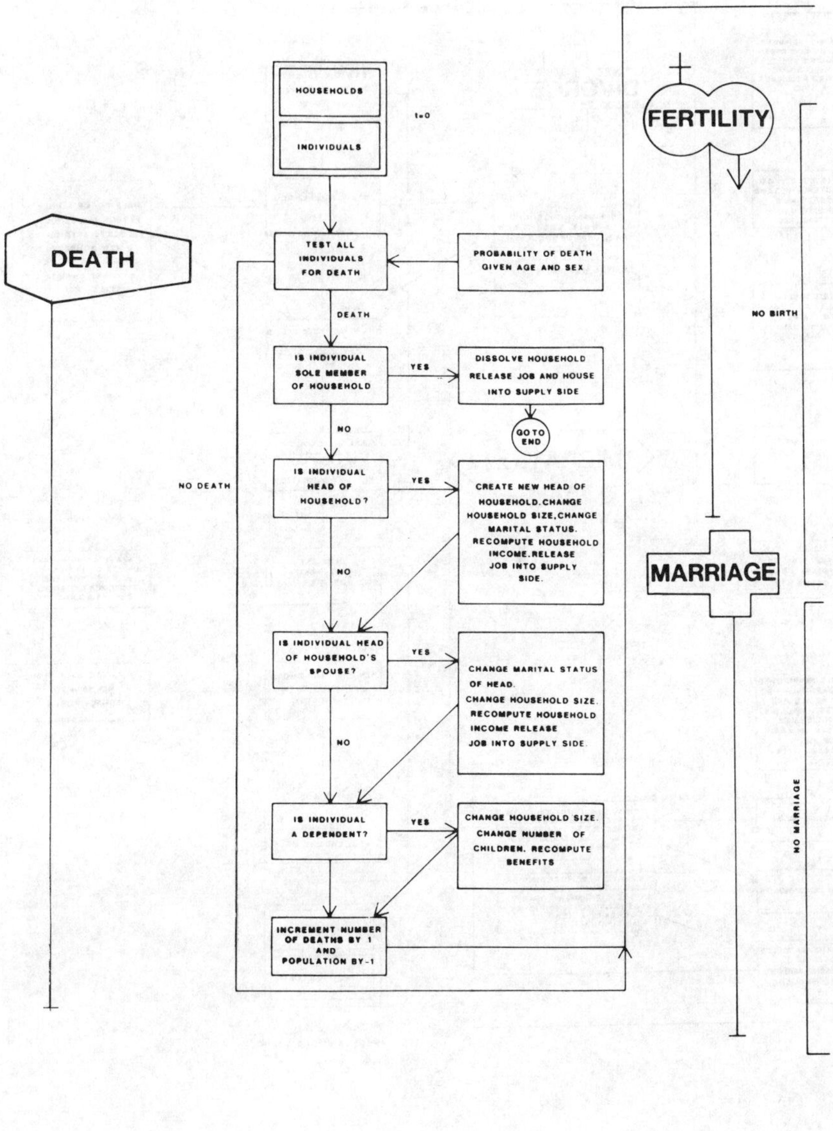

Figure 3.1: The structure of the demographic model.

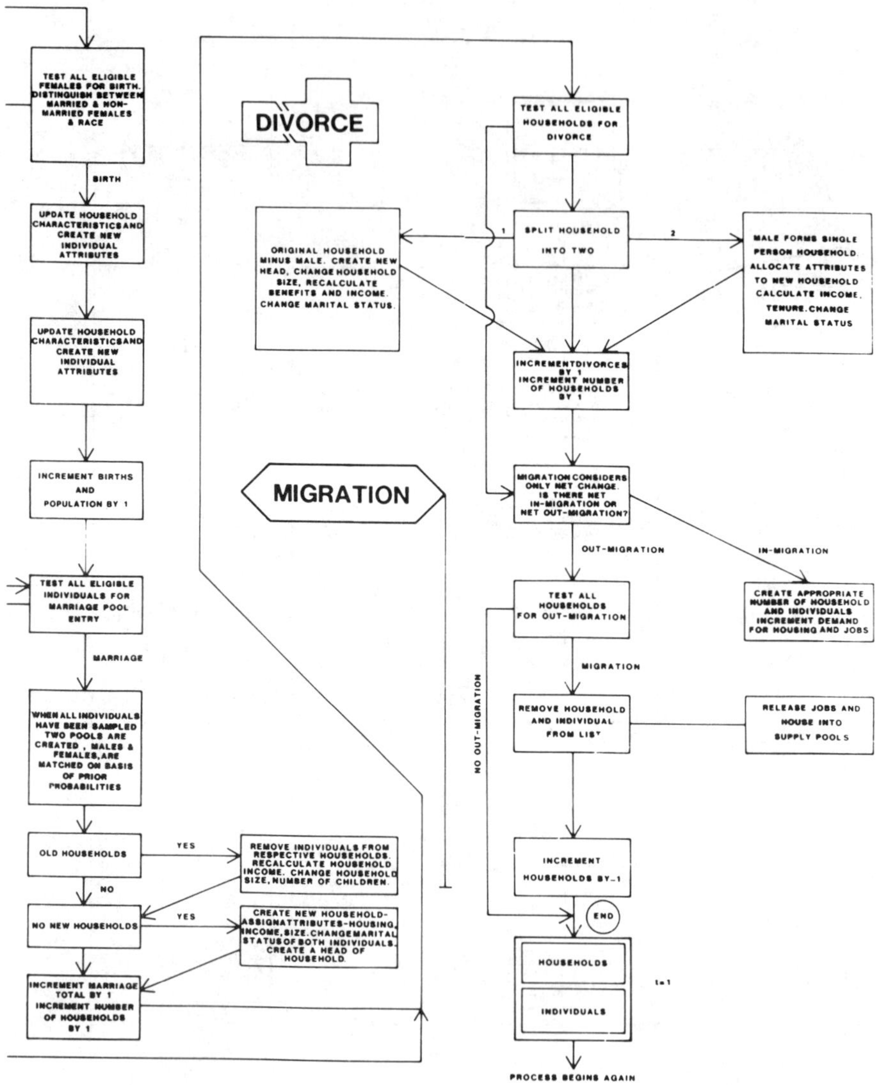

Figure 3.1: Continued

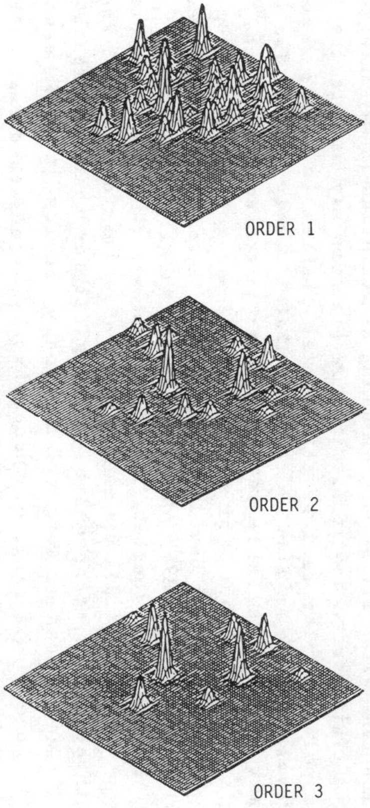

Figure 3.2: Equilibrium plots, β=(0.5,0.4,0.3)

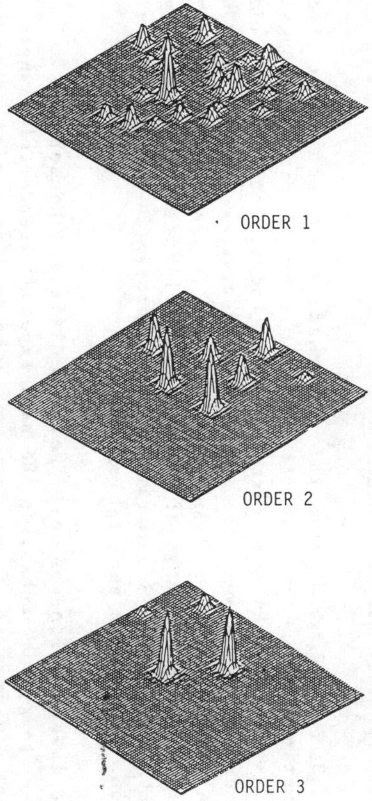

Figure 3.3: Equilibrium plots, β=(0.4,0.3,0.2)

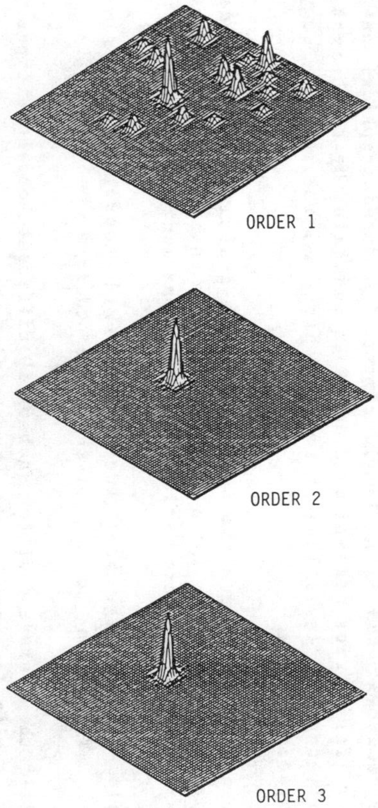

Figure 3.4: Equilibrium plots, β=(0.3,0.2,0.1)

superimposed over the city. In the computer produced graphical output the amount of facilities present at a point is indicated by the height of the peak, although due to scaling factors the plots are not strictly comparable in an absolute sense.

In addition we have the output from the associated micro-model. For illustration we present the following tables, while emphasising that many more cross-tabulations are available for reasons mentioned earlier:

(i) the average cost incurred by all households in a zone in travelling to each of the three orders;

(ii) the average cost for each of 10 different household types, in the city for each of the three goods; and,

(iii) the average cost incurred in travel by different age groups in the city for each order.

Tables 3.3, 3.4 and 3.5 present this information for each of the three sets of parameter values. The most striking yet expected pattern amongst these results is that the average distance travelled by households in different zones increases rapidly as the value of β^g decreases, namely by 277%, 317% and 296% between the highest and lowest values for each order, respectively. More specifically the range of distances travelled varies enormously as β^g decreases, and this is illustrated by the increase in variance and standard deviation, as shown on each table.

The distributions of costs between different households and between different age groups emphasises the same expected trend--trip lengths become longer as β decreases. There is not a great deal of variation between the different groups in these distributions--it would require a more detailed analysis of activity patterns and constraints to account for mobility differences between these groups. Some suggestions on this score are made in Section 4.

We now can switch attention to looking at how planners may intervene in our system of interest, and to examine how planners may measure the impact of their policies. The task we set for the planner is how to encourage retail development in an area of the city that is badly served. The policy instrument that is available to the planner is to subsidise development costs in a zone. This course of action may allow a zone to move from a no-development-possible to a development-possible state. The zone we select for analysis is supply Zone 294, which is located within demand zone 12 and is an outer area zone. The equilibrium model first was run with no variation in k_j, with β^g values of 0.35, 0.3 and 0.25 and α^g values of 1.1, 1.15 and 1.25 for the three orders, respectively. The plots for this run are shown in Figure 3.5, and it can be observed that only the lowest order centre is present in Zone 294. A range of successive k_j values then was tested for this zone; these were factors of 0.95, 0.9, 0.8, 0.5 and 0.1 of the original

TABLE 3.3

IMPACT ANALYSIS FROM RETAIL MODEL--AVERAGE COST OF TRAVEL BY ZONE

	β = (0.5, 0.4, 0.3) Order			β = (0.4, 0.3, 0.2) Order			β = (0.3, 0.2, 0.1) Order		
Zone	1	2	3	1	2	3	1	2	3
1	1.92	7.90	5.42	2.93	13.22	9.80	7.41	24.67	22.42
2	0.70	8.72	5.87	1.50	12.23	10.93	3.92	21.77	21.48
3	2.12	1.35	1.57	2.60	16.08	2.04	1.70	30.08	27.50
4	1.34	0.67	1.60	0.88	0.67	3.02	0.94	0.67	2.60
5	3.34	8.93	9.46	4.82	8.26	10.32	4.40	17.78	15.33
6	2.75	11.94	12.14	3.88	14.13	12.68	9.80	18.11	16.47
7	2.86	8.04	9.33	4.24	8.26	9.67	10.00	15.09	12.23
8	3.53	7.01	7.63	7.08	6.77	8.59	7.82	6.19	4.12
9	0.60	0.60	1.80	0.66	20.03	7.38	1.42	36.30	33.27
10	2.14	5.54	6.84	3.83	13.71	7.90	7.16	14.27	15.56
11	2.94	5.24	5.55	5.78	5.14	6.85	5.75	4.83	5.86
12	1.76	3.11	4.77	2.45	5.12	6.87	4.14	15.93	18.75
13	2.71	5.10	6.22	5.77	5.60	7.79	6.15	4.81	2.03
14	1.23	0.81	1.31	1.35	0.92	2.11	1.78	16.95	13.94
15	4.30	4.70	6.26	5.72	15.80	7.13	6.63	15.41	14.74
16	1.76	6.55	8.89	2.54	11.14	9.71	6.82	26.09	23.26
17	1.46	1.83	2.71	1.88	17.32	13.65	3.00	24.49	25.40
18	1.22	1.15	2.21	1.63	17.32	7.28	2.30	28.50	25.75
19	2.09	7.52	7.98	4.21	7.17	9.02	8.08	7.01	10.01
20	1.62	7.86	8.17	4.26	7.91	9.32	8.14	7.53	9.65
21	1.58	8.17	2.91	1.92	14.18	4.91	3.06	12.93	12.30
22	1.41	1.06	2.54	1.75	14.34	5.70	2.82	14.07	15.86
23	3.17	6.86	7.73	4.97	6.05	8.50	8.58	11.09	8.02
24	2.01	5.43	5.92	2.61	19.65	7.21	6.06	36.01	33.47
25	1.54	1.27	3.05	2.02	15.91	12.24	3.33	16.71	14.79
26	1.48	10.17	5.45	2.19	10.80	12.97	4.04	26.60	23.59
27	1.23	0.72	1.12	1.52	13.84	1.81	2.17	13.23	13.57
28	1.26	7.25	8.56	2.16	15.95	9.91	7.81	15.82	16.52
29	2.23	9.30	10.23	3.15	9.22	10.23	8.02	12.21	9.88
30	1.32	1.37	4.11	1.75	1.42	13.16	2.59	30.05	28.54
Mean	1.98	5.52	5.58	3.07	10.82	8.51	5.19	17.51	16.56
S. D.	0.87	3.53	3.04	1.66	5.55	3.39	2.71	9.28	8.50
Var	0.73	12.03	9.13	2.68	29.79	11.12	7.12	83.27	69.76

TABLE 3.4

IMPACT OF RETAIL MODEL ON DIFFERENT HOUSEHOLD TYPES
AVERAGE TRANSPORT COSTS BY HOUSEHOLD

Household Type	β = (0.5, 0.4, 0.3) Order 1	2	3	β = (0.4, 0.3, 0.2) Order 1	2	3	β = (0.3, 0.2, 0.1) Order 1	2	3
1	1.93	5.29	5.27	2.99	10.85	8.12	5.03	17.77	16.90
2	1.84	4.86	5.20	3.00	10.10	8.38	4.80	17.22	17.52
3	2.04	5.41	5.17	3.01	11.33	8.70	5.12	18.34	17.52
4	2.13	5.94	5.89	3.15	10.18	8.95	5.35	16.96	16.06
5	1.83	5.32	4.97	2.68	11.30	8.86	4.62	18.48	17.58
6	1.53	6.21	5.82	2.79	10.62	11.70	3.83	16.69	16.33
7	1.78	5.49	5.33	2.75	11.58	8.77	4.88	18.41	17.61
8	N.A.	N.A.	N.A.	N.A.	N.A.	N.A.	N.A.	N.A.	N.A.
9	1.67	5.33	5.05	2.79	11.02	8.22	4.66	18.60	17.71
10	1.83	5.42	5.52	2.70	11.11	8.68	4.89	17.97	17.16

NOTE: N.A. designates 'not available.'

TABLE 3.5

AVERAGE TRANSPORT COSTS FOR DIFFERENT AGE GROUPS

Age Group	β = (0.5, 0.4, 0.3) Order 1	2	3	β = (0.4, 0.3, 0.2) Order 1	2	3	β = (0.3, 0.2, 0.1) Order 1	2	3
< 25	1.80	6.22	5.55	2.98	11.01	8.88	5.14	17.66	16.69
25-34	1.63	5.23	5.16	2.67	11.59	8.86	4.64	19.10	18.22
35-44	1.87	5.44	5.30	2.81	11.40	8.77	4.88	18.33	17.55
45-54	1.97	5.39	5.18	2.86	11.54	8.98	4.99	18.55	17.72
55-64	1.94	5.41	5.30	2.87	11.19	8.58	4.96	18.59	17.72
65+	2.08	5.35	5.33	3.13	10.73	8.39	5.15	17.62	16.80

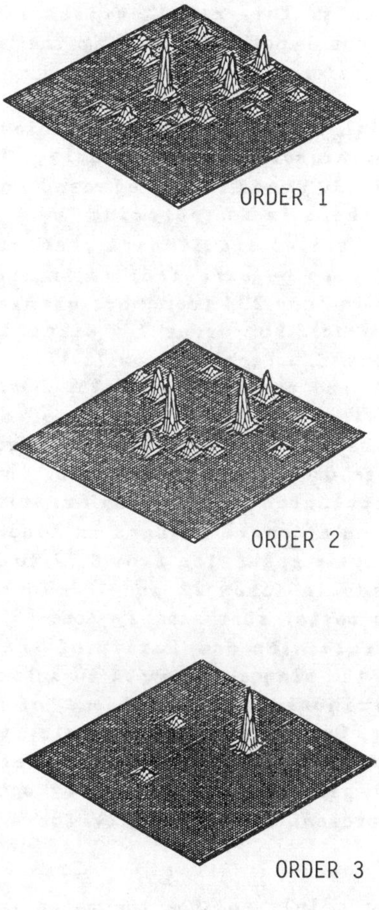

Figure 3.5: Equilibrium with no intervention.

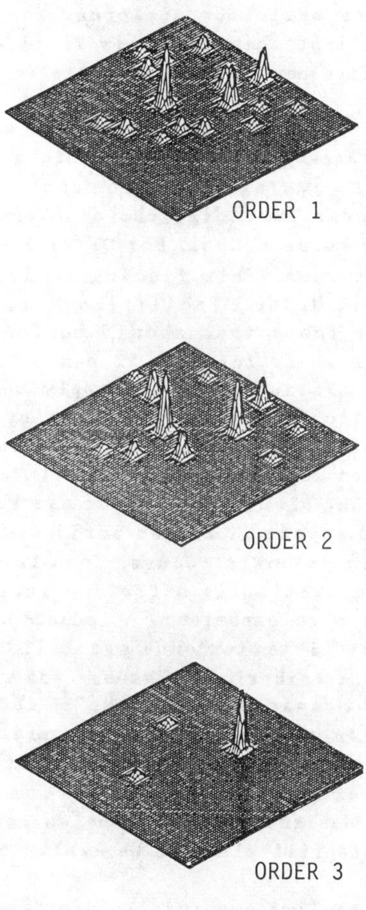

Figure 3.6: Zone 294 factored by 0.95.

value.

It can be observed that a second order centre emerges when the k_j factor is 0.95 (Figure 3.6) and a third order centre emerges when the k_j factor is 0.8 (Figure 3.7). This suggests that higher order centres (of which there are naturally fewer) require a greater amount of 'persuasion' to establish facilities in a zone that does not already contain some facilities of that order. As the k_j factor becomes much smaller, Zone 294 begins to dominate the pattern of structural activity, as can be observed from Figure 3.8. Perhaps a more important feature to observe from these plots is that when a centre is 'encouraged' to locate in a certain area, its locating almost certainly has ramifications for the provision of that facility in the rest of the system. As can be seen from the plots, when a centre does emerge and grows in Zone 294, it does so at the expense of already existing centres which either shrink or disappear completely. This result arises simply through the fact that demand is fixed and our hypothesis stating that only normal profits can be made prevents an expansion of supply.

If we now move attention to the spatial impacts of these policies, in terms of average costs we will note a similar set of events arising. Table 3.6 gives the average costs for each of the 30 zones for a selected number of k_j factors, including the case where there is no factoring (k_j = 1.0), for each of three orders. For Order 1 there are no significant changes for each of the runs. This finding is largely to be expected; as in the 'no policy' case, Order 1 facilities do exist in Zone 294 (remember changes in Zone 12 are those that should be focused on). For Order 2 a significant reduction in costs for Zone 12 can be observed, changing from 11.47 to 5.92 just for a 0.95 factoring of supply costs, and continues to fall. However, this reduction in costs is offset by an increase in the costs of other zones, particularly Zone 25 which already had fairly high costs. For Order 3 the situation is more complex. Little change occurs for k_j factorings of 0.9 and 0.95 (what change does occur can be attributed to sampling error), and as we would expect it is not until 0.8, when a centre emerges in Zone 294, that a shift in costs occurs. Zone 12 enjoys a reduction from 8.72 to 1.74; but, this reduction is offset by increases in Zones 22 and 19. However, other zones also experience a reduction in costs, most notably Zones 11, 21 and 25. This situation once again illustrates the complexity of spatial equity and distributional issues, and when the planner attempts to intervene in a market system to effect a redistribution, the outcome of that intervention can be difficult to predict. On a more positive note, these methods do allow the analyst to investigate the decision--or parameter--space that exists, but the methods do not guarantee some globally optimal plan to be designed. Rather the methods present an opportunity for a wide range of potential plans to be evaluated.

These preliminary results are designed mainly to show the potential of the approach rather than to justify it. As we noted earlier, many of the tables produced here could be generated by a corresponding aggregate model. When we proceed to examine more complex systems, the benefits of using this

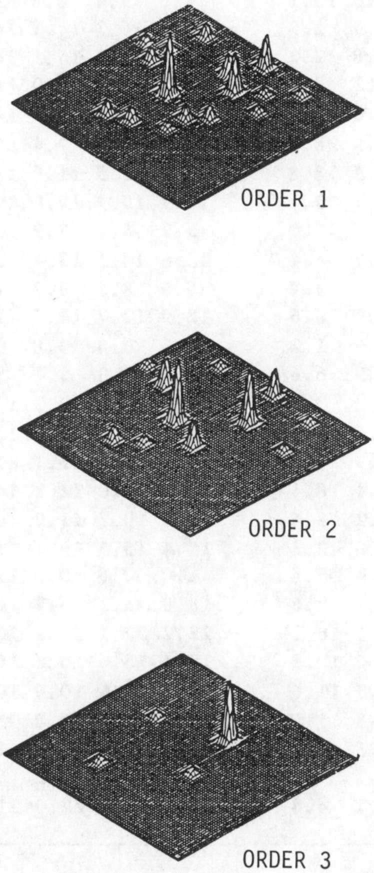

Figure 3.7: $k_{294} = 0.8$

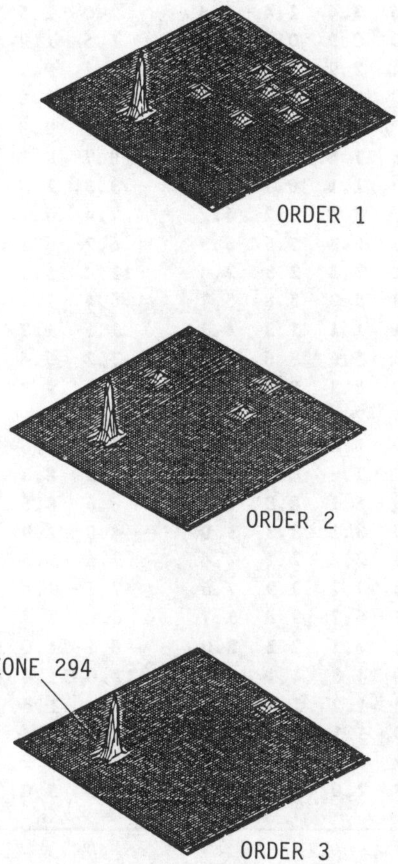

Figure 3.8: $k_{294} = 0.1$.

TABLE 3.6

IMPACT ANALYSIS FROM RETAIL MODEL--AVERAGE COST OF TRAVEL BY ZONE

Zone	Order 1 K Factor				Order 2 K Factor				Order 3 K Factor			
	1.00	0.95	0.90	0.85	1.00	0.95	0.90	0.85	1.00	0.95	0.90	0.85
1	6.2	6.5	6.6	0.8	9.4	9.5	10.1	9.7	14.2	14.1	14.1	14.2
2	2.8	2.5	2.8	3.3	9.4	9.8	10.0	10.1	8.8	9.4	8.8	9.1
3	1.6	1.4	1.4	1.4	2.0	2.5	1.9	1.8	16.3	16.2	16.1	16.3
4	1.0	0.9	0.9	0.8	1.5	1.2	1.8	2.2	15.2	15.1	15.1	14.4
5	2.4	2.9	2.7	3.1	9.4	9.7	9.2	9.0	8.0	8.0	8.0	8.1
6	8.9	8.7	8.7	8.4	13.0	12.8	13.0	12.9	12.5	12.6	12.5	12.1
7	8.8	9.0	8.8	8.6	9.3	9.8	9.5	8.8	7.4	7.4	7.4	7.7
8	7.5	7.4	7.0	7.3	8.7	8.3	8.3	8.8	11.5	11.5	11.6	11.6
9	0.8	1.1	0.7	0.9	3.3	2.5	2.7	3.1	19.9	19.9	19.9	20.0
10	6.7	5.8	6.2	6.2	7.4	6.7	6.5	7.0	5.7	5.6	5.9	5.8
11	5.3	5.3	5.5	5.5	6.2	6.1	6.7	6.4	13.6	14.2	13.4	9.9
12	2.8	2.3	2.3	2.4	11.4	5.9	5.7	4.4	8.6	8.6	8.7	1.7
13	5.4	5.9	5.8	5.7	6.4	7.1	7.2	6.6	12.9	12.7	12.7	12.0
14	1.1	1.1	1.3	1.2	1.5	1.7	1.5	1.2	0.8	0.8	0.8	0.9
15	5.3	5.6	5.6	5.2	7.2	6.4	5.9	6.6	5.9	5.0	5.4	5.0
16	6.3	8.3	8.3	8.4	9.2	9.3	9.2	9.4	10.6	10.6	10.7	10.6
17	2.5	2.7	3.0	2.2	5.2	6.1	5.5	6.1	12.7	13.0	12.5	12.3
18	1.3	1.6	1.6	1.6	3.7	3.6	3.9	4.8	17.1	17.1	17.1	17.3
19	7.7	7.6	7.8	7.7	8.2	8.3	8.4	8.4	12.1	12.0	12.2	14.1
20	7.9	8.1	8.0	8.1	8.6	8.5	8.2	8.6	11.1	10.7	11.2	10.8
21	2.9	3.1	3.3	3.0	4.0	2.9	2.8	2.3	15.4	15.5	15.6	1.3
22	2.5	2.2	2.1	2.4	2.9	3.8	4.4	5.4	1.0	0.8	0.9	11.2
23	7.2	7.7	7.9	7.6	7.5	7.4	7.5	7.6	6.0	6.1	6.2	6.3
24	5.7	6.2	5.8	5.7	6.2	5.8	6.3	6.7	22.2	22.1	22.1	22.3
25	2.2	2.1	2.2	2.0	8.1	11.0	10.6	10.5	15.4	15.6	15.6	10.0
26	3.9	4.3	4.1	4.1	12.0	11.4	11.5	11.6	10.6	10.6	10.6	10.8
27	1.2	1.2	1.6	1.3	2.2	1.4	1.8	1.5	0.9	1.0	0.8	1.1
28	4.0	3.1	2.9	3.0	8.5	8.4	8.6	8.5	7.3	7.4	7.3	8.0
29	8.9	8.7	9.0	8.8	8.7	9.9	9.7	9.9	8.3	8.4	8.2	8.3
30	2.3	2.0	2.1	2.4	5.0	5.0	5.1	4.4	22.2	21.8	22.0	21.7

approach should become more self-evident. We now examine how this framework can be further developed for a comprehensive urban system.

4. COMPREHENSIVE FRAMEWORKS

In Sections 1.2 and 1.3 we reviewed certain developments in model building at the meso- and macro- scales. Then an example was provided that demonstrated the method of integration for the retail sub-model. In this section we will examine some of the implications of applying this type of approach within a comprehensive modelling framework. The discussion attempts to explain the significance of Figure 4.1, wherein a schematic illustration of the type of system we propose is outlined.

For the purposes of illustration we have assumed a model with four subsystems that are reflected in the supply-side structure variables. $Z_j^k(t)$ represents the distribution of industrial activities producing goods type k in location j at time t; $W_j^g(t)$ is retail and service floorspace for good g; $H_i^l(t)$ is the stock of housing type l in zone i; and, $c_{ij}^\theta(t)$ is the transport cost between locations i and j, where the superscript θ represents the union of all supply-side activities that interface with one another through the transport network. We have assumed the demand-side to be represented by a single array of variables, $P_i^\phi(t)$, being the population in zone i at time t, with characteristics in the set ϕ, embracing attributes such as age, sex, occupation, and marital status, to an appropriate degree of detail.

Both the supply- and demand- side structure variables are assumed to be 'slow,' and form the inputs to a series of 'fast' interaction models. These interactions are taken to be generated instantaneously with respect to the structure variables, and are of three kinds. The classical journey-to-work and shopping models arise directly at the interface between population distributions and the suppliers; we therefore refer to these as 'supply-demand interactions.' At the same time, many interdependencies exist between the supply sectors (e. g., there are many product flows within the industrial subsystem, while all suppliers must compete for scarce resources such as land and capital--these are 'supply-side interactions'). Similar processes are operative on the demand-side; for example, to determine labour supply and the demand for goods (to satisfy various utilities in consumption). These latter components are 'demand-side interactions'. For a given distribution of activities, the instantaneous allocation of interactions allows a meso-scale dynamic to be generated. Thus summing the flows from a retail model at the supply-side generates a measure of revenue attracted, and this in turn may be coupled with the cost of supply to generate a slow dynamic in which the structure variables respond to disequilibrium.

With respect to the demand-side, the first observation that can be made is that the interaction matrices to be generated can be very large and sparse. For example, even a coarsely specified journey-to-work array would be characterised as $\{T_{ij}^{kmn}\}$, where k is the type of industry in which the individual works, m is the mode of the trip, and n is the occupation of the

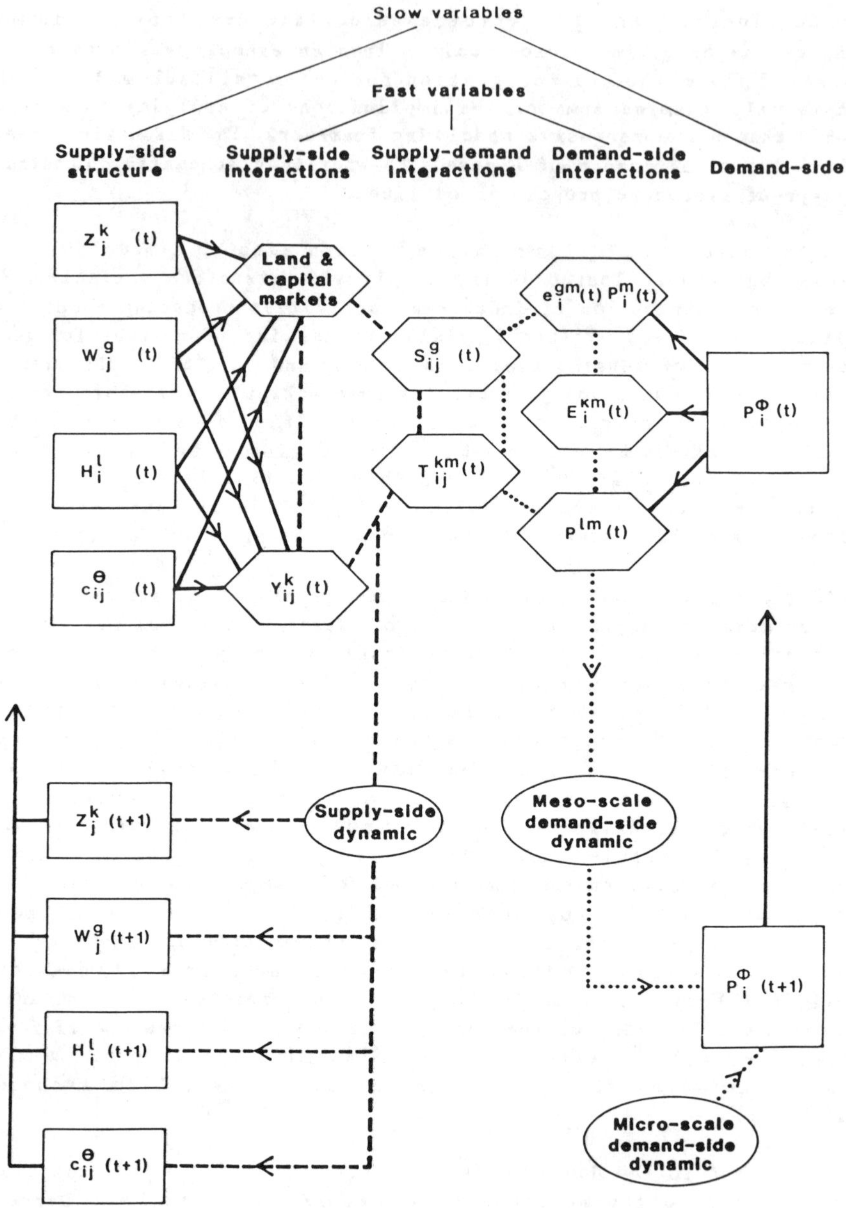

Figure 4.1: Structure of a comprehensive and integrated model.

individual. With 100 origin and destination zones, 10 occupation groups, 10 industry types and 5 modes of transport, an array with 5 million elements is implied. The techniques of micro-simulation and list-processing allow us to sidestep the need for storing and manipulating such huge arrays. Instead, the interactions may be modelled and then converted to conditional trip-making probabilities for input to the simulation process. As we have observed above, this perspective allows us to bring in much finer levels of detail at the demand-side.

The micro-simulation framework is efficient not only in terms of storage, but also with respect to accounting. Since the characteristics of each individual are treated explicitly, the appropriate conservation laws must be obeyed, and global transitions always may be traced back to their original components. Furthermore, by processing the individual lists in a variety of ways, numerous types of aggregate variable may be examined for the same system. This approach may be particularly useful in the calculation of appropriate PERFORMANCE INDICATORS to examine various policy issues (G. Clarke and Wilson, 1986).

Once the structure variables have been updated through the 'slow' dynamic, the whole process is repeated for the next time period or iteration. Notice that with respect to the structure variables, increments always are applied at the end of the time period, so the sub-models always are solved with respect to lagged variables. In this case a crucial determinant of system evolution is the relative speeds of change operating with respect to the different subsystem activity levels. However certain problems may arise in relation to the ORDER of solution of the 'fast' interaction models, since the outcome of one submodel may be dependent on the outcomes of one or more others. To take a simple example, the expenditure on goods might be calculated as a residual once costs of tripmaking are deducted from disposable income. Thus inputs to a retail submodel clearly would depend upon journey-to-work expenditures. Once again, however, the obvious solution is to adopt an appropriate system of lags, such that

$$S^g_{ij}(t) = A^m_i(t)\, e^{mg}_i(t-1)\, P^m_i(t)\, [W^g_j(t)]^{\alpha g(t)} \exp[-\beta^g(t)\, c^g_{ij}] \quad , \quad (4.1)$$

$$e^{mg}_i(t) = f[S^g_{ij}(t), T^{km}_{ij}(t), \ldots] \quad . \quad (4.2)$$

The parameters of the model (4.1), namely α^g and β^g, could be projected completely independently (e. g., through a time series extrapolation).

5. CONCLUSIONS

In this paper we have tried to concentrate our attention on significant theoretical developments, although there remain many non-trivial technical problems to be resolved. On the activity side this involves consideration of factors such as disaggregation (by mode, by trip purpose), incorporation of more sophisticated measures of the cost matrix, and the treatment of problems of closure. In relation to the supply-side, we need to concern ourselves again with more sectorally disaggregate specifications, taking into account store size, the way in which choices are perceived by consumers, and whence trip patterns are generated. These problems are discussed in more detail by Clarke (1985, Chapter 6).

The kind of integrated framework introduced in this paper represents a potentially important route into the linkage of such consumer choice (or discrete choice) problems within the spatial interaction-activity-structure framework. However, it should be emphasised, as we observed in Section 3, that the main use of the micro-simulation device at present is as a solution technique, which allows efficient storage and manipulation of micro-data that are essentially synthetic. Thus one outstanding research task is the problem of interfacing behaviouralist and activity oriented work (e. g., Lenntorp, 1976; Jones, 1979) with micro-simulation methodology (cf. Thrift, 1981).

The second major theme in applied research in this area is the extent to which methods being developed are taken up by planners. The construction of a micro-simulation package that would serve a wide variety of needs, ranging from demographic and household forecasting, to labour market analysis and housing system analysis, would be a useful addition to the small range of models in practical use at the moment. In view of the potential integration of micro- and macro-level approaches in a multi-subsystem context, which we have tried to explore here, together with ongoing developments in information storage and processing, we believe the potential is greater now than ever before for breakthroughs from model-building in theory to model use in actual planning practice.

6. REFERENCES

Alonso, W., 1964, Location and Land-use, Cambridge, Mass.: Harvard University Press.

Anas, A., 1983, Discrete Choice Theory, Information Theory and the Multinomial Logit and Gravity Models, Transportation Research B, 17: 13-23.

Birkin, M., and G. Clarke, 1985, On the Design and Implementation of an Information System for Urban Modelling and Spatial Planning, Working Paper 441, School of Geography, University of Leeds.

______, and A. Wilson, 1985a, Some Properties of Spatial-Structural-Economic-Dynamic Models, Working Paper 440, School of Geography, University of Leeds.

______, and ______, 1986b, Industrial Location Models I: A Review and an

Integrating Framework, Environment and Planning A, forthcoming.

_____, and _____, 1986c, Industrial Location Models II: Weber, Palander, Hotelling and Extensions Within a New Framework, Environment and Planning A, forthcoming.

Bonsall, P., 1979, Micro-simulation of Mode Choice: A Model of Organised Car Sharing, PTRC Proceedings, Summer Annual Meeting, University of Warwick, July 4-7.

Campbell, D., J. Crutchfield, D. Fermer and E. Jen, 1985, Experimental Mathematics: The Role of Computation in Non-linear Science, Communications of the Association for Computing Machinery, 28: 374-384.

Clarke, G., 1985, Retail Centre Usage and Structure: Empirical, Theoretical and Dynamic Explorations, unpublished doctoral dissertation, School of Geography, University of Leeds.

_____, and A. Wilson, 1986, A Review of Indicators, Planning and System Performance, Working Paper 446, School of Geography, University of Leeds, forthcoming.

Clarke, M., 1981, A Note on the Stability of Equilibrium Solutions of Production-Constrained Spatial Interaction Models, Environment and Planning A, 17: 175-183.

_____, 1985, Integrating Dynamical Models of Urban Structure and Activity: An Application to Urban Retailing Systems, unpublished doctoral dissertation, School of Geography, University of Leeds.

_____, P. Keys and H. Williams, 1981, Microanalysis and Simulation of Socio-economic Systems, in Quantitative Geography: A British View, edited by N. Wrigley and R. Bennett, London: Methuen, pp. 248-256.

_____, and H. Williams, 1985, Micro-Analysis in Urban and Regional Systems, Cambridge: Cambridge University Press.

_____, and A. Wilson, 1983a, The Dynamics of Urban Spatial Structure: Progress and Problems, Journal of Regional Science, 23: 1-18.

_____, and _____, 1983b, Exploring the Dynamics of Urban Housing Structure: A 56 Parameter Residential Location and Housing Model, Working Paper 363, School of Geography, University of Leeds.

_____, and _____, 1985, The Dynamics of Urban Spatial Structure: The Progress of a Research Programme, Transactions of the Institute of British Geographers, 10: 387-416.

_____, _____ and M. Birkin, 1985, The Dynamics of Urban Spatial Structure, unpublished manuscript, School of Geography, University of Leeds.

Couclelis, H., 1985, Cellular Worlds: A Framework for Modelling Micro-Macro Dynamics, Environment and Planning A, 17: 585-596.

Dendrinos, D., and H. Mullally, 1985, Urban Evolution: Studied in the Mathematical Ecology of Cities, Oxford: Oxford University Press.

Department of Employment, 1981, New Earnings Survey, 1981, London: Government Statistical Service.

_____, 1982, Family Expenditure Survey. Report for 1981 Giving the Results for the United Kingdom, London: HMSO.

Feinberg, S., 1970, An Iterative Procedure for Estimation in Contingency Tables, Annals of Mathematical Statistics, 41: 907-917.

Fisher, I., 1922, The Making of Index Numbers, Boston: Houghton.

Griffith, D., and A. Lea (eds.), 1983, Evolving Geographical Structures, The Hague: Martinus Nijhoff.

Harris, B., and A. Wilson, 1978, Equilibrium Values and Dynamics of Attractiveness Values in Production-Constrained Spatial Interaction Models, Environment and Planning A, 10: 371-388.

Havemann, R., and K. Hollenbeck (eds.), 1980, Micro-economic Simulation Models for Public Policy Analysis, New York: Academic Press.

Jones, P., 1979, HATS: A Technique for Investigating Household Decisions, Environment and Planning A, 11: 59-70.

Kain, J., and W. Apgar, 1977, Simulation of the Market Effects of Housing Allowances, Department of City and Regional Planning, Harvard University.

Lenntorp, B., 1976, Paths in Space-time Environments, Lund Studies in Geography, Series B, Human Geography, 44.

Lombardo, S., and G. Rabino, 1984, Non-linear Dynamic Models for Spatial Interaction: The Results of Some Empirical Experiments, Papers of the Regional Science Association, 55: 83-101.

McFadden, D., S. Cosslett, G. Duguay and W. Jung, 1977, Demographic Data for Policy Analysis: The Urban Travel Demand Forecasting Project, Phase 1, Final Report Series, Vol 8, Institute of Transportation Studies, University of California, Berkeley.

Orcutt, G., S. Caldwell and R. Wertheimer, 1976, Policy Exploration Through Micro-Simulation, Washington, D. C.: The Urban Institute.

OPCS, 1980, 1981 Census: Small Area Statistics. Prospectus. OPCS Monitor CEN 80/8, London: HMSO.

_____, 1983, General Household Survey, 1981. OPCS Social Survey Division, Series GHS11, London: HMSO.

Shannon, C., 1948, A Mathematical Theory of Communication, Bell System Technical Journal, 27: 379-423 and 623-656.

Thrift, N., 1981, Behavioural Geography: A Paradigm in Search of a Paradigm, in Quantitative Geography: A British View, edited by N. Wrigley and R. Bennett, London: Methuen, pp. 352-365.

Wegener, M., 1983, A Simulation Study of Movement in the Dortmund Housing Market, Tijdschrift voor Economische en Sociale Geografie, 74: 267-281.

Williams, H., P. Keys and M. Clarke, 1986, Vacancy Chain Models for Housing and Employment Systems, Environment and Planning A, 18: 89-105.

Wilson, A., 1981, Catastrophe Theory and Bifurcation: Applications to Urban and Regional Systems, London: Croom Helm.

_____, 1985, Structural Dynamics and Spatial Analysis, Working Paper 431, School of Geography, University of Leeds.

_____, and M. Birkin, 1984, Models of Agricultural Location in a Spatial Interaction Framework, Working Paper 399, School of Geography, University of Leeds.

_____, and M. Clarke, 1979, Some Illustrations of Catastrophe Theory Applied to Urban Retailing Structures, in Developments in Urban and Regional Analysis, edited by M. Breheny, London: Pion, pp. 5-27.

_____, and C. Pownall, 1976, A New Representation of the Urban System for Modelling and for the Study of Micro-level Interdependence, Area, 8: 246-254.

NEW DEVELOPMENTS OF A DYNAMIC URBAN RETAIL MODEL WITH REFERENCE TO CONSUMERS' MOBILITY AND COSTS FOR DEVELOPERS

Silvana T. Lombardo
University of Rome 'La Sapienza'
Italy

1. INTRODUCTION

In the context of locational problems, the location of shopping centres is certainly the one most easily modelled by simulation of the economic mechanisms of demand and supply formation, as well as the clearing of markets (Anas, 1986; Wilson, 1985). In the case of shopping centres, we find that on the consumers' side a large number of operators can make a free choice on the basis of their evaluations of supply conditions (for example, they can take into account the access time to the various centres, and/or the advantages connected with the presence of a number of shops of the same or different kind in the same shopping centre). On the developers' side, we also have a large number of operators who will make their locational choices on the basis of an evaluation of costs (e. g., the costs of setting up and running a business, or of transport goods), and revenues (from sales, taking into account competition with other shops) relative to a given location. Naturally the construction of a model to deal with this general problem is far from simple, considering the complexity of those processes involved in representing these two sides of the market.

A number of studies have been carried out, such as those by the Leed's School (e. g., Clarke, Clarke and Wilson, 1985; Fotheringham, 1983), mainly on the role played by the presence of different kinds and levels of services in the choice behaviour of consumers. Much less attention has been paid to the representation of costs for developers. Though some interesting insights are provided by Birkin and Wilson (1985) and by Haag (1986), there is yet much work to be done. If we were to take into account only the aspects relating to 'location costs,' that is the costs of installation of a shopping centre, keeping all other factors constant, we already would be able to generate a whole family of models. There is a need to consider the behaviour of the developer in relation to a commercial plan. There are the questions of technological indivisibilities of the size of the centre, positive externalities (e. g., use of existing infrastructures) and negative externalities (e. g., absence of carparks). But there are also 'interaction costs' related to accessibility--the costs of transporting goods from wholesalers--which depend upon both the location of wholesalers and on the structure of the transport network. This aspect has been neglected so far.

Another aspect of the problem that we wish to emphasize here is the fact that all the above phenomena involve a time dimension. The choice of shopping centre by a consumer, for example, will vary over time according to prevailing supply conditions, by comparison among centres, or simply by random choice. Similarly, on the side of the developer the size of the shopping centre will be the result of a sequence of decisions taken over a period of time on the basis of expectations, costs, and the such. These factors lead us to think in terms of a model made up of two sets of dynamic equations (Haag, 1986), one relating to the behaviour of developers and the other to that of the consumers. These equations would need to be constructed in such a way that they would take into account the dependence of the behaviour of each developer on the behaviour of other developers and the consumers, and the dependence of the behaviour of each consumer on that of other consumers as well as the developers.

2. THE GENERAL MODEL FRAMEWORK

To model these dynamic processes there is a wide choice of alternative mathematical formulations. The dynamic equations can be written as difference equations, as differential or integral equations, as deterministic or stochastic, and can treat space as continuous or discrete. Without wishing to suggest that this is necessarily the best choice, it should be emphasized that the differential form highlights mechanisms of change in the dynamic behaviour of the model more successfully than do integral forms, and that discrete space is more manageable than is continuous space. A deterministic treatment certainly is more approximate than is a stochastic treatment; but, on the other hand, determinism avoids many analytical and computational complications. The dynamic behaviour of actors can be interpreted in two different ways:

(a) we can see the variation in behaviour as a transition from one state to another; or,

(b) we can see the variation in behaviour as the response to a stimulus.

The former of these two options leads to the so-called compartmental approach (Godfrey, 1983; Weidlich and Haag, 1983). Let P_i be the variable that describes the i-th state (and therefore a certain kind of behaviour) of the actors. The changes in behaviour (i. e., in states) are described by the following differential equations:

$$\dot{P}_i = \sum_j P_j r_{ji}(\alpha,P,Q,t) - P_i \sum_j r_{ij}(\alpha,P,Q,t) \quad . \tag{2.1}$$

The right hand side of equation (2.1) enumerates all possible transitions between different states of the system, and the variable r_{ji} represents the rate of transition (i. e., the probability of a transition per unit of time) between state j and state i. In equation (2.1) we can see that the transition rate is a function of parameters (α), of the state of the system itself (P), of other variables (Q), and of time (t). The specification of the transition rate, as a function of all market factors, is the aim of the construction of a shopping model.

The second option mentioned above leads to the so-called logistic approach (Volterra, 1926). Changes in this system are described by the following differential equations:

$$\mathring{P}_i = f_i[K_i(\alpha,P,Q,t) - G_i(\alpha,P,Q,t)] \quad . \qquad (2.2)$$

The term in square brackets represents the stimulus that caused variation, and consists of the difference between a certain entity that is a function of the above quantities, and another entity that is a function of these same quantities. The function f_i specifies how this stimulus is translated into a variation in the state of the system. The specification of functions f_i, K_i and G_i in relation to all factors is the aim of the construction of the shopping model here, too.

As we said previously, a general model can be constructed as two sets of equations of type (2.1) or (2.2), which respectively relate to consumers and developers. As far as spatial aspects are concerned, we can observe that on the developers' side their behaviour can be considered either from the viewpoint of (a) or (b) described above. If we denote a given location with index i, we can see the problem of location of shopping centres either as a move from one zone to another in response to variations in demand [approach (a)], or as the opening, expansion or closure of a retail centre in a given zone, in response to variations in demand [approach (b)]. Although both phenomena occur in reality, it is reasonable to assume that the latter is more important than the former. For this reason, in the model proposed here the behaviour of developers is treated as being of the second type (logistic). The behaviour of consumers also can be considered from two different points of view. The demand in a given zone can be seen as spatial mobility in response to a variation in supply [approach (a)], or as an increase/decrease in the number of consumers in a given zone in response to a variation in supply, as well as other factors, such as demographic factors [approach (b)]. Although both phenomena occur in reality, the elasticity of demand with respect to supply is far less important than the spatial mobility of consumers; therefore, their behaviour is treated here in terms of the first approach. Basically we are attempting to model the dynamics of the mobility of an activity (consumer behaviour) in the context of the dynamics of a stock (developer behaviour).

3. THE TREATMENT OF CONSUMERS' MOBILITY

Given that consumer mobility can be described by an equation of type (2.1), and shopping centre dynamics by an equation of type (2.2), we need to see what forms the functions f_i, K_i, G_i, and the transition rates r_{ji} may take. For the model of consumer mobility, we make two assumptions. First, the process of mobility between the various shopping centres depends upon the place of residence of the consumers (i. e., we must consider the different distances of the various shopping centres from the place of residence). This assumption implies that mobility must be disaggregated for each zone of residence, with the total mobility being obtained by summing the results for all the zones of residence. Second, as well as the factor considered above, the choice of shopping centre is made by comparing the supply in terms of size, possibly taking into account the sensitivity of the consumer to the effect of economies of scale. Consequently, the transition rate is expressed as

$$r^k_{ji} = \lambda W^{\alpha}_i \exp(-\beta c_{ki}) / W^{\alpha}_j \exp(-\beta c_{kj}) \quad , \tag{3.1}$$

where W_i is the size of the shopping centre in zone i, c_{ki} is the cost of travel from the zone of residence k to shopping zone i, λ is a constant that measures the general level of mobility of consumers, and α and β are parameters that define the weight of economies of scale and the deterrence of the cost of travel, respectively. It should be observed that these assumptions represent a rather simple set of hypotheses that serves for the first stage of formulation of the model, but which easily could be substituted by more complex hypotheses, both in relation to the choice process (e. g., the choice of the nearest shopping centre) and the evaluation of supply (e. g., retail prices, the possibility of access to other kinds of services, or time lag in perception of supply).

Two assumptions are made for the stock model, too. First, the stimulus for the increase or decrease in the size of shopping centres in a particular zone is given by the difference between revenues and costs for the sale of goods. In order to obtain a rough estimate, instead of keeping a detailed sales account we assume revenue to be proportional to the number of customers. Similarly, instead of calculating all the costs (purchase of goods, staff expenses, owner's income, management costs and costs of changing the size of the centre), costs are assumed to be proportional to the size of the centre itself. Second, the variation in the size of a shopping centre in a given zone is directly proportional to the difference between revenues and costs. Therefore, the functions f_i, K_i, and G_i can be expressed as follows:

$$f_i = \varepsilon[K_i - G_i] \quad , \tag{3.2}$$

$$K_i = \sum_k e\, P^k_i \quad , \tag{3.3}$$

$$G_i = g_i\, W_i \quad , \tag{3.4}$$

where ε is the ratio between the speed of variation in shopping centre size and the difference between costs and revenues, e is the revenue (as defined above) per consumer, g_i is the unit cost (as defined above) of the production of the service in zone i, and P^k_i is the number of consumers living in zone k who use the shopping centre in zone i. If these four preceding assumptions are combined (remembering that they consist of the simplest hypotheses and can be replaced by more complex ones), then the following model is obtained:

$$\mathring{W}_i = \varepsilon[\sum_k e\, P_i^k - g_i\, W_i] \quad , \tag{3.5a}$$

$$\mathring{P}_i^k = \sum_j P_j^k\, r_{ji}^k - P_i^k \sum_j r_{ij}^k \quad , \tag{3.5b}$$

$$\text{with } r_{ji}^k = \lambda W_i^\alpha \exp(-\beta c_{ki}) / W_j^\alpha \exp(-\beta c_{kj}) \quad .$$

The stationary states of the model (3.5) can be obtained from the conditions

$$\mathring{W}_i = 0 \quad , \text{ and} \tag{3.6}$$

$$\mathring{P}_i^k = 0 \quad . \tag{3.7}$$

These conditions imply, respectively, that

$$\sum_k e\, P_i^k = g_i\, W_i \quad , \text{ and} \tag{3.8}$$

$$\sum_j P_j^k\, r_{ji}^k = P_i^k \sum_j r_{ij}^k \quad . \tag{3.9}$$

That is, for every zone i, costs are equal to revenues and the number of new consumers in the zone is equal to the number who choose other zones. In addition, if we insert into equation (3.9) the rates of transition r_{ij}^k given by equation (3.1), and bearing in mind that at equilibrium the W_i are constant, we obtain, after completing algebraic manipulations, the following expression for the P_i^k at equilibrium:

$$P_i^k = (\sum_i P_i^k) W_i^{2\alpha} \exp(-2\beta c_{ki}) / \sum_i W_i^{2\alpha} \exp(-2\beta c_{ki}) \quad , \tag{3.10}$$

where $\sum_i P_i^k = P_*^k$ is the (constant) total number of consumers living in zone k. We observe immediately that in equilibrium the distribution of consumers among the shopping zones corresponds to a spatial interaction model whose attraction and impedance factors coincide with the transition rate factors, except for a ratio of 1 to 2 in the relative parameters. If we then substitute equation (3.10) into (3.8) we obtain, for the equilibrium W_i,

$$W_i = (1/g_i) \sum_k e\, P_*^k\, W_i^{2\alpha} \exp(-2\beta c_{ki}) / \sum_i W_i^{2\alpha} \exp(-2\beta c_{ki}) \quad . \tag{3.11}$$

In equation (3.11) we can recognize with no difficulty the equilibrium solution of the Harris and Wilson model (Harris and Wilson, 1978). This result is immediately comprehensible if we observe that this last model consists of equation (3.5a) in which the P_i^k are defined by equation (3.11), which is the equilibrium solution of (3.5b). In other words, the Harris and Wilson model is a special case of our proposed model. Moreover, the Harris and Wilson model is a special case in which the consumer dynamics are many times faster than the shopping centres dynamics, so that in the model a stationary solution can be used instead of the transitory states of the former dynamics.

In reality, however, the entrepreneurial dynamics appear to be capable of a relatively rapid response in relation to demand. Therefore, it would seem more reasonable to utilize the more general model, and interesting to compare, using numerical simulations, the results obtained from these two different models, especially with regard to transitory states, as the stationary states already have been shown to be the same.

3.1. Simulation Results

A set of simulation experiments was carried out using a hexagonal grid of the Christallerian type already used for previous simulations with the Harris and Wilson model. The behaviour of this latter model, therefore, already has been explored both for these specific conditions (Lombardo and Rabino, 1985) and for general conditions (Clarke and Wilson, 1981; Rijk and Vorst, 1983; Lombardo and Rabino, 1984, 1986). It should be noted that even for a limited number of zones, the number of equations to be integrated in order to solve the proposed new model can be very large. In fact, whereas in Harris and Wilson's model the number of equations is equal to the number of zones with possible shopping centres, in our model it is equal to the number of zones with shopping centres, multiplied by the number of residential zones plus one. Considering the effort of computation involved, it was decided for the initial model tests to use a simpler situation in which only two zones were considered to be residential. There are good reasons, however, for believing that this simplification does not restrict the generality of the results, except for the number and type of equilibria obtained (which are equal in any case to those obtained with the Harris and Wilson model). For the integrations of the differential equations, the Runge-Kutta method was used.

In order to ensure complete comparability between Harris and Wilson's model and our own, it was necessary for the simulation conditions to be the same. Therefore, besides adopting the same territorial system (equal travel costs matrix and resident population), we also used the same initial values of W_i, and the same initial values for service demand coming from each zone. In Harris and Wilson's model this demand is calculated through a spatial interaction model that is always in a stationary state. In the proposed model the quantity $\sum_k P_i^k$ corresponds to the same spatial interaction model, and is pre-calculated for given values of W_i and of the other parameters. Also, in the Harris and Wilson model, the values α and β have been multiplied by a factor of 2. In Figure 1 the numbering of the zones, the distribution of consumers, and also a number of equilibrium solutions for different values of α and β are shown. We should remember that these equilibrium solutions coincide in the two models ($\alpha \to 2\alpha$ and $\beta \to 2\beta$, for the Harris and Wilson model). In addition, for $2\alpha < 1$ a positive solution is obtained (non-zero revenues in all zones) whereas for $2\alpha > 1$ solutions are not necessarily positive, except in the case of Figure 1.d, where the symmetry of the structure, the initial condition (W_i = 100 in all zones) and

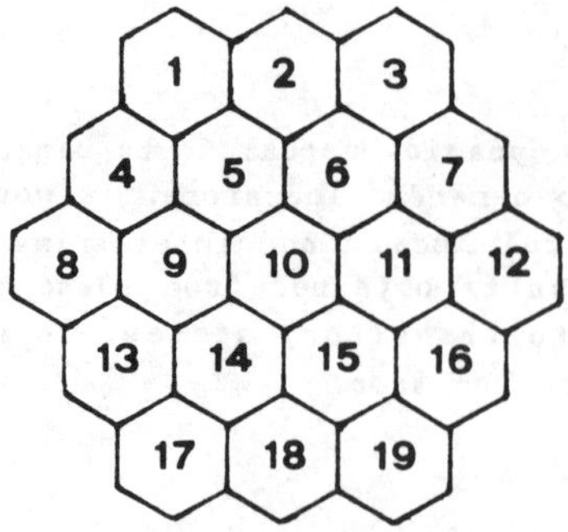

a. Zone numbering

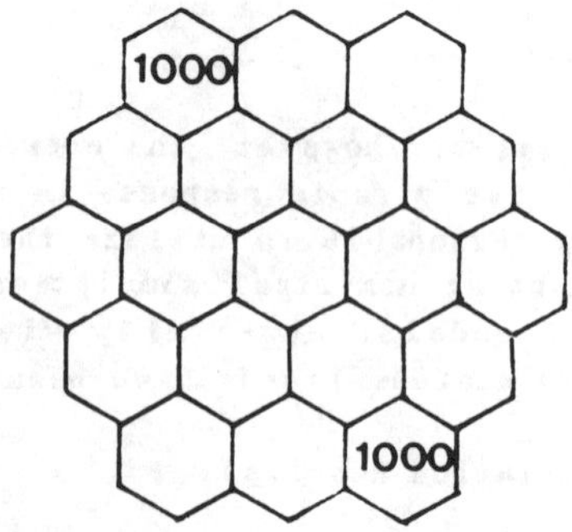

b. Consumers by place of residence

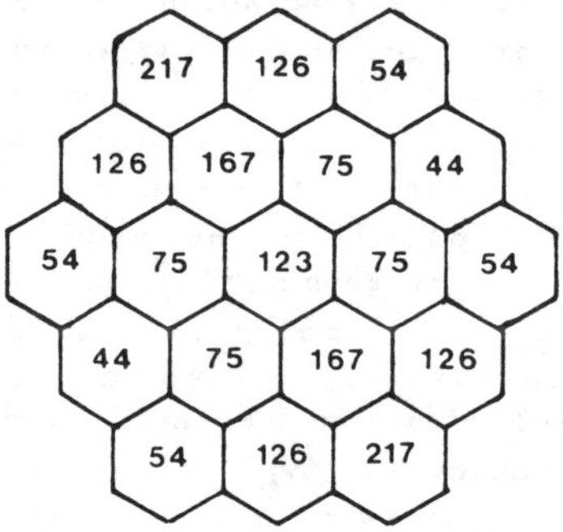

c. Shopping centres distribution with $\alpha = 0.35$, $\beta = 0.25$

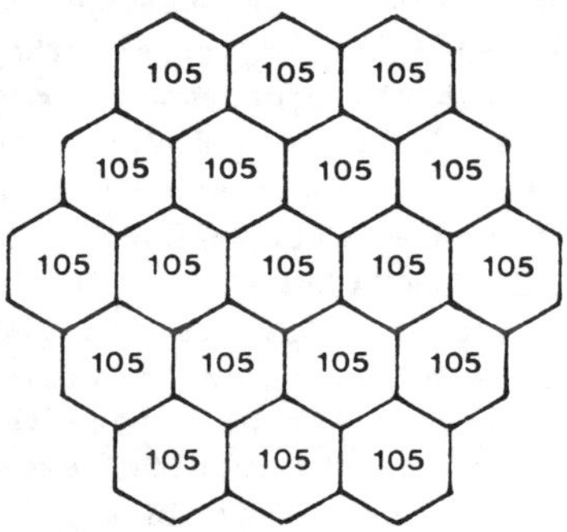

d. Shopping centres distribution with $\alpha = 0.65$, $\beta = 0.0025$

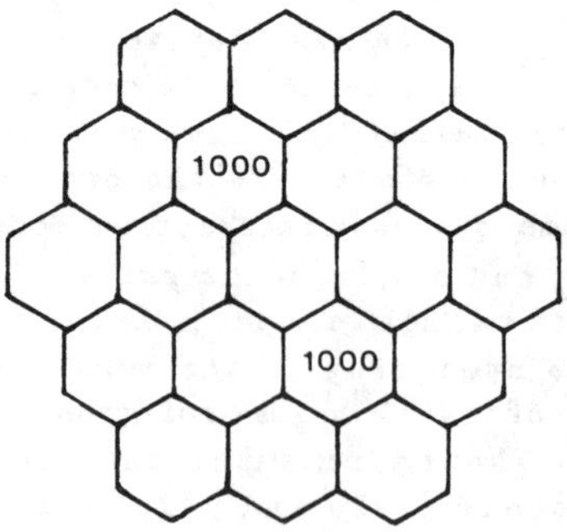

e. Shopping centres distribution with $\alpha = 0.65$, $\beta = 0.025$

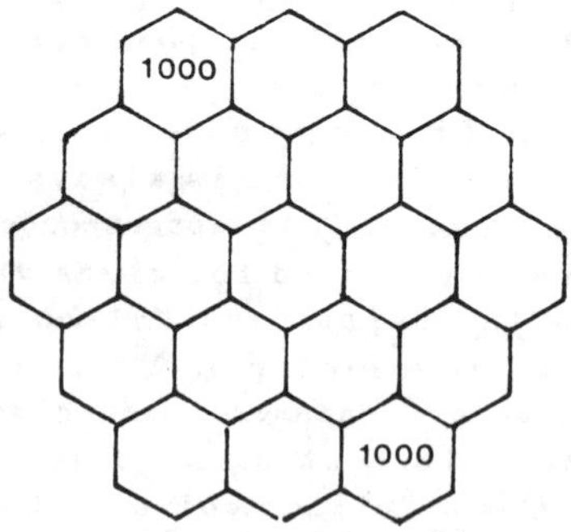

f. Shopping centres distribution with $\alpha = 0.65$, $\beta = 0.25$

Figure 1: Equilibrium solutions for different values of α and β.

the lack of importance of travel cost (β practically zero) permit a uniform distribution of revenue. In Table 1 the flows from residential Zone 1 to the various shopping zones in the initial state and the final state (see Figure 1.c and 1.d) are shown. For the final state Table 1 shows the flows calculated both with the spatial interaction model and with the dynamic mobility equations. We observe that, as predicted, in equilibrium these two flows coincide (within rounding error).

Figure 2 illustrates a number of significant cases of the evolution from the initial state to an equilibrium for shopping facilities in Zone 1, with $\alpha = 0.35$ and $\beta = 0.25$ (see Figure 1.c) for different values of ε and λ. There are, in fact, two anomalous cases, namely $\varepsilon = 0$ and $\lambda = 0$. In the

TABLE 1

FLOWS FROM ZONE 1 TO THE SHOPPING AREAS

Shopping Area	$\alpha = 0.35$, $\beta = 0.25$ t = 0	t = equilibrium proposed model	t = equilibrium Harris/Wilson model	$\alpha = 0.65$, $\beta = 0.025$ t = 0	t = equilibrium proposed model	t = equilibrium Harris/Wilson model
1	117	175	178	55	0	0
2	93	100	100	54	0	0
3	58	39	38	53	0	0
4	53	100	100	54	0	0
5	53	118	119	54	512	512
6	58	44	46	53	0	0
7	38	23	22	52	0	0
8	58	39	38	53	0	0
9	58	46	46	53	0	0
10	58	62	62	53	0	0
11	38	30	30	52	0	0
12	25	18	17	51	0	0
13	38	23	22	52	0	0
14	38	30	30	52	0	0
15	38	47	47	52	489	489
16	25	26	26	51	0	0
17	25	18	17	51	0	0
18	25	26	26	51	0	0
19	25	35	35	50	0	0

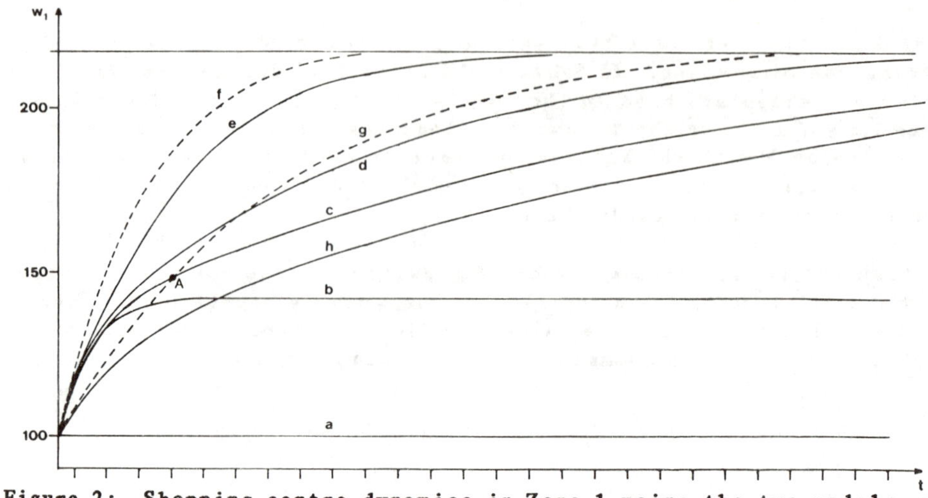

Figure 2: Shopping centre dynamics in Zone 1 using the two models, with α=0.35 and β=0.25, and for different values of ε and λ.

first case, as $\dot{W}_i = 0$, W_i does not change (curve a). In the new model P_i^k could vary up to an equilibrium pattern of flows corresponding to the given distribution of W_i, but as the initial situation adopted is assumed to be in equilibrium already, P_i^k does not vary. In the second case, as $\dot{P}_i^k = 0$, P_i^k does not vary but W_i adapts itself to the constant demand. The state of equilibrium obtained, therefore, is not the same as that obtained with the two models since demand no longer is a function of supply [curve (b)]. In all other cases (λ ≠ 0, ε ≠ 0) we find that the two models converge to the same equilibrium states.

Curves (c), (d) and (e) of Figure 2 represent the evolutionary path for a given value of ε (0.5) and different values of λ. For the purpose of comparison, curve (f) shows the path of the Harris and Wilson model for the corresponding values of ε. The first feature we observe is that the higher the value of λ the more the curves of the two models tend to coincide. The second feature is that the effect of λ is to increase the time necessary to reach equilibrium, and that the amount of time needed to reach equilibrium is inversely related to the mobility of consumers. The main characteristic, however, is the great difference between these two types of curves. The curve relating to the Harris and Wilson model is very similar to an exponential curve, which also corresponds with the hypothesis that the speed of adjustment is proportional to the distance from the equilibrium state. For the new model, while an initial section of the curve looks like an exponential one, the curve takes on a more linear shape. This result is due to the initial dominance of the adjustment of supply to demand, whereas in the final part of the curve the demand-supply imbalance is reduced, and hence the curve is influenced more by demand mobility. To confirm this contention we can try to see if the effects of demand mobility can be reproduced by the Harris and Wilson model with a lower ε value; that is, by slowing down the process of adjustment of supply.

In order to achieve this goal, we can compare that curve obtained with the new model for a given value of ε (0.5) and λ [see curve (c)] with that curve obtained from the Harris and Wilson model for a lower value of ε [0.2; see curve (g)] so that the same condition is reached after the same period of time [point A, at the intersection of curves (c) and (g)]. We observe that although the initial pick-up of the Harris and Wilson model is slower (the adjustment speed adopted being substantially slower), in the second part of the curve convergence to equilibrium is faster than in curve (c). This difference between the two models is a significant one, since if one has available only a sufficiently precise series of data on the stock dynamics of real systems (it is not easy to obtain time series data for flows). Nevertheless, it is possible to calibrate both supply behaviour and the mobility of demand.

A final analysis to be reported here concerns comparisons between the behaviour of the two models with respect to changes in the speed of adjustment of supply to demand ($\varepsilon = 0.2$). In the case of the Harris and Wilson model, as is easily demonstrated and can be seen by comparing curves (f) and (g) in Figure 2, this variation causes a change in time scale equal to the ratio between the two speeds. A comparison of curves (c) and (h), on the other hand, shows that this feature is not present in the new model, where only the first part of the curves behaves in this way, while the upper part of the curves tends to change by a decreasing time factor. This conclusion once again confirms that the system dynamics is strongly influenced by demand mobility.

4. THE TREATMENT OF COSTS FOR DEVELOPERS

In Section 3 a simple form of the cost function was adopted, [i. e., the costs were assumed to be proportional to the size of a shopping centre, the term $g_i W_i$ in equation (3.5)]. Let us now investigate what kind of results can be obtained when one uses a more complex and realistic cost function. It must be noted that the following considerations concerning the use of a different cost function can be couched both in terms of the Harris and Wilson model, and in terms of the model presented in Section 3, since they concern the equilibrium solutions.

The form of the cost function proposed here introduces the consideration of space, in terms of accessibility of developers to their suppliers as well as to their consumers. Hence, this form concerns the transport of goods from wholesalers to shops. The transport of goods can be considered as a choice process that depends mainly on the location of wholesalers, the size of the shops, and the cost of transportation.

Therefore, this phenomenon can be modeled by a spatial interaction model of the form

$$M_{ki} = Q_k B_k W_i^{\gamma}\exp(-\delta v_{ki}) \quad , \tag{4.1}$$

where M_{ki} is the flow of goods distributed from wholesalers in zone k to the retailing of zone i, Q_k is the quantity of goods distributed by wholesalers in zone k, W_i is the size of the shops in zone i, v_{ki} is the cost of transport of a unit of good from zone k to zone i, and γ and δ are parameters measuring, respectively, economies of scale and the deterrence of transport cost. In addition,

$$B_k = [\sum_i W_i^{\gamma}\exp(-\delta v_{ki})]^{-1} \quad , \tag{4.2}$$

which ensures that all the goods considered are distributed. We now can calculate that element of the cost function G_i which depends upon transport costs, namely

$$S_i = \sum_k (q + v_{ki}) Q_k B_k W_i^{\gamma}\exp(-\delta v_{ki}) \quad , \tag{4.3}$$

where q is the price of a unit of good for the retailer. If v_{ki} increases, the quantity of goods transported decreases, due to $\exp(-\delta v_{ki})$, but the cost of a unit of good increases. Assuming a high mobility of goods (i. e., a relatively low value of δ), the latter phenomenon becomes dominant, so that the cost S_i can be viewed as a positive function of v_{ki}. We can assume that the other elements of the cost function depend on the size of a shop, or on both its size and location, and that they are proportional to the size of the shop. Accordingly, we can write, for the former case,

$$U_i = u W_i \quad , \tag{4.4}$$

and for the latter case,

$$Z_i = z_i W_i \quad . \tag{4.5}$$

Combining these two results yields

$$R_i = (u + z_i) W_i \quad . \tag{4.6}$$

Now, rather than equation (3.4), the cost function assumes the form

$$G_i = \sum_k (q + v_{ki}) Q_k B_k W_i^{\gamma}\exp(-\delta v_{ki}) + R_i \quad . \tag{4.7}$$

4.1. Considerations About Equilibrium Solutions

By comparing this model with the Harris and Wilson model, we can see immediately that the equilibrium solutions will not be the same, because of the different and more complex form of the cost function [equation (4.7)]. Here we analyze this point heuristically, while simulations of the dynamic behaviour of the whole model will be carried out next.

In the Harris and Wilson model the form of the revenue curve depends upon the value of the parameter α (see Figure 3), while the cost curve is a straight line. The possible equilibrium solutions are well known and are represented in Figure 3 (Wilson, 1981). In the new model, we have a revenue curve that takes the same shape as in Figure 3, while the cost curve is the sum of a curve similar to the revenue one and a line [see equation (4.7)]. The shape of the S_i curve [see equation (4.3)] depends upon the value of the parameter γ. This result is shown in Figure 4, where different contributions of the transport cost and of the other costs to the resulting cost curve also are indicated.

Many interesting situations can be uncovered, depending on the values of the exogenous variables. An example is given here, formulated by keeping constant the revenue curve, the 'other costs' line and the parameters α and γ, and varying the cost of transport of goods (and consequently S_i), which is the new element introduced. In Figure 5.a we can see that for a certain

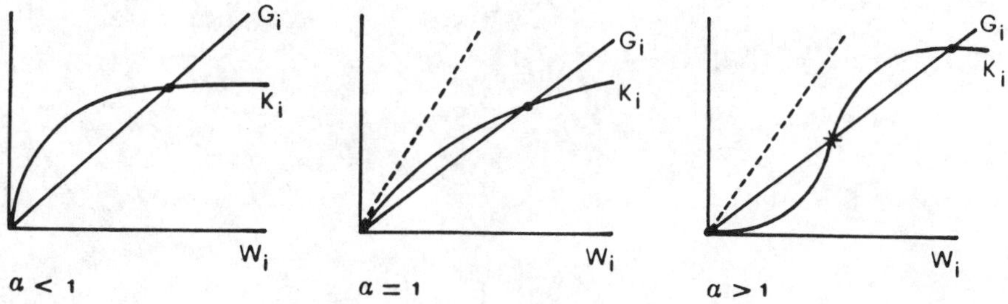

Figure 3: Harris and Wilson model; revenue curve and cost curve with stable and unstable equilibrium points.

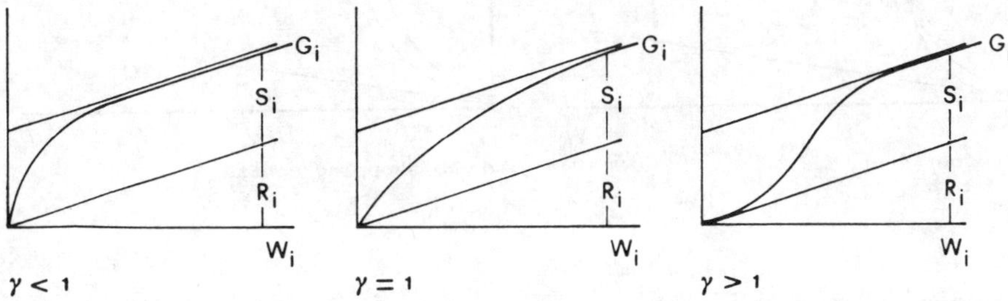

Figure 4: The new model; cost curve.

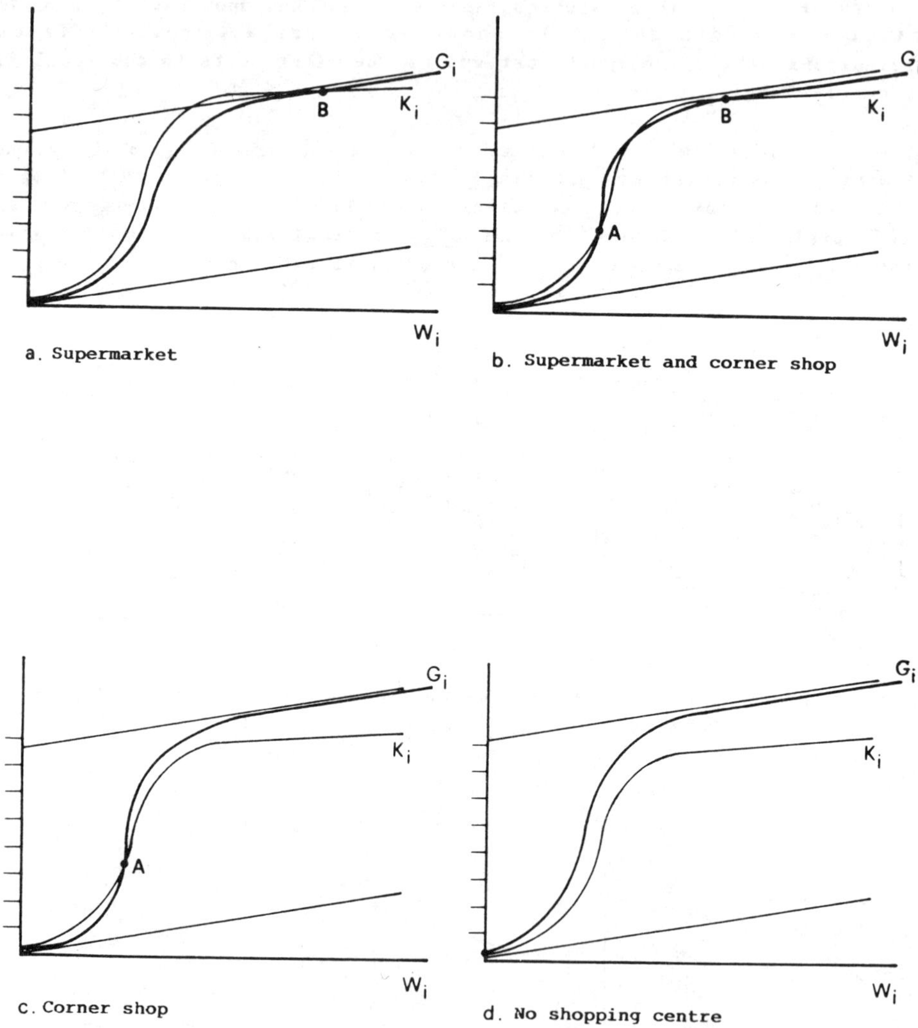

Figure 5: The new model; effects of varying cost of transporting goods.

value of S_i we obtain one stable equilibrium point, which is different from zero. Increasing the value of S_i, two stable equilibrium points different from zero appear (see Figure 5.b). Further increases of the value of S_i lead first to a unique non-zero equilibrium solution (see Figure 5.c), and then to a solution equal to zero (see Figure 5.d). As the points A and B represent quite different sizes of shopping centres, they can be interpreted, respectively, as the corner shop solution and the supermarket solution. Therefore, in this example (where $\alpha > 1$ and $\gamma > 1$), other things being equal, a low cost of transport of goods allows the existence of a large shop (supermarket), while a high cost leads to a small corner shop or to the absence of shops.

The most interesting case, though, is that represented in Figure 5.b, where two stable equilibrium solutions appear. In this case both the supermarket solution and the corner shop solution are possible. Analyzing this case, it is possible to see that the transition from one kind of solution to another is not continuous; rather, a slight variation of S_i can correspond to the disappearing of one of the two solutions. This result can be shown by plotting the values taken by W_i for different values of S_i (see Figure 6.a). When the value of S_i is high (say $S_i > H$), only the corner shop solution is possible. Further, a small decrease of S_i suddenly permits the supermarket solution to be possible, together with the corner shop (see also Figure 6.b). For a further decrease of S_i (say $S_i < E$) the corner shop solution disappears, and only the supermarket solution is possible. Consequently, the values of S_i, such that $E \leq S_i \leq H$, are a set of critical values for which zone i can jump from one solution to another, because of slight fluctuations of the system.

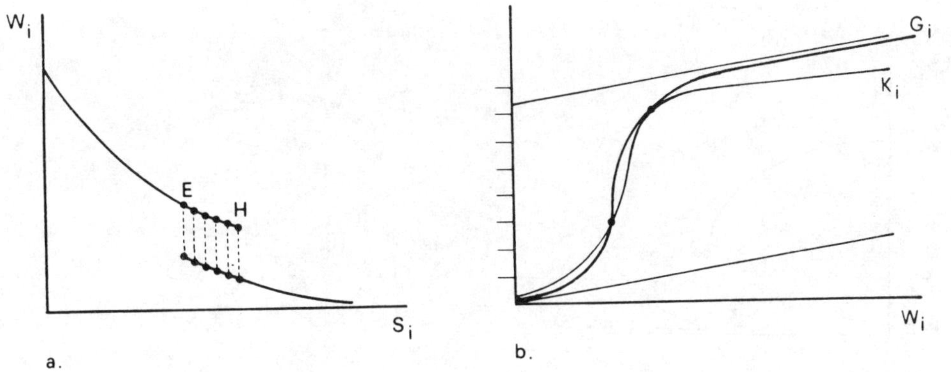

Figure 6: The new model; (a) shopping centre size for different values of S_i, and (b) revenue curve and cost curve for S_i=H.

5. CONCLUSIONS

In this paper new developments in the dynamic modeling of retail services location are presented. The new formulations can be compared with the Harris and Wilson model, which provides a basic point of reference in this field. The development proposed in Section 3 concerns the introduction of dynamic behaviour for demand, and its relations with the dynamics of supply. By investigating the results of a set of simulations that were carried out using both the new model and the Harris and Wilson model, we can see that despite the similarities in the formulations of the two models, which means that many results are the same (in particular the equilibrium solutions), an important difference in the transitory behaviour of the system emerges. Since in a real system the surrounding conditions within which the demand and supply for retail services operate are extremely variable (e. g., population changes, and changes in the transport network), transitory states become a normal condition. From this fact we draw the conclusion that simulating real systems with the model proposed here produces significantly different results from the Harris and Wilson model. Moreover, the new model, having greater generality, leads towards a more satisfactory representation of real systems.

The model development proposed in Section 4 concerns a more accurate description of the costs for developers, with particular focus on 'interaction costs' relating to problems of accessibility for developers. Comparing the behaviour of this model with the behaviour of the Harris and Wilson model, we found that in this latter case the transition from a large shopping centre to a small one cannot be represented, and is provided as an interpretation of the transition between a solution different from zero to a solution equal to zero (Wilson, 1981, p. 81). By means of the new formulation, this transition is more realistically represented as a movement between a large structure and a small, but measurable one. Moreover, in the

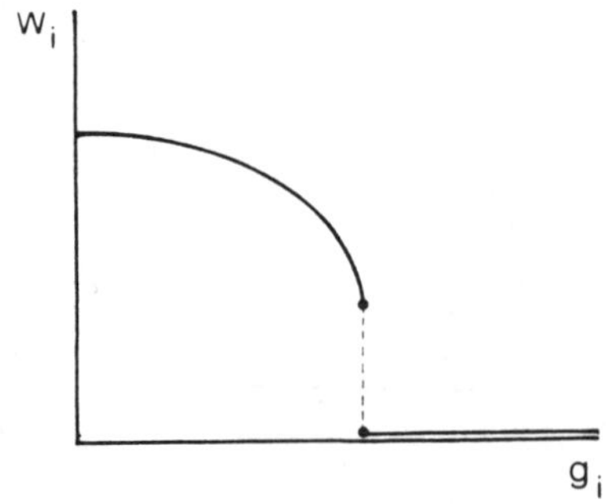

Figure 7: Harris and Wilson model; shopping centre size for different values of g_i.

Harris and Wilson model the bifurcation set is one point that separates the two solutions (see Figure 7), while in this model it is, more realistically, a set of values that describes the concomitance of the two solutions (see Figure 6a). Thus, this new model allows for the coexistence of a corner shop and a supermarket in the same zone. Therefore, the new model version can be considered as a substantial improvement.

Given the foregoing considerations, two main directions for future research can be identified. First there is a need for a more accurate modeling of the dynamic behaviour of demand by means of the identification of factors that determine such behaviour (i. e., in the direction of giving a more appropriate formulation to the transition rate r^k_{ji}). Second, a more comprehensive analysis is needed of the cost function, such that this function is derived from a more detailed study of the economic behaviour of developers. Developments in these directions will be useful in the provision of this kind of model perspective with both a more sound economic and behavioural basis, and a higher reliability when simulating real systems.

6. REFERENCES

Anas, A., 1986, Economic, Urban and Regional Simulation, Encyclopedia of Economics, forthcoming.

Birkin, M., and A. Wilson, 1985, Some Properties of Spatial Structural Economic Dynamic Models, paper presented at the IV European Colloquium of Theoretical and Quantitative Geography, Veldhoven, The Netherlands, September 9-13.

Clarke, G., M. Clarke and A. Wilson, 1985, Multiple Bifurcation Effects with a Logistic Attractiveness Function in the Supply Side of a Service System, Working Paper 397, School of Geography, University of Leeds.

Clarke, M., and A. Wilson, 1981, The Dynamics of Urban Spatial Structure: Progress and Problems, Working Paper 313, School of Geography, University of Leeds.

Fotheringham, A., 1983, A New Set of Spatial Interaction Models: The Theory of Competing Destinations, Environment and Planning A, 15: 15-36.

Godfrey, K., 1983, Compartmental Models and Their Application, London: Academic Press.

Haag, G., 1985, A Stochastic Dynamical Theory of Residential and Labor Mobility, Including Travel Networks, paper presented at the NATO Advanced Studies Institute entitled 'Transformations Through Space and Time,' Hanstholm, Denmark, August 3-11.

______, 1986, Services, in Urban Systems: Design for an Integrated Dynamic Model, edited by C. Bertuglia, G. Leonardi and A. Wilson, London: Croom Helm, forthcoming.

Harris, B., and A. Wilson, 1978, Equilibrium Values and Dynamics of Attractiveness Terms in Production-constrained Spatial Interaction Models, Environment and Planning A, 10: 371-388.

Lombardo, S., and G. Rabino, 1984, Non-linear Dynamic Models for Spatial Interaction: The Results of Some Empirical Experiments, Papers of the Regional Science Association, 55: 83-101.

______, and ______, 1985, A Note on the Convergency Rate of the Harris and Wilson Model, unpublished paper, Dipartimento di Pianificasione Territoriale e Urbanistica, Universita di Roma.

______, and ______, 1986, Calibration Procedures and Problems of Stability in Nonlinear Dynamic Spatial Interaction Modeling, Environment and Planning A, 18: in press.

Rijk, F., and A. Vorst, 1983, Equilibrium Points in an Urban Retail Model and Their Connection With Dynamical Systems, Regional Science and Urban Economics, 13: 383-399.

Volterra, V., 1926, Variazioni e Fluttuazioni Nel Numero di Individui in Specie Animali Conviventi, Memorie Accademia Nazionale Lincei, 2(s/6): 31-113.

Weidlich, W., and G. Haag, 1983, Concepts and Models of a Quantitative Sociology: The Dynamics of Interacting Populations, Berlin: Springer-Verlag.

Wilson, A., 1981, Catastrophe Theory and Bifurcation: Applications to Urban and Regional Systems, London: Croom Helm.

______, 1985, Structure Dynamics and Spatial Analysis From Equilibrium Balancing Models to Extended Economic Models for Both Perfect and Imperfect Markets, Working Paper 431, School of Geography, University of Leeds.

SECTION 3

TRANSFORMATIONS INVOLVING INTERACTIONS OVER SPACE

The term 'interaction' carries two quite distinctive meanings in the theoretical literature in geography--interactions between subsystems (such as between consumers and producers, workers and firms) and interactions or movements over space, including transfers of households and labour, or flows of income. The specification of subsystem interaction and the ways in which spatial separation, spatial relationships and areal differentiation influence these system interactions are central issues in the papers in this section, which exemplify the range of problems in this area and the approaches that can be made. In each case space enters explicitly in model formulation, for the operation of the process mechanism is in part a function of the relative positioning of the entities coupled with the spatial metric in which the entities are located.

In the first paper Kanaroglou and Papageorgiou are interested in an evolutionary model of population systems in which special emphasis is placed upon disequilibria. The aim of this paper is to test ideas about disequilibria through a preliminary analysis of the annual Canadian regional population system for the time period of 1961 thru 1983. Linear and non-linear modeling results are contrasted. The nested logit model furnishes the kernel of the analysis, and migration is the phenomenon embraced by spatial autoregressive structures. Cyclical behavior of temporal disequilibria is disclosed whose patterns suggest the presents of bifurcation points. Certainly the non-linear relations and non-zero spatial autocorrelation necessary for bifurcation to occur are present. The notable finding in this paper is that the modelled system behavior provides a reasonably close approximation to observed behavior, with the non-linear model performing considerably better than its linear counterpart.

Dejon's paper is a theoretical examination of a production and exchange economy in which the geographical location and differentiation of actors (firms, landlords, households) is explicitly included in the model. Stochastic choice theory, where individuals can indulge a certain randomness in responding to alternatives, was used by Hildenbrand (1971) for the analysis of large exchange economies, but assuming independence between the actors both in their preferences and their endowments. He demonstrated the existence of price equilibria in such circumstances, a situation Follmer (1974) later showed will break down if one introduces local (nearest neighbour) interaction between individuals (so called 'waves of imitation'). Dejon discusses the situation where prices as well as quantities are included in the choice sets. The approach is developed and spatialised by introducing search costs into the model, with economic agents migrating between markets in response to different levels of attractiveness defined in terms of such things as wage rates and price levels.

Amrhein and MacKinnon develop a model of a spatial labour market that explicitly recognizes the positions of firms and workers in the market area as well as employee characteristics. Individual firm and worker contact fields (in the tradition of Hagerstrand) bind the labour market area together, providing the spatial framework for workers to receive local information about possible job assignments and for firms to make local comparisons relating to profit levels. It is through such local awareness and local comparisons that firms and workers adjust, and the system is transformed through time as firms seek higher profits and labour higher wages. One of the interesting possibilities opened up by a model of this sort is the ability not to identify and model the internal spatial attributes of labour markets, but also to explore in more detail the emergence of segmented and partitioned spaces. Such spatial structures have been highlighted and mathematical generators explored elsewhere and in different contexts by, among others, Curry (1985), Haining (1985), and Bennett, Haining and Wilson (1985). These are deep properties of spatial systems, hardly yet explored. But enough is known already to show how they may arise from a variety of often quite simple spatial interaction mechanisms.

The paper by Fotheringham and Knudsen, like the one by Lombardo in the preceding section, deals with discontinuous change in urban retailing, developing the ideas of applied catastrophe theory that have emerged from work at Leeds University and Brussels. The key idea here is that smooth changes in independent variables can give rise to sudden discontinuous change in a dependent variable--here interpreted in terms of urban retailing and the size and spacing of shopping areas. Mathematically there is a common, if yet unexplored, link between the bifurcations of these urban retailing models and the segmented and partitioned spaces generated in the phase transitions of certain spatial interaction models (Stewart, 1981). The paper by Fotheringham and Knudsen presents a framework for explaining discontinuities in both the size and relative location of retail facilities. Space enters explicitly into the model as consumers have different potential levels of expenditure at different sites, depending upon, among other things, relative location. The familiar gravity model is replaced by a competing destinations model that recognizes the importance of agglomeration economies and the benefits that may accrue from the spatial proximity of retail stores. The authors motivate their study by drawing on empirical work from Gainesville, Florida.

REFERENCES

Bennett, R., R. Haining and A. Wilson, 1985, Spatial Structure, Spatial Interaction and Their Integration: A Review of Alternative Models, Environment and Planning A, 17: 625-645.

Curry, L., 1985, Inefficiencies in the Geographical Operation of Labour Markets, Regional Studies, 19: 203-216.

Follmer, H., 1974, Random Economies with Many Interacting Agents, Journal of Mathematical Economics, 1: 51-62.

Haining, R., 1985, The Spatial Structure of Competition and Equilibrium Price Dispersion, Geographical Analysis, 17: 231-242.

Hildenbrand, W., 1971, Random Preferences and Equilibrium Analysis, Journal of Economic Theory, 3: 414-429.

Stewart, I., 1981, Applications of Catastrophe Theory to the Physical Sciences, Physica, 2D: 245-305.

DISEQUILIBRIUM IN THE CANADIAN REGIONAL SYSTEM: PRELIMINARY EVIDENCE, 1961-1983

P. S. Kanaroglou and Y. Y. Papageorgiou
McMaster University
Canada

1. INTRODUCTION

Disequilibrium seems to be a natural concept for dynamical systems. To fix ideas, consider the following situation (Kanaroglou, et. al., 1985, pp. 26-27):

> Suppose that you hold in your hands an invisible spherical bowl with a visible ball in it. As you walk, the ball moves inside the bowl. At any moment, given the characteristics of your walk, there is a point around the bottom of the bowl on which the ball would rest. This is the <u>steady-state</u> of the moment for the system ball-bowl. As it changes during your walk, the ball 'tries' to follow it, never catching it except by chance, and then for a fleeting moment. You could, of course, describe in three dimensions the trajectory of the ball during your walk without any reference to the bowl: after all, the only thing you see is the ball. Yet a study of the relationships between the ball, the bowl and the characteristics of your walk help you to understand the object of your concern. In particular, suppose that you could model this system and represent the characteristics of your walk by a set of parameters. At any moment, you can monitor the position of the ball and estimate the set of parameters associated with that moment. Since the bowl is spherical you know that, for any particular feasible set of parameters, there is a single position of the ball which represents a steady-state. Introducing the estimated set of parameters in your model, you can compute the steady-state of the moment. Comparing the difference between the observed and the steady-state positions of the ball, you could derive a measure of <u>disequilibrium</u> for the system, which corresponds to that particular moment. For example, the distance between the observed and steady-state positions of the ball could serve as a simple measure of disequilibrium. You may repeat the same procedure several times to learn something about the evolution of

> disequilibrium in your system, i. e. about the evolution of the difference between the observable trajectory of the ball and the unobservable trajectory of the corresponding steady-state.

Now transport these ideas to a migratory system consisting of a population distributed over a fixed number L of well-defined regions. A particular state of the system here is a particular distribution of the population over the regions. This corresponds to a particular position of the ball relative to the bowl for the system ball-bowl. The three-dimensional space of the system ball-bowl becomes the L-dimensional space of the migratory system. The laws of physics operating over the inner surface of the bowl as you move are replaced by another set of relations operating over the domain of feasible states as the socio-economic profile of the migratory system unfolds. Since the fundamental decision entity in the migratory system is an individual, the corresponding set of relations stems from the desire of individuals for higher utility that directs their migratory behaviour at the micro-level, and which, in aggregation, creates evolution at the macro-level. At any moment, given the socio-economic profile of the migratory system, there is a steady-state in the sense that the observed population distribution would remain invariant if the system found itself precisely in that state under the given socio-economic profile. Nevertheless, matters now become more complicated. In the system ball-bowl, further to the existence of a steady-state at every moment of its evolution, you knew that the steady-state was globally stable--hence it also was unique. This was ascertained by the perfectly spherical shape of the bowl. Were the shape of the bowl more complicated, global stability, hence uniqueness of steady-states, might not have been retained. Under these more general circumstances multiple steady-states could co-exist, each with different stability characteristics. Similarly, global stability of a steady-state represents only a special case for migratory systems; and even if only one steady-state existed at a particular moment of the system's evolution, individuals, in their quest to achieve higher utility, would not necessarily behave so as to bring the system inevitably closer to its steady-state under that particular socio-economic profile. In other words, under that particular socio-economic profile, the system could evolve toward a limit cycle. In general, as the socio-economic profile changes, particular steady-states appear, may evolve, and then disappear. At any moment, barring limit cycles, the desire of individuals for higher utility will direct the migratory system toward one of its steady-states. This tendency may persist over a certain period, beyond which the system comes to interact with another steady,-state. It is in this sense that both systems 'try' to follow their mutual steady-state, never catching it except by chance, and then for a fleeting moment.

The relationship between the observed state of a migratory system and the corresponding unobservable steady-state, as well as the evolution of this interaction, has drawn the interest of several scholars (see, for example, Rogers, 1967; Vining, 1975; Haag and Weidlich 1984). Empirical questions feeding such interest include: how clearly does the model of macro-interaction between observed and steady-states, which is deduced from the premise that individuals are directed toward higher perceived utility, account for the observed evolution of a migratory system? Does, for example, an observed regional population above (below) the corresponding steady-state over some time-period necessarily imply that the population decreases (increases)? Does it imply that these trends will continue in the near future? Furthermore,

> Is it or is it not the case that the arrangement of humans across space, say an integrated system of regions, is tending toward some stable form in all countries? Or is it rather the case that such patterns or arrangements are inherently unstable, alternating between two or more such 'equilibrium' states? (Vining, 1975, p. 157)

A fundamental attribute complicating answers to all these questions is that systems such as the ball-bowl and the migratory ones are typically _non-linear_. For the system ball-bowl, the dynamics of the ball change continuously in the sense that, for given motion characteristics corresponding to a particular instant, the curvature of the bowl imposes a distinct configuration of forces corresponding to every position of the ball relative to the bowl; the dynamics of the ball at every moment fully depend upon the corresponding state of the system. Similarly, given a socio-economic profile, the dynamics of a migratory system fully depend upon the corresponding population distribution. This dependency emerges because the distribution of utilities itself, through returns to scale and the like, must be sensitive to the population distribution. Thus, a particular socio-economic profile applied to an evolving population distribution will produce an evolving dynamical structure. Beyond these similarities, it is perhaps useful to discuss a further complication related to the non-linear nature of the two systems. At every instant during its motion, the ball 'knows' little about where it is actually heading, in fact infinitesimally little, which, together with the boundary conditions of the moment, determines the choice of the immediate future. In other words, over an infinitesimal time-period, the motion of the ball can be approximated by a _linear_ system determined over the joint characteristics of your walk and the relative position of the ball at the beginning of the infinitesimal time-period. During this period, the current state of the system seems to be directed toward the globally stable steady-state of the corresponding linear system (i. e., toward its _linear_ steady-state). Of course, at the beginning of the next infinitesimal time-period, the linear system approximating the motion of the ball will change even though the original characteristics of your walk remain invariant. This happens because, now, the initial conditions of the linear system include the position of the ball at the beginning of the new infinitesimal time-period--which has changed unless the system is in its steady-state. Hence, given the characteristics of your walk, the trajectory of the ball can be decomposed and described by an infinite series of linear systems leading to a <u>non-linear</u> steady-state associated with those characteristics. At every moment, given the characteristics of your walk, one may distinguish between the linear steady-state that directs the system in the short-run, and a non-linear steady-state that the system attains in the long-run.

Now consider the migratory system. It seems reasonable to assume that individuals in the system cannot be fully aware of the non-linear intricacies of this system. Further, if the system is large enough, every individual appears insignificant relative to the entire system. Hence no individual decision to migrate can, by itself, have a significant impact on the evolution of the system. Therefore it seems reasonable to assume that individuals take the current distribution of utilities in the system as parametric. In other words, these individuals do not take into account the effect of their potential move on the current distribution of utilities. That is, individuals move as if the system were linear. In the short-run, similar to the system ball-bowl, this would imply that during its evolution the observed state adjusts toward the linear steady-state--a testable hypothesis. In the long-run, however, given a particular socio-economic profile, the move of many individuals adjusting to present circumstances will force non-linearities in the migratory system precisely as the changing curvature of the bowl did in the system ball-bowl. Clearly, the myopic behaviour of individuals combined with the non-linear nature of the migratory system produce a major spatial externality, which provides another example to Schelling's (1971) ecology of micro-motives. This externality renders the interaction between observed and non-linear steady-states complex enough to preclude any hypotheses about its nature.

In this paper our objective is to test these ideas in a preliminary analysis of the Canadian regional population system for the time-period 1961 to 1983. The theoretical foundations of the methodology used is elucidated in Kanaroglou, Liaw and Papageorgiou (1985).

2. THEORETICAL FOUNDATION

2.1. The System

Consider a system of N individuals distributed over L areas. A state of this system is a random vector R[t] with realizations

$$\underline{x} \equiv (x_1, \dots, x_L) \quad \text{, with } 0 \leq x \leq 1 \text{ for } i = 1, \dots, L,$$

$$\sum_i x_i = 1 \quad , \tag{1}$$

where x_i denotes the proportion of N in region i. Let $\text{Prob}\{R[t]=\underline{x}\} \equiv P[\underline{x};t]$ represent the probability that the system is in state $\underline{x}$ at time t. Given that the Markov assumption holds, and given that the decisions of individuals to migrate are stochastically independent, the evolution of this probability is determined by the <u>master equation</u>

$$\frac{d}{dt} P[\underline{x}; t] = \sum_{i,j} (E_i^{+1} E_j^{-1} - 1)\, q_{ji}[t]\, x_i\, P[\underline{x}; t] \quad , \tag{2}$$

where $E_i^{+1} E_j^{-1} f[\underline{x}] \equiv f[x_1, \ldots, x_i+1, \ldots, x_j-1, \ldots, x_L]$, and $q_{ji}[t]$ is the <u>migration rate</u> from i to j at time t (see Haag and Weidlich, 1984). In general, migration rates can be written as

$$q_{ji}[t] = f_{ji}(\underline{x}[t]; \underline{\theta}[t]) \quad , \tag{3}$$

where $\underline{\theta}[t]$ represents all factors, other than the state of the system, that influence the decision of individuals to migrate. Cast into the terminology of the introduction, $\underline{\theta}[t]$ stands for the socio-economic profile of the migratory system at time t.

In principle, the master equation contains all information about the system. However, it consists of C(N+L-1, L-1) coupled linear differential equations, each one corresponding to a distinct state. Thus, excepting special cases, a direct handling of the master equation appears technically prohibitive. A feasible alternative is to examine the evolution of a particular state, rather than simultaneously the evolution of all states, as does the master equation. Under these circumstances one moves from evolutionary to dynamical modelling. The question now becomes one about the choice of state to be examined. Intuitively, the mean-value state which, for j = 1, ..., L, satisfies

$$\frac{d}{dt} \bar{x}_j[t] = \sum_i \sum_{\underline{x}} (q_{ji}[t]\, x_i - q_{ij}[t]\, x_j)\, P[\underline{x}; t] \tag{4}$$

is a reasonable choice. If the probability distribution over states is unimodal and symmetric, the mean-value state is the most probable, and equation (4) determines the most probable evolution of the migratory system. If unimodality or symmetry, or both, do not hold, however, then (4) may mislead in the sense that it may not represent the most probable distribution. A more general alternative is to use one of the several approximation results that have been proposed, in order to study the evolution of the most probable state. All of these equations stem from the observation that, as the population of the system becomes arbitrarily large, the variance of the corresponding population distribution around the most probable state becomes arbitrarily small. Perhaps the most powerful of these results belongs to Kurtz (1978) (see de Palma and Lefevre, 1983). Kurtz studies a sequence of dynamical systems in which the population increases without bound. Let $\underline{x}^N[t]$ denote the most probable state at time t belonging to the system with total population N. Consider the system of differential equations

$$\begin{aligned} \frac{d}{dt} x_j[t] &= \sum_i (q_{ji}[t]\, x_i[t] - q_{ij}[t]\, x_j[t]) \\ &\equiv F(\underline{x}[t]; \underline{\theta}[t]), \text{ for } j = 1, \ldots, L \quad , \end{aligned} \tag{5}$$

with initial conditions $\underline{x}[0]$ and solution $\underline{x}^*[t]$ at time t. Then if F is continuous and if

$$\lim_{N\to\infty} \underline{x}^N[0] = \underline{x}[0] \qquad , \text{ for every } \tau < \infty ,$$

then

$$\lim_{N\to\infty} \sup_{t\leq\tau} |\underline{x}^N[t] - \underline{x}^*[t]| = 0 \quad \text{almost everywhere.} \tag{6}$$

That is, for arbitrarily large migratory systems, the solution to equation (5) describes the evolution of the most probable state--provided that the corresponding initial conditions match, and provided that the time horizon is finite. As already mentioned, the essence of the intuition is to be found in that, for arbitrarily large systems, the most probable state completely dominates in the sense that

$$\lim_{N\to\infty} P[\underline{x}^N; t] = 1 \quad . \tag{7}$$

Then, of course, the most probable state is necessarily the mean-value state because it is the only possible state. Substituting equation (7) into (4) yields equation (5), as required for consistency.

We now are ready to define more precisely the concepts of steady-state discussed in the introduction. Along the observed trajectory of the system, take any given instant τ as a reference point. Relative to this point, the socio-economic profile is fixed and given by $\underline{\Theta}|_\tau$. Steady-states exist for both the linear and non-linear system, and are characterized by $d\underline{x}/dt = \underline{0}$. Using equation (3), migration rates explicitly depend on the state of the system. Hence equation (5) is non-linear. However, for any $t \geq \tau$, the linear migration rates are given by $q_{ji}|_\tau = f_{ji}[\underline{x}|_\tau; \underline{\Theta}|_\tau]$, which corresponds to individuals adjusting myopically at time τ as if the spatial distribution of utilities in the system remained fixed during an adjustment process within fixed socio-economic environment (i. e., as if their potential moves, along with others', did not affect the state of the system observed at τ). Under these circumstances, the steady-state associated with τ is given by the solution to the <u>linear</u> system

$$0 = \sum_i q_{ji}|_\tau \, x_i[t] - q_{ij}|_\tau \, x_j[t] \qquad , \text{ for } j = 1 ,\ldots, L \quad , \tag{8}$$

using equation (5). Being linear, this system has a unique solution that represents the globally stable linear steady-state, and which can be found analytically. Solving equation (8) over a sequence of instances, τ, along the system's trajectory, one always can define the corresponding unique steady-state trajectory. On the other hand, for any $t \geq \tau$, the non-linear migration rates are given by

$$q_{ji}[t] = f_{ji}(\underline{x}[t]; \underline{\Theta}|_\tau) \equiv g_{ji}(\underline{x}[t]; \tau) \quad .$$

Using equation (5) once again, the steady-state associated with τ is given by the solution to the non-linear system

$$0 = \sum_i g_{ji}(\underline{x}[t];\ \tau)\ x_i[t] - g_{ij}(\underline{x}[t];\ \tau)\ x_j[t] \quad ,$$

$$\text{for } j = 1\ ,\dots,\ L \quad . \tag{9}$$

It is known that a solution to this system always exists--hence a non-linear steady-state always exists (see Kanaroglou, et. al., 1985). However, being non-linear, this system might have several solutions with varying stability characteristics; and unlike the linear system, it is not generally possible to solve equation (9) analytically. Thus, for any τ, one must use an iterative solution algorithm, beginning with the observed $\underline{x}[\tau]$ that simulates the adjustment process described by equation (5), and given the socio-economic profile $\underline{\theta}|_\tau$. If the algorithm converges to $\underline{x}^*|_\tau$, this value can be interpreted as the steady-state solution for the system (5) which corresponds to $\underline{\theta}|_\tau$ with initial conditions $\underline{x}[\tau]$. If the algorithm converges to more than one solution, then $\underline{x}[\tau]$ is in the vicinity of a bifurcation point along the trajectory of the system equation (5) that corresponds to the socio-economic profile $\underline{\theta}|_\tau$. And, if the algorithm does not converge, then we may conclude that either the equation system (5), which corresponds to $\underline{\theta}|_\tau$ with initial conditions $\underline{x}[\tau]$, is indeed on a limit cycle, or that the algorithm operating from $\underline{x}[\tau]$ to $\underline{x}^*|_\tau$ encounters numerical difficulties. In any case, solutions to equations (8) and (9) are taken to approximate the most probable linear and non-linear steady-states of the migratory system. Estimating and studying these solutions for the Canadian regional system is our main concern.

2.2. The Structure of Migration Rates

Let $p_{ij}[t,\Delta t]$ be the probability that an individual migrates from j to i during time period $[t,t+\Delta t]$. Then,

$$q_{ij}[t] = \lim_{\Delta t \to 0} p_{ij}[t,\Delta t]/\Delta t \quad . \tag{10}$$

Consequently, the structure of migration rates may be determined through the corresponding probabilities to migrate. This capability is convenient because the probabilities to migrate can arise from first principles using discrete choice theory (see for example Ben-Akiva and Lerman, 1985).

Suppose that the decision of an individual to migrate from j to i can be partitioned into two levels, namely the decision to leave j, and the decision to select i as a destination given that the individual has decided to move out of j. Thus,

$$p_{ij}[t,\Delta t] = p_{i|j}[t]\ p_{\cdot j}[t,\Delta t] \quad , \tag{11}$$

where $p_{\cdot j}[t,\Delta t]$ is the probability at time t of an individual leaving j during the time-interval Δt, and $p_{i|j}[t]$ is the probability at time t of an individual selecting i as a destination given that he has decided to move out of j. Notice that, according to equation (11), the probability of leaving j does not only depend upon the distribution of utilities in the system, which is a fundamental assumption of discrete choice theory, but also on the time interval, Δt. In order to distinguish between these two effects, we assume that the <u>odds of out-migration</u>, $p_{.j}/(1 - p_{.j})$, as a function of utilities and the time interval, is separable in the following sense:

$$p_{.j}[t,\Delta t]/(1 - p_{.j}[t,\Delta t]) = h[t,\Delta t]\ p_{.j}[t,l(t)]/\{1 - p_{.j}[t,l(t)]\}\ , \quad (12)$$

where $h[t,\Delta t]$ is a continuous, increasing function of Δt such that $h[t,0] = 0$ and $h[t,l(t)] = 1$. The function l represents an <u>intrinsic time-interval</u> which, at time t, renders the function h equal to one.

Using discrete choice theory, we can derive

$$p_{i|j}[t] = \exp(v_{ij}[t]/\mu[t] - I_j[t]) \quad , \text{and} \quad (13)$$

$$p_{.j}[t,l(t)] = \{\exp(\mu[t]\ I_j[t] - v_{jj}[t])/\{1 + \exp(\mu[t]\ I_j[t] - v_{jj}[t]) \quad . \quad (14)$$

(see Kanaroglou, et. al., 1985). Here, $v_{ij}[t]$ and $v_{jj}[t]$ represent the <u>advantage to migrate</u> from j to i at time t, and the <u>advantage to stay</u> in j at time t, respectively. $\mu[t]$ is an index of inter-correlation of the stochastic parts of the advantages to migrate from j to i, $i \neq j$ (see Ben-Akiva and Lerman, 1985). And, I_j is the <u>inclusive value</u> expressed as

$$I_j[t] = \ln\{\textstyle\sum_{k\neq j} \exp(v_{kj}[t]/\mu[t])\} \quad . \quad (15)$$

Substituting equation (14) into (12), we obtain

$$p_{.j}\ [t;\Delta t] = \frac{\exp(\alpha_0[t,\Delta t] + \mu[t]\ I_j[t] - v_{jj}[t])}{1 + \exp(\alpha_0[t,\Delta t] + \mu[t]\ I_j[t] - v_{jj}[t])} \quad , \quad (16)$$

where $\alpha_0[t,\Delta t] \equiv \ln(h[t,\Delta)t]$. We call equation (13) the <u>destination choice model</u>, and equation (16) the <u>departure model</u>. These two equations constitute a <u>nested logit model</u>.

Now it is possible to determine the migration rates that drive the evolution of a migratory system through the adjustment process described by equation (5). Expanding h, as a Taylor series, about $\Delta t = 0$, we get

$$h[t,\Delta t] = h[t,0] + \partial h/\partial \Delta t\ |_{\Delta t=0}\Delta t + (1/2)\partial^2 h/\partial \Delta t^2\ |_{\Delta t=0}\Delta t^2 + \ldots$$
$$\equiv \omega[t]\ \Delta t + \omega_2[t]\ \Delta t^2 + \ldots , \quad (17)$$

because the first term of the series is zero by our assumptions about equation (12). Retaining only the linear term in approximation yields

$$h[t,\Delta t] \;=\; \omega[t]\,\Delta t \quad . \tag{18}$$

Since $h[t,l(t)] = \omega[t]\,l[t] = 1$, an approximation to the intrinsic time interval $l[t]$ is the inverse of $\omega[t]$. From now on, for simplicity, we shall omit the explicit time notation. Using equations (10), (11), (13), (15), (16) and (18),

$$\begin{aligned} q_{ij} &= p_{i|j} \lim_{\Delta t \to 0} p_{\cdot j}[\Delta t]/\Delta t \\ &= \exp(v_{ij}/\mu - I_j) \lim_{\Delta t \to 0} \frac{\omega\,\Delta t \exp(\mu\, I_j - v_{jj})}{\Delta t[1 + \omega\,\Delta t \exp(\mu\, I_j - v_{jj})]} \\ &= \exp(v_{ij}/\mu - I_j)\;\omega \exp(\mu\, I_j - v_{jj}) \\ &= \omega \exp[(\mu - 1)\, I_j + v_{ij}/\mu - v_{jj}] \quad . \end{aligned} \tag{19}$$

Since the inclusive value depends on the entire distribution of advantages to migrate, and since these must depend on both the state of the migratory system and the corresponding socio-economic profile, the migration rates given by equation (19) indeed are represented by equation (3). Notice that in the special case, when the advantages to migrate from j to i, $i \neq j$, are perfectly uncorrelated, $\mu = 1$ and $q_{ij} = \omega \exp(v_{ij} - v_{jj})$ which agrees with Haag and Weidlich (1984). It perhaps is useful to emphasize that, in general, migration rates within the present framework are constructed from first principles based on the behavioural assumptions of standard discrete choice theory. Therefore, through the adjustment process described by equation (5), these rates provide an explicit link between the micro-level of individual decisions to migrate and the evolution of the migratory system at the macro-level.

3. EMPIRICAL ANALYSIS

3.1. The Nested Logit Model

We begin with conceptual arguments about the structure of the advantage to stay v_{jj}, and the advantage to migrate, v_{ij}. Table 1 is designed as a reference for these arguments. In what follows, we shall classify the explanatory variables used by Ledent and Liaw (1985) employing the scheme of Table 1. The subsequent interpretation is strictly ours, and in no way is it meant to convey the spirit of Ledent and Liaw.

In our framework, decisions to migrate depend on comparisons of utility levels in the various regions of the migratory system. Thus the fundamental components of the advantage to stay and the advantage to migrate must be utilities. It seems reasonable to assert that utility in a region does not only depend on the level presently attainable by an individual there, but

TABLE 1

THE STRUCTURE OF ADVANTAGES TO STAY AND TO MIGRATE

	ORIGIN	DESTINATION
	Propensity to Stay $s = \underset{\sim}{\gamma}^{o}\underset{\sim}{z}$	Migration Cost $c_{ij} = \underset{\sim}{\gamma}^{1}\underset{\sim}{d}_{ij}$
	Origin Size Effect $(\alpha^{o} x_j^{o})$	Destination Size Effect $(\alpha^{1} x_i^{1})$
	Origin Utility $(u_j^{o} = \underset{\sim}{\beta}^{o}\underset{\sim}{y}_j^{o})$	Destination Utility $(u_i^{1} = \underset{\sim}{\beta}^{1}\underset{\sim}{y}_i^{1})$
Attractivity	Origin Attractivity $(v_j^{o} = \alpha^{o} x_j + u_j^{o})$	Destination Attractivity $(v_i^{1} = \alpha^{1} x_i + u_i^{1})$
Advantage	Advantage to Stay $(v_{jj} = v_j^{o} + s)$	Advantage to Migrate $(v_{ij} = v_i^{1} - c_{ij})$

also upon his expectations about future levels of utility. Present levels of utility depend on the consumption of private and public goods. Expectations, on the other hand, could arise from recent past as well as present indicators, such as capital accumulation, unemployment, and the like, which are expected to affect utility formation in the future. Given this classification of explanatory variables for utilities, it now becomes necessary to distinguish between <u>origin utility</u>, u_j^0, and <u>destination utility</u>, u_i^1 (see Table 1). On empirical grounds, this separation is supported by several studies, which found that destination characteristics tend to be more important than origin characteristics in the decision to migrate (e. g. Lowry, 1966). Such a distinction is clearly made in Ledent and Liaw (1985), whose estimation of the associated nested choice model for the Canadian regional system has provided the basis for our empirical analysis. In general, Ledent and Liaw adopt a standard linear-in-parameters form for both v_{jj} and v_{ij}. First consider their specification of origin utilities in the context of our previous remarks. For current levels, since individuals aim to maximize utility, it is theoretically justifiable to invoke the indirect utility function and estimate utility gained from the consumption of private goods through corresponding expenditure in real terms. Expenditure in their study is represented by the <u>wage level</u>, expressed as the ratio of the provincial to national average wage rates. An explanatory variable associated with the consumption of public goods is the <u>cultural attachment</u> of individuals to a region, defined as the percentage share of British and French in the provincial population. The rationale is that a longer settlement history may cause a stronger attachment to places. On the other hand, for expectations, Ledent and Liaw have chosen regional <u>unemployment benefits</u> which, in their estimation, proved to be statistically insignificant. Consider next their specification of destination utilities. Expenditure here is given by the <u>average wage rate</u>, measured as the average weekly salary, in real terms. The quality of the environment is represented by the <u>annual number of degree-days below 18° C</u> averaged over 1951-1980. Finally, expectations are approximated by the <u>average unemployment benefit</u>, which reflects the lack of employment opportunities at a given destination.

Utility does not represent the only factor influencing the decisions of individuals. Population size also is significant in the determination of $p_{.j}$ and $p_{i|j}$. The <u>origin size effect</u> in Ledent and Liaw is given by the <u>population share</u>, expressed as a percentage of national population; the <u>destination size effect</u>, on the other hand, is given by the <u>population size</u>. The estimated coefficients of these effects represent the deep impact of population size on factors such as regional returns to scale and the quality of the environment. Because of the latter, part of the size effects may belong to the corresponding utilities. In any case, the linear combination of origin utility and size effects is defined here as <u>origin attractivity</u>, v_j^0; whereas, the linear combination of destination utility and size effects is defined as <u>destination attractivity</u>, v_i^1 (see Table 1). Attractivities, in turn, are distorted by the <u>propensity to stay</u>, s, and <u>migration costs</u>, c_{ij}. The former is associated with origin attractivities, and reflects the truth that provincial departure rates strongly depend on fluctuations in the national economy. The latter is associated with destination attractivities, and arises from the friction of distance and the attachment of individuals to experienced social environments. In Ledent and Liaw (1985), the propensity to stay has been represented by the <u>national average wage rate</u>, which proved to be statistically insignificant, and by the <u>national unemployment rate</u>; hard times (i. e., times of low wages and high unemployment) are times of low mobility. In contradistinction, migration costs have been represented by the <u>logarithm of distance</u> and by an index of <u>ethnic dissimilarity</u> between origin and destination. Thus, in Ledent and Liaw,

$$v_{jj} = \alpha^0 x_j^0 + \underline{\beta}^0 \underline{y}_j^0 + \underline{\gamma}^0 \underline{z} , \tag{20}$$

where $\underline{y}_j^0$ includes the wage level, cultural attachment and unemployment benefits, and $\underline{z}$ includes the national average wage rate and the national unemployment rate. In addition,

$$v_{ij} = \alpha^1 x_i^1 + \underline{\beta}^1 \underline{y}_i^1 + \underline{\gamma}^1 \underline{d}_{ij} , \tag{21}$$

where $\underline{y}_i^1$ includes the average wage rate, the annual number of degree-days below 18° C, and the average unemployment benefit, and $\underline{d}_{ij}$ includes the logarithm of distance and ethnic dissimilarity.

The migration data set used by Ledent and Liaw consists of annual matrices for place-to-place flows spanning the years 1961-1983, a total of 22 observations. Each 10-by-10 migration matrix has off-diagonal elements m_{ij} representing the number of individuals who migrated from j to i during the corresponding year. The population distribution over the ten provinces of Canada, $\underline{n} = (n_1,\ldots,n_{10})$, for that year also is known. An observation that corresponds to $p_{i|j}$ is given by $m_{ij} / \sum_{k\neq j}^{10} m_{kj}$. On the other hand, an observation that corresponds to $p_{\cdot j}$ is given by For $m_{kj} / \sum_{k\neq j}^{10} n_j$. For each time period [t,t+1], t = 1961,...,1982, the estimation problem is defined by equations (13), (15), (16), (20) and (21). We first estimate the parameters of the destination choice model in the form α^1/μ, $\underline{\beta}^1/\mu$ and $\underline{\gamma}^1/\mu$. Next we calculate the inclusive value. This then is used to estimate the parameters α_0, μ, α^0, $\underline{\beta}^0$, and $\underline{\gamma}^0$ the departure model. In Ledent and Liaw (1985), the destination choice model was estimated on an annual basis. The departure model, on the other hand, was estimated for all years simultaneously. Ledent and Liaw performed the estimation procedure several times, with various combinations of independent variables, in order to examine complementarity and latent collinearity, and in order to achieve the best goodness of fit. For the purposes of our paper, we adopted their best run at the time our research was carried out. This run corresponds to the independent variables discussed in conjunction with Table 1.

3.2. The Adjustment Process

Once the parameters of the destination choice and departure models become available, it is possible to estimate the migration rates given by equation (19) for each time-period. These rates, in turn, determine an estimate of the adjustment process described by equation (5) for the Canadian regional system between 1961 and 1983. Holding these rates constant for each time-period, the steady-state equations (8), which correspond to the linear system, are readily obtained. One then could calculate the evolution of steady-states for both the linear and the non-linear systems over the period under consideration. The difference between observed and linear (non-linear) steady-states will provide measures of disequilibrium as discussed in the introduction.

The evolution of observed states for the Canadian regional system between 1961 and 1983 appears the Appendix. In our context, a state consists of ten entries, each one corresponding to the proportion of total population in one of the ten Canadian provinces. To determine the evolution of linear steady-states, instead of estimating the rates $q_{ij}|_\tau$ for equation (8), we elected an alternative which allows computation directly from the migration data. Moreover, using equation (10),

$$p_{ij}[t,dt] = q_{ij}[t]\, dt \quad . \tag{22}$$

Integrating both sides of equation (22), and assuming that the rates are constant over the observation interval Δt, we have

$$p_{ij}[\tau,\Delta t] = \int_{\tau}^{\tau+\Delta t} q_{ij}[t]\, dt = q_{ij}|_\tau \, \Delta t \quad . \tag{23}$$

Upon substituting equation (23) into equation (8), we get

$$0 = \Delta^{-1} t(\textstyle\sum_i p_{ji}|_\tau \, x_i - x_j) \quad , \text{ for } j = 1, \ldots, 10 \quad 0. \tag{24}$$

An observation that corresponds to p_{ji} is given by m_{ji}/n_i. Consequently, the linear steady-state is the unique non-trivial solution to the system

$$\textstyle\sum_i p_{ji}|_\tau \, x_i = x_j \quad , \text{ for } j = 1, 0\ldots,10 \quad , \tag{25}$$

where the quantities p_{ji} are known. This solution is given by the left eigenvector of the probability matrix defined in equation (23) and corresponding to its unit eigenvalue. The resulting evolution of linear steady-states for the Canadian regional system between 1961 and 1983 appears in the Appendix. A visual comparison between observed and linear steady-states involving four of the ten Canadian provinces is shown in Figure 1. It is evident that our hypothesis concerning the close interaction between observed and linear steady-states is fully substantiated in this figure. Furthermore, upon comparison of the first and second tables in the Appendix,

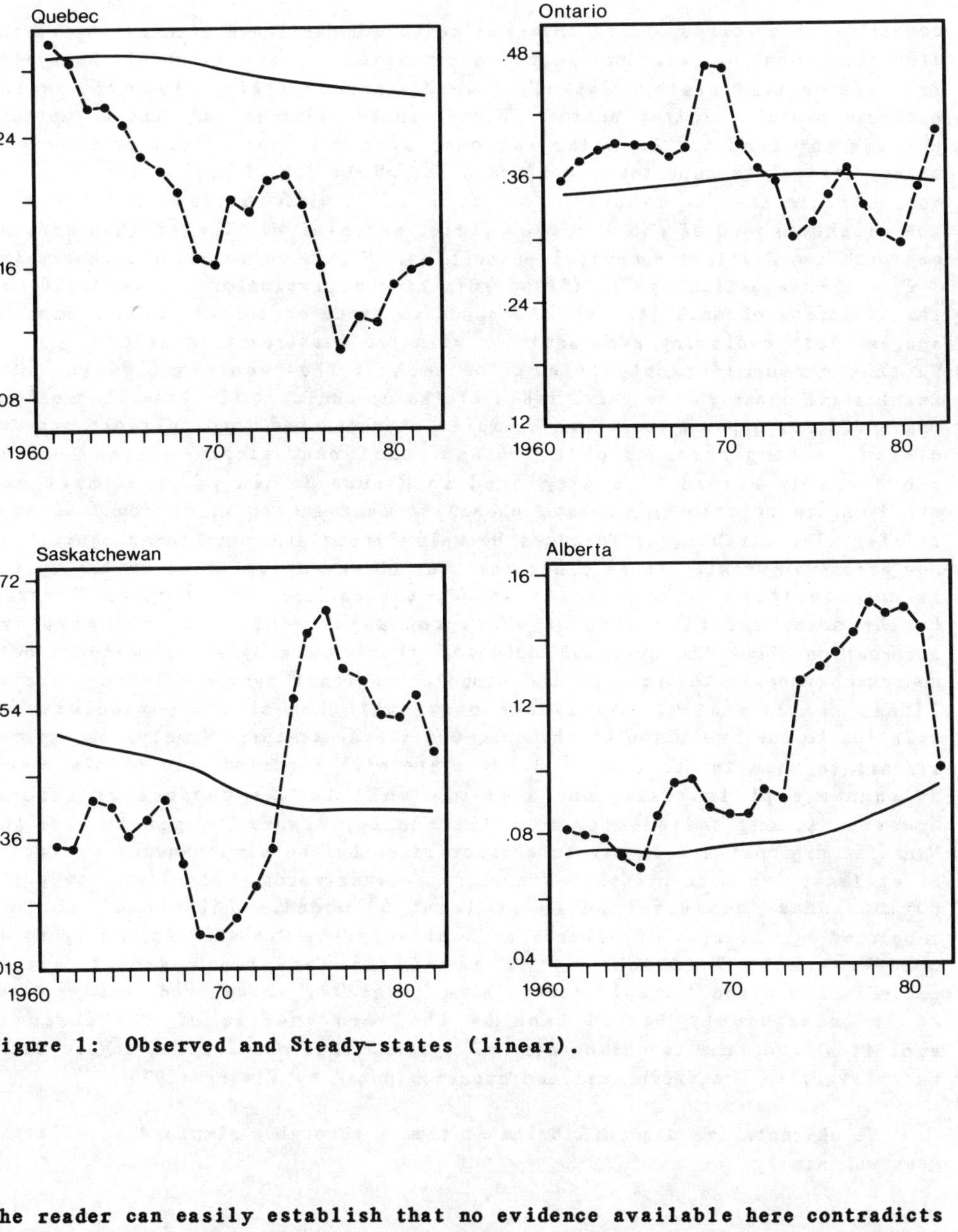

Figure 1: Observed and Steady-states (linear).

the reader can easily establish that no evidence available here contradicts our hypothesis. Thus it seems that the myopic behaviour of individuals in their migration decisions is consistent with the evidence reported here.

In order to determine the evolution of non-linear steady-states, we employed an iterative solution algorithm (IMSL subroutine ZSPOW) on the non-linear equation system (5). For each time-period, we used as initial

conditions the corresponding observed state. Under these circumstances, the iterative solution procedure actually simulates the evolution of the system from its present state, possibly toward a steady-state, given the socio-economic profile of that period. Since, in any time-period, the adjustment process may lead the Canadian regional system to none, one, or multiple steady-states, we repeatedly perturbed the evolution of the system in order to ascertain that no solution to equation (9), with initial conditions in the neighbourhood of the observed state, was missed. Toward this end, we employed two distinct numerical principles. First, we forced the constraint $\sum_i x_i = 1$ on equation system (5) by adjusting a particular x_i, i=1,...,10, at the beginning of each iteration. Second, we executed ten additional runs by successively replacing each equation with the constraint equation $_i\, x_i = 1$. We thus performed twenty trials for each of the twenty-two years. Our results are shown in the third table of the Appendix. All three theoretical possibilities appear; there are examples of none, one, and multiple steady-states. A comparison of observed and non-linear steady-states for the provinces in Figure 1 is displayed in Figure 2. In years of multiple steady-state solutions, we have chosen to show the solution found in most trials. The striking differences between linear and non-linear equilibria now become apparent. It is clear that the observed evolution of the system is not sensitive to non-linear steady-states (see, for example, Alberta). Furthermore, the direction of short-run adjustments does not give any information about the eventual outcome; the two steady-state patterns bear no resemblance in the case of the Canadian regional system. In contrast to linear steady-states, non-linear ones exhibit a strong periodicity in relation to the evolution of the observed steady-states. Namely, the system appears to move in and out of steady-state with a periodicity of six years. We cannot explain this; indeed it may well be the product of chance. However, strong periodicity notwithstanding, Figure 2 suggests that the Canadian regional system may be characterized by the simultaneous evolution of at least two most probable states. In other words, that $P[\underline{x};t]$ over the period under consideration is at least bi-modal. This conclusion is supported by the case of Alberta and, clearly, by Quebec. According to an observation by Weidlich, it is as if the Canadian regional system periodically directs itself toward some 'parallel,' unrealized steady-state, to be subsequently forced back by the mere inertia of its observed evolution. For the Canadian regional system, this observation may be taken as relevant to the aforementioned question posed by Vining (1975).

We characterize disequilibrium at time t through a simple dissimilarity measure, namely,

$$D[t] = \sum_{i=1}^{10} |x_i[t] - x_i^*[t]| \quad , \tag{26}$$

where $\underline{x}^*$ denotes a steady-state. That is, disequilibrium is defined as the total distance between the observed state and the corresponding steady-state. Figure 3 describes the evolution of disequilibrium for both the

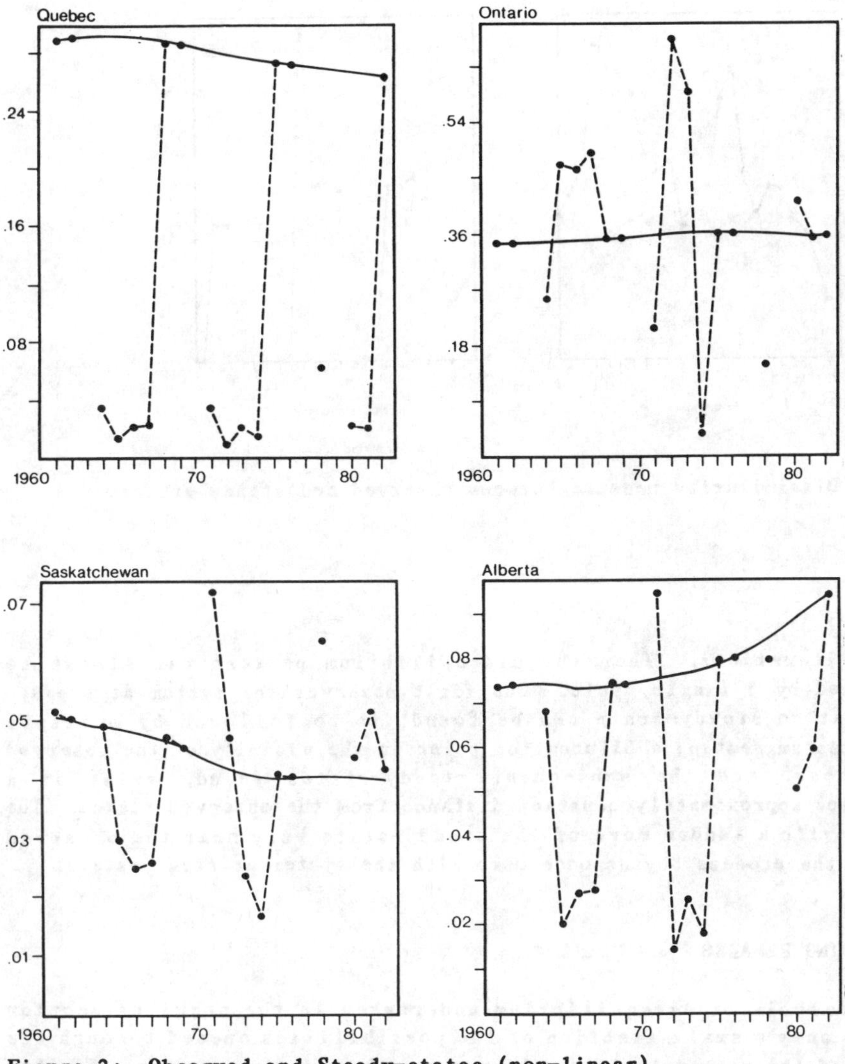

Figure 2: Observed and Steady-states (non-linear).

linear and the non-linear case. For the latter, as in Figure 2, we chose those steady-state solutions found in most trials. Although disequilibrium in the short-run exhibits a definitely increasing trend, the pattern becomes strongly regular when the estimated non-linear nature of the adjustment process is fully taken into account. This tendency is especially strong if one discounts the non-linear solution that corresponds to the year 1974-75, which was found very far away from the observed state and after a large

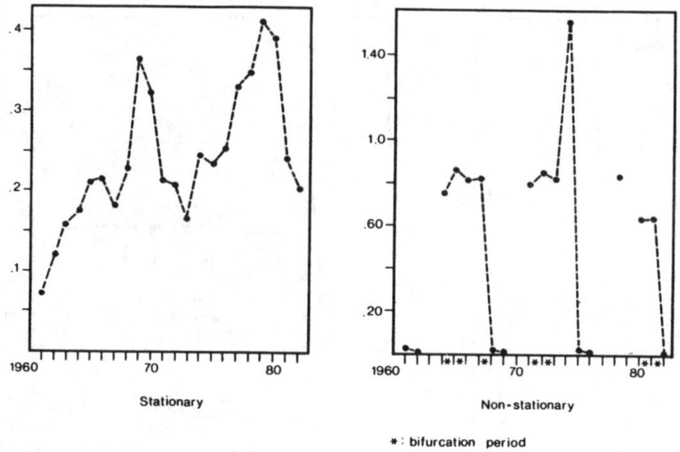

Figure 3: Dissimilarity measure between observed and steady states.

number of iterations. Then the disequilibrium pattern can almost be characterized by a single cycle. One first observes the system at steady-state. Next no steady-state can be found, to be followed by multiple steady-states suggesting a bifurcation point in the vicinity of the observed state. These, and the subsequent steady-states found, exist at a considerable, approximately constant distance from the observed states. The cycle ends with a sudden move of the steady-state very near the observed state, and the process begins once more with the system at steady-state.

4. CONCLUDING REMARKS

The analysis of disequilibrium undertaken in the previous section represents only a small fraction of the possibilities opened through the estimation of the nested logit model over a number of time-periods. Similar interesting research questions are raised in the study of other elements, such as the origin and destination utilities displayed in Table 1. It seems that a wealth of information can be ordered and understood in the context of the present dynamical framework. These issues will be the object of a more detailed study in the future.

5. REFERENCES

Ben-Akiva, M., and S. Lerman, 1985, Travel Behaviour: Theories Models and Prediction Methods, Cambridge, Mass.: MIT.

de Palma, A., and C. Lefevre, 1983, Individual Decision-making in Dynamic Collective Systems, Journal of Mathematical Sociology, 90: 103-124.

Kanaroglou, P., K.-L. Liaw and Y. Papageorgiou, 1985, An Operational Framework for the Analysis of Migratory Systems, QSEP Research Report No. 136, McMaster University.

Kurtz, T., 1978, Strong Approximation Theorems for Density-Dependent Markov Chains, Stochastic Processes and Their Applications, 6: 223-240.

Haag, G., and W. Weidlich, 1984, A Stochastic Theory of Interregional Migration, Geographical Analysis, 16: 331-357.

Ledent, J., and K.-L. Liaw, 1985, Interprovincial Migration Outflows in Canada, 1961-83: Characterization and Explanation paper presented at the IUSSP Seminar On Internal Migration and Regional Development, Montreal.

Lowry, I. 1966, Migration and Metropolitan Growth, San Francisco: Chandler.

Rogers, A., 1967, Matrix Analysis of Interregional Population Growth and Distribution, Papers of the Regional Science Association, 18: 177-196.

Schelling, T., 1971, On the Ecology of Micromotives, The Public Interest, 25: 61-98.

Vining, D., 1975, The Spatial Distribution of Human Populations and its Characteristic Evolution Over Time: Some Recent Evidence from Japan, Papers of the Regional Science Association, 35: 157-178.

6. APPENDIX: Observed and Steady-states for the Canadian Regional System Over the Period 1961-1983.

Provinces are represented as: 1 - Newfoundland, 2 - Prince Edward Island, 3 - Nova Scotia, 4 - New Brunswick, 5 - Quebec, 6 - Ontario, 7 - Manitoba, 8 - Saskatchewan, 9 - Alberta, and 10 - British Columbia.

1. Observed States

Year	1	2	3	4	5	6	7	8	9	10
1961-62	0.025	0.006	0.040	0.033	0.288	0.342	0.051	0.051	0.073	0.089
1962-63	0.025	0.006	0.040	.0033	0.289	0.342	0.050	0.050	0.074	0.089
1963-64	0.025	0.006	0.040	0.032	0.290	0.342	0.050	0.049	0.074	0.090
1964-65	0.025	0.006	0.039	0.032	0.289	0.344	0.050	0.049	0.074	0.090
1965-66	0.025	0.006	0.038	0.031	0.289	0.346	0.049	0.048	0.074	0.091
1966-67	0.025	0.005	0.038	0.031	0.289	0.348	0.048	0.048	0.073	0.094
1967-68	0.024	0.005	0.037	0.030	0.288	0.350	0.047	0.047	0.073	0.095
1968-69	0.024	0.005	0.037	0.030	0.286	0.351	0.047	0.046	0.074	0.097
1969-70	0.024	0.005	0.037	0.030	0.285	0.352	0.047	0.046	0.074	0.098
1970-71	0.024	0.005	0.037	0.029	0.282	0.355	0.046	0.044	0.075	0.100
1971-72	0.024	0.005	0.037	0.029	0.279	0.357	0.046	0.043	0.075	0.101
1972-73	0.024	0.005	0.036	0.029	0.278	0.358	0.045	0.042	0.076	0.103
1973-74	0.024	0.005	0.036	0.029	0.276	0.359	0.045	0.041	0.077	0.104
1974-75	0.024	0.005	0.036	0.029	0.274	0.360	0.045	0.040	0.077	0.106
1975-76	0.024	0.005	0.036	0.029	0.272	0.360	0.045	0.040	0.078	0.107
1976-77	0.024	0.005	0.036	0.029	0.271	0.359	0.044	0.040	0.080	0.107
1977-78	0.024	0.005	0.036	0.029	0.270	0.359	0.044	0.040	0.082	0.107
1978-79	0.024	0.005	0.036	0.029	0.268	0.359	0.044	0.040	0.084	0.108
1979-80	0.024	0.005	0.035	0.029	0.267	0.358	0.043	0.040	0.086	0.109
1980-81	0.024	0.005	0.035	0.029	0.266	0.356	0.043	0.040	0.089	0.111
1981-82	0.023	0.005	0.035	0.029	0.264	0.354	0.042	0.040	0.092	0.113
1982-83	0.023	0.005	0.035	0.028	0.263	0.354	0.042	0.040	0.094	0.113

2. Linear Steady-States

Year	1	2	3	4	5	6	7	8	9	10
1961-62	0.022	0.007	0.036	0.027	0.297	0.352	0.048	0.035	0.080	0.093
1962-63	0.018	0.004	0.031	0.023	0.284	0.370	0.047	0.035	0.079	0.107
1963-64	0.016	0.004	0.030	0.022	0.256	0.382	0.045	0.041	0.078	0.123
1964-65	0.013	0.003	0.027	0.023	0.258	0.390	0.041	0.041	0.073	0.129
1965-66	0.015	0.004	0.029	0.023	0.248	0.389	0.035	0.036	0.068	0.153
1966-67	0.016	0.004	0.030	0.022	0.228	0.390	0.037	0.038	0.082	0.150
1967-68	0.020	0.005	0.035	0.027	0.219	0.377	0.042	0.041	0.094	0.136
1968-69	0.021	0.005	0.035	0.024	0.206	0.385	0.041	0.034	0.096	0.150
1969-70	0.012	0.003	0.032	0.020	0.166	0.465	0.035	0.023	0.088	0.151
1970-71	0.019	0.006	0.034	0.029	0.161	0.464	0.036	0.022	0.085	0.139
1971-72	0.027	0.006	0.038	0.031	0.201	0.391	0.036	0.024	0.085	0.157
1972-73	0.025	0.007	0.047	0.034	0.192	0.365	0.040	0.028	0.093	0.165
1973-74	0.020	0.006	0.039	0.031	0.211	0.353	0.047	0.035	0.090	0.165
1974-75	0.024	0.007	0.041	0.038	0.216	0.298	0.043	0.053	0.126	0.151
1975-76	0.025	0.007	0.045	0.040	0.200	0.314	0.046	0.064	0.130	0.126
1976-77	0.022	0.007	0.038	0.033	0.165	0.341	0.049	0.068	0.134	0.142
1977-78	0.021	0.007	0.040	0.031	0.107	0.367	0.045	0.060	0.140	0.178
1978-79	0.021	0.005	0.040	0.030	0.129	0.332	0.038	0.059	0.151	0.191
1979-80	0.022	0.006	0.035	0.028	0.126	0.304	0.033	0.053	0.147	0.243
1980-81	0.021	0.004	0.034	0.025	0.148	0.292	0.038	0.052	0.149	0.233
1981-82	0.019	0.005	0.036	0.028	0.157	0.350	0.049	0.056	0.142	0.155
1982-83	0.030	0.006	0.041	0.035	0.163	0.403	0.047	0.048	0.098	0.127

3. Non-linear Steady-States

Year	1	2	3	4	5	6	7	8	9	10
1961–62	0.026	0.000	0.041	0.033	0.289	0.345	0.051	0.051	0.074	0.090
1962–63	0.025	0.006	0.040	0.033	0.289	0.344	0.050	0.050	0.074	0.089
1963–64	No solution found.									
1964–65	0.024	0.048	0.106	0.093	0.037	0.240	0.040	0.049	0.068	0.295
	0.017	0.032	0.061	0.070	0.582	0.108	0.023	0.026	0.030	0.050
	0.026	0.063	0.274	0.133	0.037	0.281	0.030	0.034	0.041	0.080
1965–66	0.023	0.087	0.117	0.182	0.012	0.469	0.039	0.029	0.020	0.021
	0.022	0.088	0.110	0.145	0.007	0.051	0.476	0.049	0.028	0.023
1966–67	0.024	0.119	0.160	0.080	0.021	0.481	0.024	0.025	0.027	0.038
1967–68	0.025	0.115	0.147	0.080	0.022	0.493	0.025	0.026	0.028	0.040
	0.012	0.458	0.441	0.037	0.006	0.015	0.006	0.007	0.008	0.010
1968–69	0.025	0.000	0.038	0.031	0.287	0.354	0.047	0.047	0.074	0.097
1969–70	0.024	0.005	0.037	0.029	0.285	0.354	0.047	0.046	0.074	0.098
1970–71	No solution found.									
1971–72	0.032	0.058	0.141	0.119	0.035	0.204	0.052	0.072	0.095	0.194
	0.033	0.064	0.187	0.142	0.035	0.188	0.049	0.067	0.085	0.148
1972–73	0.019	0.062	0.060	0.049	0.007	0.678	0.047	0.047	0.015	0.017
	0.026	0.058	0.154	0.413	0.007	0.037	0.044	0.051	0.016	0.016
	0.029	0.169	0.139	0.192	0.007	0.045	0.088	0.379	0.029	0.024
1973–74	0.023	0.091	0.106	0.066	0.021	0.577	0.023	0.024	0.027	0.042
1974–75	0.023	0.411	0.375	0.071	0.014	0.032	0.014	0.016	0.018	0.027
1975–76	0.025	0.000	0.037	0.030	0.273	0.363	0.045	0.041	0.079	0.108
1976–77	0.024	0.005	0.036	0.029	0.271	0.362	0.044	0.040	0.080	0.107
1977–78	No solution found.									
1978–79	0.040	0.055	0.115	0.118	0.062	0.151	0.047	0.063	0.080	0.268
1979–80	No solution found.									

1980-81	0.040	0.110	0.130	0.077	0.022	0.412	0.042	0.044	0.051	0.074
	0.040	0.106	0.122	0.073	0.019	0.077	0.048	0.055	0.071	0.389
1981-82	0.043	0.124	0.150	0.079	0.021	0.353	0.050	0.052	0.060	0.068
	0.042	0.109	0.124	0.072	0.019	0.078	0.058	0.068	0.087	0.343
1982-83	0.024	0.000	0.035	0.029	0.264	0.357	0.043	0.040	0.095	0.114

MODELLING AN ECONOMY IN SPACE AND TIME: THE DIRECT EQUILIBRIUM APPROACH WITH ATTRACTION-REGULATED DYNAMICS

Bruno Dejon
University of Erlangen-Nuernberg
Federal Republic of Germany

1. INTRODUCTION

This paper provides a brief exposition of the direct equilibrium approach and related attraction-regulated dynamics, as applied to a spatial production and exchange economy. The approach is specified as direct because after introduction of utility functions in price-quantity space, economic equilibrium is established by appeal to the individual rationality principle without the intermediary of market demand functions. The type of dynamics adequate to this direct equilibrium approach was introduced by Dejon and Graef (1983), and by Dejon (1983), at the San Miniato NATO Institute entitled 'Evolving Geographical Structures,' as attraction-regulated dynamics. The so-called passive attraction-regulated form of migration laws is particularly suited to revealing the relationship that exists between a direct equilibrium and stationarity.

2. THE DIRECT EQUILIBRIUM APPROACH TO SPATIAL PRODUCTION AND EXCHANGE

We consider finitely many commodity markets $g \varepsilon \mathbf{G}$. Any two of the markets possibly may differ only by their geographical locations, since the physical nature of their respective goods will be held constant. In addition, we consider finitely many behaviorally homogeneous (finite) populations $n \varepsilon \mathbf{N}$, and let P^n denote the number of economic agents in population n. $\mathbf{N}^{gd}$ denotes those populations from $\mathbf{N}$ that are on the demand side of market g, and $\mathbf{N}^{gs}$ denotes those that are on the supply side. On the supply side of the labor market are populations of households, possibly segmented by age, by size and by education of the head of household, for example. On the demand side of the labor market are populations of firms, in general segmented according to economic sector and possibly by technological standard.

A discrete price-quantity space is associated with each market g, namely $(\mathbf{P}^g x \mathbf{Q}^g) \cup \{0^g\}$. The prices $p \varepsilon \mathbf{P}^g$ are non-negative numbers, while the quantities $q \varepsilon \mathbf{Q}^g$ are positive numbers, and 0^g designates the null-quantity of good g, which does not require any associated price. Each population n has, formally, the same choice set $\mathbf{A} := \pi(\mathbf{P}^g x \mathbf{Q}^g) \cup \{0^g\}$, where π denotes the Cartesian product over all markets g, and := denotes that the expression on the left-hand side is posited to equal the expression on the right-hand side. The fact that actual choice sets of different populations may well be different may be accounted for by attributing a utility value of $-\infty$ to alternatives $a \varepsilon \mathbf{A}$ that are not feasible for population n, be this for technological or for budgetary reasons.

At any time t, let $P^n(a;t)$, or more briefly $P^n(a)$, designate the number of agents of population n associated with alternative $a \varepsilon \mathbf{A}$. $\underline{P} := [P^n(a)]_{n \varepsilon \mathbf{N}, a \varepsilon \mathbf{A}}$ then is the state vector of the system.

At any market g, supply from current production at price p, $SP^g(p)$ say, is $\sum_{n \varepsilon \mathbf{N}^g s} \sum_{a \varepsilon \mathbf{A}'} q \cdot P^n(a)$, where $\mathbf{A}' := \mathbf{A}(p^g{=}p)$ stands for the set of all those alternatives $a \varepsilon \mathbf{A}$ that associate a non-zero quantity $q \varepsilon \mathbf{Q}^g$ of good g with the specified price $p \varepsilon \mathbf{P}^g$. Satisfied demand, that is the physical volume of transactions, at market g, is obtained in an analogous fashion. Total supply $S^g(p)$ of good g at price level p is the sum of supply from current production and supply from stock. The latter quantity always equals zero if good g is instantaneously perishable. Finally, relative (satisfied) demand, $r^g(p)$, of good g at price level p is obtained as the quotient of satisfied demand to total supply, at price level p.

Each population n is assumed to attribute some utility $U^n(a|\underline{r})$ to alternative $a \varepsilon \mathbf{A}$, the value of $U^n(a|\underline{r})$ being dependent upon the vector $\underline{r}$ of all relative demands. This formulation permits spillover effects to come into play. The so-called deterministic or strict utilities $U^n(a|\underline{r})$ serve as parameters to stochastic utilities in such a way that after utility maximization by each agent--according to the individual rationality principle (McFadden, 1978; Hildenbrand, 1971)--some normalized distribution $d^n(a)$ of the agents of population n over the choice set $\mathbf{A}$ is obtained. Here 'normalized' means $\sum_{a \varepsilon \mathbf{A}} d^n(a) = 1$. The specific expression for $d^n(a)$ naturally depends on the probability distribution function of the stochastic utilities; if, for example, these are independent and identical Gumbel distributions, with concentration parameters γ^n, the $d^n(a)$ are proportional to $\exp[\gamma^n U^n(a|\underline{r})]$. This is the well-known multinomial logit model (McFadden, 1978). Subsequent theoretical results will not depend on this specific expression for $d^n(a)$; in simulation work, however, so far we have worked exclusively with the multinomial logit model.

A state $\underline{P}$ of the system is called a direct equilibrium state if, for each population n, $P^n(a) = P^{n \cdot} d^n(a)$ for all $a \varepsilon \mathbf{A}$. This definition will prove useful only if equilibria defined in this way have essential features in common with equilibrium states as defined in standard general equilibrium theories. An important feature of direct equilibrium proven by Dejon, Gueldner and Wenzel (1985a, 1985b) under rather general conditions concerning the strict utilities $U^n(a|\underline{r})$ and the distributions $d^n(a)$ is as follows. As randomness of the stochastic utilities diminishes, the distributions $d^n(a)$, in particular their marginal distributions along the price axes of the various goods g that are actually of concern to population n, grow more peaked. The locations $pp^g \varepsilon \mathbf{P}^g$ of these peaks may be conceived of as non-Walrasian and Walrasian equilibrium prices, respectively, depending upon whether or not economic agents at these price levels experience quantity rationing (Benassy, 1982). Quantity rationing will be rendered perceptible by the occurrence of a search cost on the demand side, coupled with a cost of uncertainty of demand on the supply side.

A salient feature of the direct equilibrium approach resides in the fact that no market demand functions are required. Equilibrium is calculated directly from strict utilities. However, these have to be made dependent on the state $\underline{P}$ of the system, more precisely, on relative demand $\underline{r}$.

3. SELECTED POPULATIONS, MARKETS AND CHOICE SETS

Let hr designate the population of all households of any region r (for the sake of formal simplicity, no further disaggregation of households is introduced). Let lr denote the (aggregated) labor market in region r. Again, for simplicity, we assume that each household has only a single quantity of labor to offer at any regional market, which is 50 man-hours per week. Firms, however, will be assumed to ask for multiples of 50 man-hours per week. Therefore, $\mathbf{Q}^{lr} := \{50k|k=1,2,\ldots,K\}$. The price axes $\mathbf{P}^{lr}$ are constituted by (finitely many) wage rates p^{lr}. Any household in region r' may choose to offer its labor services at any labor market lr, and at any wage rate p^{lr}. Thus, the choice set $\mathbf{A}$ of population hr' consists of the Cartesian product $\pi(\mathbf{P}^{lr} x \mathbf{Q}^{lr}) \cup \{0^{lr}\}$, formed over all regions r. As an aside, the alternative '$q^{lr} = 0^{lr}$ for all regions r' stands for voluntary unemployment.

Landlords in region r will constitute another population, llr. Without seriously restricting generality one may assume that each landlord owns one unit of land. Households and firms will ask for entire multiples of the land unit. Therefore, $\mathbf{Q}^{lmr} := \{1,\ldots,LM\}$ forms the quantity axis of the land market lmr of region r. Similarly, as before, $\mathbf{P}^{lmr}$ designates the (discrete) price axis of the land market lmr. Although each landlord has only one unit of land to offer at the land market, formally one may set the choice set $\mathbf{A}$ of population llr to equal $(\mathbf{P}^{lmr} x \mathbf{Q}^{lmr}) \cup \{0^{lmr}\}$, where 0^{lmr} represents withdrawal from the market.

Yet another population, fr say, is constituted by the firms in region r (assuming, for the sake of simplicity, that there is only one good gr, the composite good, that is being produced). Firms are endowed with a comparatively more complex choice set. Among other things, a firm in region r chooses the quantity q^{gr} of the composite good gr that it is going to produce and offer at market gr (located in region r), as well as the offering price p^{gr} (in case q^{gr} is different from 0^{gr}). In addition, a firm chooses a price-quantity pair (p^{lmr},q^{lmr}), or 0^{lmr}, at its regional land market lmr, as well as a price-quantity pair (p^{lr},q^{lr}), or 0^{lr}, at its regional labor market. All of these selections, of course, are made in accordance with a firm's technological constraints.

For any given set of occupation numbers $P^{fr}(p^{lr},q^{lr},p^{lmr},q^{lmr},p^{gr},q^{gr})$ [for firms producing and selling in region r] and $P^{hr'}(p^{lr},50)$ [for households living in region r' and offering their labor services in region r at the wage rate p^{lr}], one can calculate total supply and (satisfied) demand of labor services in region r at the wage rate p^{lr}, and subsequently obtain relative demand $r^{lr}(p^{lr})$ at wage rate p^{lr} in region r. One then can calculate, for any region of residence r', the number of households that achieve wage income class yw by summing over all products $r^{lr}(p^{lr})\cdot P^{hr'}(p^{lr},50)$ for which $p^{lr} - c^{rr'}$ lies in wage income class yw, $c^{rr'}$ denoting the cost of commuting between r and r'. Adding non-wage income (i. e., land rent and distributed profits of the firms), one obtains total income and then can calculate the numbers $P^{yr'}$ of households in region r' that belong to income class y. These households constitute a population, yr' say, whose agents act as demanders at the various commodity markets gr and at the land market lmr' (since they consume residential space). Their choice set A is the Cartesian product $[\pi(P^{gr} x Q^{gr}) \cup \{0^{gr}\}] \times [(P^{lmr'} x Q^{lmr'}) \cup \{0^{lmr'}\}]$.

Given the formulation presented thus far, the numbers P^{hr} and P^{fr} for households and firms respectively in regions r would have to remain constant over time. The resulting direct equilibrium would be one that results in an imaginary country where households and firms are not permitted to relocate. In order to permit relocation, one has to introduce an additional population f comprising all firms (nationwide), and another population h comprising all households. Both of these populations may have identical choice sets, that is, the set of all regions r=1,2,...,R. As households and firms migrate between regions r, the population totals P^{hr} and P^{fr} change over time. This change instantaneously induces related changes of the occupation numbers $P^{hr}(a)$ and $P^{fr}(a)$. The dynamic laws of these changes are discussed in Dejon and Gueldner (1985) under the heading of 'splitting rules', and so will not be discussed here.

4. STRICT UTILITIES

The dependence of strict utilities $U^{n}(a|\underline{r})$ on relative demands $r^{g}(p)$ is of two kinds, depending on whether population n is on the supply or on the demand side of market g. At any labor market lr, for instance, households hr' are on the supply side. Strict utility $U^{hr'}_{lr}(p^{lr},50|\underline{r})$ of offering labor services in region r at the wage rate p^{lr} could be posited to equal $[r^{lr}(p^{lr})]^{\beta hr'}\cdot p^{lr} - k^{hr'}(c^{rr'})$. The first term in this sum may be interpreted as wage income p^{lr} diminished by the cost of uncertainty of demand for labor, which here is specified by $[r^{lr}(p^{lr})]^{\beta hr'}$. The second term accounts for the cost of commuting between r and r', which may be computed in a transportation submodel. The exponent $\beta hr'$ is an empirical constant, non-negative on a priori grounds, and presumably also not larger than one. A value for $\beta hr'$ near zero means there exists a low cost of uncertainty of demand for labor. In computer simulation work such values lead to households shifting sluggishly to lower (claimed for) wage rates in response to weak demand for labor at higher wage rates.

Firms in region r are on the demand side of their labor market lr. Assuming that they are profit maximizers, strict utility $U^{fr}(p^{lr},q^{lr},p^{lmr},q^{lmr},p^{gr},q^{gr}|\underline{r})$ could be set equal to

$$[r^{gr}(p^{gr})]^{\beta fr}q^{gr}p^{gr}-\{p^{lr}+s^{lr}[r^{lr}(p^{lr})]\}q^{lr}-\{p^{lmr}+s^{lmr}[r^{lmr}(p^{lmr})]\}q^{lmr} \quad ,$$

where s^{lr} and s^{lmr} stand for search costs at the labor market lr, and the land market lmr, respectively. One should note that search cost depends on relative demand, both increase simultaneously.

Before introducing utilities of consumption plans, as perceived by households in population yr (living in region r and incurring income y), we make the following simplifying assumption, which does not seriously restrict the generality of our results: any single household shops only in a single region, so that alternatives consist merely in choosing some shopping region r', the quantity q of a good to be purchased and the shopping price p, in addition to choosing some quantity q^{lmr} of land in region r (for residential purposes) as well as the related rental price p^{lmr}. A corresponding utility function then might be written as

$$U^{yr}(r',p,q,p^{lmr},q^{lmr}) \quad :=$$

$$u_o^y(q,q^{lmr}) - k^y(r',r) - s^g[r^{gr'}(p)] - s^{lm}[r^{lmr}(p^{lmr})] -$$

$$s(y-cc^{r'r}-pq-p^{lmr}q^{lmr}),$$

where $u_o^y(q,q^{lmr})$ = the utility associated (in the more classical sense) with consuming the quantities q and q^{lmr},
$k^y(r',r)$ = a measure of the encumbrance of travelling between regions r' and r for shopping purposes,
$s^g[r^{gr'}(p)]$ = the search cost in market gr', at price level p,
$s^{lm}[r^{lmr}(p^{lmr})]$ = the search cost in market lmr, at price level p^{lmr},
$cc^{r'r}$ = the monetary cost of shopping trips between regions r' and r, and,
$s(y-cc^{r'r}-pq-p^{lmr}q^{lmr})$ = +∞ or 0, depending on whether or not the budget constraint '$y-cc^{r'r}-pq-p^{lmr}q^{lmr} \geq 0$' is violated.

The preceding utility function does not contain a term accounting for the utility of saving, in other words of $y-...-p^{lmr}q^{lmr}$. In future work such a term may have to be incorporated.

The expressions for strict utilities given above are, of course, not compulsory. Dejon, Gueldner and Wenzel (1985a, 1985b) have formulated rather general properties of strict utilities that together with certain properties of the distributions $d^n(a)$ secure the desirable features of those direct equilibria briefly alluded to in Section 2. Three of the more salient of these general properties are

(1) monotonicity: (i) $U^n(a|\underline{r})$ does not decrease (increase) as the price p^g of any good g increases, provided population n is on the supply side (demand side) of market g, (ii) $U^n(a|\underline{r})$ does not decrease (increase) as relative demand $r^g(p)$ for any good g, at any price level p, increases, provided population n is on the supply side (demand side) of market g, and (iii) $d^n(a1)$ is larger than $d^n(a2)$ if $U^n(a1|\underline{r})$ is larger than $U^n(a2|\underline{r})$;

(2) concentrability: the distributions $d^n(a)$ are controlled by some parameter ρ [measuring randomness of the random utilities associated with the strict utilities $U^n(a|\underline{r})$], in such a way that, for each population n and for ρ tending to zero, the distribution $d^n(a)$ grows more and more peaked at that alternative that exhibits the largest value $U^n(a|\underline{r})$ of strict utility; and,

(3) complementarity of search cost and of uncertainty of demand: there exists a search cost threshold α, $0 \leq \alpha \leq 1$ such that search cost, as a function of relative demand, sets in only above α, while the cost of uncertainty of demand sets in only below α. The foregoing examples of utility functions only approximately exhibit that kind of complementarity described here, which does not alter our results to any significant extent.

From these and other selected properties, the following characteristics can be obtained for equilibrium states $\underline{P}$:

(a) at each market g relative demand $r^g(p)$ is a non-increasing function of p; and, hence,

(b) at any market g, the lowest price level, say p^g_+, at which the cost of uncertainty of demand sets in, is higher than that price level, say p^g_-, at which search cost begins; and,

(c) for a sufficiently small randomness parameter ρ, economic agents concentrate at the equilibrium price level, and there, in particular, at quantity levels that optimize the agents' respective utility functions.

For a sufficiently small randomness parameter ρ, the following additional propositions hold: p^g_+ and p^g_- are neighboring price levels in the discrete set P^g, or at worst separated by on additional level $\bar{p}\varepsilon P^g$. Meanwhile at $\bar{p}$ there is neither a search cost nor a cost of uncertainty of demand. Furthermore, if $\bar{p}$ exists, practically all supply and demand at market g is concentrated at price level $\bar{p}$. If there is no $\bar{p}$, practically all supply and demand at market g is concentrated at price levels p^g_+ and/or p^g_-. Therefore,

$\bar{p}$, p_+^g and p_-^g will be called equilibrium price levels. For the discussion of the Walrasian or non-Walrasian character of price levels p_+^g and p_-^g it is expedient to notice that in a situation where p_+^g and p_-^g both exist (i. e., neither is p_-^g the highest price occurring in $\mathbf{P}^g$ nor p_+^g the lowest one), noticeable supply at price level p_+^g is possible only if the cost of uncertainty of demand at level p_+^g is negligible, and similarly, noticeable demand at price level p_-^g is possible only if the search cost at level p_-^g is negligible. Therefore, and because of the above attribute (c), if p_+^g and p_-^g both exist at all markets g, then the equilibrium state under consideration may be qualified as Walrasian. If, however, at one or several markets g (whose collection is denoted by **GR**) only p_+^g or p_-^g exists, and a noticeable cost of uncertainty of demand or search cost, respectively, prevails, the economic agents' behavior is still one of optimizing, but under quantity rationing constraints at the markets g that form **GR**. The equilibrium state then may be viewed as non-Walrasian. Thus, in terms of rationing if only p_-^g exists, and hence equals the largest price level occurring in $\mathbf{P}^g$, it is the demand side that is rationed at market g, unless the search cost at price level p_-^g is negligible. The analogous observation, concerning p_+^g, will be left as an exercise for the reader to complete.

Vanishing (total) excess demand, at any market g, is not obtainable from direct equilibrium conditions. In fact computer simulation results--preliminary though they may be--have furnished equilibrium states that exhibit involuntary unemployment. More generally, negative excess demand may result at markets of any goods that are instantaneously perishable. In markets of non-perishable goods, vanishing excess demand--in the sense of supply from current production equaling (satisfied) demand--is a consequence of stationarity. In a stationary state, where stocks are constant, current production has to equal (satisfied) demand (as the stored good is not perishable). As a final comment here, instead of absolute prices, as indicated above in the sample expressions of utility functions, relative prices are being used. Therefore, utility functions are homogeneous of degree zero with respect to prices.

Next we shall discuss briefly conditions under which stationary states are direct equilibrium states, and vice versa.

5. ATTRACTION-REGULATION AS THE DYNAMICS GERMANE TO THE DIRECT EQUILIBRIUM APPROACH

Direct equilibrium, as introduced in Section 2, postulates, for each population n, $V^n(a) := P^n d^n(a)/P^n(a)$ to equal 1 for all alternatives $a \varepsilon \mathbf{A}$. In disequilibrium, an alternative a of population n is said to exhibit high attraction if $V^n(a)$ is larger than 1, and low attraction if $V^n(a)$ is less than 1.

Dynamics germane to the direct equilibrium approach consist, broadly speaking, of inducing changes of state $\underline{P}$ by having economic agents migrate predominantly from low to high attraction alternatives. This phenomenon is what is meant by attraction regulation. The migration rate M^n_{ba} from alternative a to alternative b (of population n) may be written trivially as the product of the so-called moverpool ,M^n_a, and the (conditional) transition probability, h^n_{ba} (conditioned on moving to occur). That is,

$$M^n_{ba} = h^n_{ba} M^n_a \quad .$$

A large class of attraction regulated dynamics is obtained by positing (Dejon, 1983; Dejon and Gueldner, 1985)

$$h^n_{ba} := V^n(b)\, h^{no}_{ba}/V_a \quad ,$$

$$\text{with } V_a := \sum_b V^n(b)\, h^{no}_{ba} \quad \text{, and}$$

$$M^n_a := \Delta^n(V^n[a]/V_a)\cdot M^{no}_a \quad ,$$

where h^{no}_{ba} are intrinsic transition probabilities, V_a is the average attraction as perceived by economic agents currently associated with alternative a, $V^n(a)/V_a$ is the relative attraction of alternative a, and Δ^n is a modifying factor--depending upon relative attraction of alternative a--that modifies the so-called intrinsic moverpool M^{no}_a. Typically, the intrinsic quantities M^{no}_a and h^{no}_{ba} are simple functions of the state vector $\underline{P}$. Without loss of generality, the function Δ^n may be assumed to be standardized such that $\Delta^n(1)$ equals 1. Thus, in an equilibrium state [where $V^n(a)$,and therefore V_a, equals 1 for all alternatives a], Δ^n equals 1, and the intrinsic moverpool turns out to prevail in states of direct equilibrium (hereafter renamed simply 'equilibrium').

Ways of designing still larger classes of attraction-regulated migration laws are indicated in Dejon (1983) and Dejon and Gueldner (1985), where it is shown how migration rates may be obtained as solutions of network equilibrium problems. A very natural extension consists of replacing $V^n(a)$ by some strictly isotone function $f[V^n(a)]$, with $f(1)=1$ in order to preserve equilibria. Further flexibility derives from the proposition that if some specific distribution law $d^n(a)$ possesses the general properties alluded to in Section 4, then this also holds for any positive power $[d^n(a)]^\gamma$, where γ is arbitrary, but fixed. In addition, equilibrium states for $V^n(a) \propto [d^n(a)]^\gamma/P^n(a)$ are the same as those for $V^n(a) \propto d^n(a)/[P^n(a)]^{1/\gamma}$. Therefore, one may adopt the more general definition of $V^n(a) \propto d^n(a)/[P^n(a)]^\alpha$, with positive α, α being allowed to vary with n.

Two specific classes of attraction-regulated migration laws are useful to single out at this time. First, in so-called Rank 1 systems the intrinsic transition probabilities h^{no}_{ba}--and so, also, h^n_{ba}--do not depend on the source alternatives a:$h^{no}_{ba}=h^{no}_b$. As a consequence, total in-migration $\sum_a M^n_{ba}$ into alternative b takes the simple form $V^n(b)\, h^{no}_b\, M^n/V^n$, where M^n stands for the total moverpool of population n. A frequent choice of h^{no}_b is

$$h_b^{no} \propto [P^n(b)]^\lambda \quad , \text{ with } 0 \leq \lambda \leq 1 \quad .$$

The second class is one in which $\Lambda^n(x) := x^z$, with z not larger than 1 for stability reasons (Gueldner, 1984). A rather frequent choice is $z = -1$. Gueldner (1984) analyses the case of positive z and outlines its implications for the context of migration in geographic space.

In the Rank 1 case the choice of $z = 0$ leads to the following migration balances: $\sum_a M_{ba}^n - M_b^n = V^n(b)\, h_b^{no}\, M^{no}/V^n - M_b^{no}$. With the rather common choice of $M_b^{no} = \varepsilon^n P^n(b)$, where ε^n denotes intrinsic mobility, the additional Rank 1 assumption $h_b^{no} \propto P^n(b)$, which implies $M^{no} = \varepsilon^n P^n$, leads to

$$\sum_a M_{ba}^n - M_b^n = \varepsilon^n P^n [d^n(b) - P^n(b)/P^n] = \varepsilon^n P^n(b)[V^n(b) - 1] \quad .$$

This result is somewhat intuitive. The occupation numbers $P^n(b)$ of high attraction alternatives are continually being increased, while those of low attraction alternatives are being decreased. The dynamics used in Allen, Sanglier and Engelen (1984) are obtained from the preceding one by replacing $V^n(b)$ by $2 - 1/V^n(b)$. Near equilibrium this substitution constitutes only a minor change, as the functions $y = x$ and $y = 2 - 1/x$ possess a common tangent at the point $x = 1$, $y = 1$.

A third class is such that attraction-regulated migration laws are said to be in passive form if the intrinsic moverpool vector (M_a^{no}) is the principle right-hand eigenvector, for the eigenvalue 1, of the intrinsic transition probability matrix (h_{ba}^{no}), irrespective of the state $\underline{P}$ of the system. In the Rank 1 case this attribute means that $M_a^{no} = w h_a^{no}$ (where w is a scalar). This equality holds for any state $\underline{P}$ of the system. With $\Lambda^n(x) := x^z$ the migration law then can be written as

$$M_{ba}^n = V^n(b)[V^n(a)]^z\, h_b^{no}\, h_a^{no}\, w \quad ,$$

where the time-dependent scalar w is not necessarily the same as above, and in general also will be different from that mentioned below.

With h_a^{no} proportional to $[P^n(a)]^\lambda$, this preceding migration law takes the form

$$M_{ba}^n = d^n(b)[d^n(a)]^z [P^n(b)]^{\lambda-\alpha} [P^n(a)]^{\lambda - z\alpha} w \quad .$$

One should remember that this is the Rank 1 case, which does not cover the full range of gravitational laws that generally contain a spatial friction term $f(c_{ba})$. Whenever this migration law is used, often z is chosen to be equal to -1, and $d^n(b) := \exp[\gamma^n U^n(b)]$, which is the multinomial logit formulation.

Interest in the passive form of migration laws derives from the fact that stationarity in the sense of $\sum_a M_{ba}^n - M_b^n = 0$, for all n and a, is conveniently discussed for the passive form. One can easily show that equilibrium states, for which, by definition, $V^n(a) = 1$ for all n and a, are stationary. Since $\Delta^n(1) = 1$ (see above), for equilibrium states the stationarity condition maintains that

$$\sum_a h^{no}_{ba} M^{no}_a - M^{no}_b = 0 \quad \text{, for all b and n} \quad .$$

This result, however, means passiveness prevails.

A final useful class consists of passive attraction-regulated migration laws, which are said to be strongly passive if the equilibrium states are the only stationary states. In such a situation, equilibrium states are calculated (approximately) by calculating (approximate) stationary states. These latter states, in the case of asymptotic stability, may be considerably easier to calculate. Gueldner (1984) shows that Rank 1 passive migration laws with $\Delta^n(x) := x^z$ and $z < 1$ are strongly passive. The case of $z := 0$ and $h^{no}_b \propto P^n(b)$ has been treated explicitly under the second class mentioned above, and is easily seen to be strongly passive, since for $V^n(b) \neq 1$, $\sum_a M^n_{ba} - M^n_a \neq 0$. Meanwhile, for the non-Rank 1 case it might be appropriate to relax the passiveness definition to the extent of postulating merely that the intrinsic moverpool vector converges to the right-hand eigenvector, with eigenvalue 1, of the intrinsic transition probability matrix.

5.1. Some Preliminary Simulation Results

Now a brief summary will be presented of some computer simulation results that were kindly provided by Mr. Koerner, and serve to illustrate the Walrasian nature of direct equilibria when there is no quantity rationing. A previously calculated equilibrium state gets disturbed by lowering, at the goods market of a bi-sectoral economy, the threshold value α of relative demand (above which search costs set in and below which the costs of uncertainty of demand begin; see Section 4) from 0.95 to 0.915. This decrease shrinks the interval of relative demand values r^g where firms are sensitive to uncertainty of demand, while, conversely, the range of values increases for which households perceive search costs. In particular, the search costs at price level p^g_- (where, in the previously calculated equilibrium, households are rather heavily concentrated) suddenly increase. As a consequence, households move from the initial p^g_- level to higher price levels with low or no search costs at all. As a result, the average nominal transaction price increases by 1.6% during the first six time units.

The average nominal wage is observed to fluctuate much less. Over a period of 16 time units, this wage increases by about 0.12%, and then keeps returning monotonically towards its initial equilibrium value. Thus, the movements of average real wage are almost exactly linked to those of average nominal price.

At first, not only average nominal price, but also average net earnings of the firms and total stocks of stored goods increase. Among these three indices, average net earnings of the firms is the first one to change direction (to wit, after four time units), then average transaction price changes (after six time units), and finally the stock of stored goods turns (after 14 time units). Essentially, this pattern of changing directions of motion is maintained until the system has settled into a stationary state. The various indices perform damped oscillations, with a period of roughly 28 time units. The values of consumption and production levels of firms are very much in phase with average transaction prices (approximately the same locations of relative maxima and minima, respectively), while real consumption is very much in opposite phase (i. e., it is minimal when prices are maximal, and vice versa).

All indices mentioned in the preceding discussion, except the stock of stored goods, settle down again at their initial equilibrium values. The stock of stored goods undergoes an increase, which is due to the stationary state supply from stocks (which is set proportional to the stock of stored goods) equaling $(1-\alpha)$ times the total supply. Before and after lowering α, all equilibrium prices and quantities--with the single exception of the stock of stored goods--are the same (within a certain degree of precision that is controllable by the number of time units carried out), which supports the conclusion reached in Section 4 that direct equilibrium prices are Walrasian ones (for negligibly small randomness in the random utility functions, and provided no quantity rationing occurs).

6. DISCUSSION AND CONCLUSIONS

Stochastic choice theory has supplied a major impetus during the past twenty years to research in the areas of housing choice, transport choice, on the demand side, and rental status choice by landlords, on the supply side (McFadden, 1978). Choices were modelled for parametrically given prices, and stochastic demand functions were obtained (Hildenbrand, 1971). The basic idea of the direct equilibrium approach consists of introducing price dispersion by incorporating prices, besides quantities, into the choice sets. The main technical device for making this idea work consists of the use of search costs and of the cost of uncertainty of demand, which are functionally dependent on relative demand. Dejon and Gueldner (1985) have used this particular device in the context of a land use and location model, and have found that numerically calculated direct equilibria exhibited those properties that Alonso (1964) had shown must hold for equilibrium solutions. That this was so by necessity was proved by Gueldner (1984), who also outlined a procedure for implementing the direct equilibrium approach in a general equilibrium model with two sectors (households and firms), yet no stocks, and has provided some preliminary computer simulation results. Dejon, Gueldner and Wenzel (1985a, 1985b) have analyzed the relationship between direct general equilibrium and equilibrium with quantity rationing. Nevertheless, numerous issues remain unresolved. In particular, the monetary sector and rational expectations need to be incorporated in those modelling strategies reported here.

Direct equilibrium modelling is neither at the micro nor the macro scale, but rather occurs at the meso scale. Macro in the sense of modern macro-economics is not adequate to describe this modelling approach, since the temporal law of change of macro variables, like average income of households or average production of firms, depends not only on macro variables, but also on the distributions of variables. To be more precise, it depends on the distribution of populations of economic agents along price and/or quantity axes of choice sets. The micro level is inadequate, too, if one is ready to accept as typically micro a theory like search theory for the labor market (Rogerson, 1983), or a micro analysis of the decision to migrate as discussed by Leonardi (1983).

A final question of interest is that of uniqueness of direct equilibria. Treating this question in any detail would have been outside the scope of this paper. Only a cursory remark may be in order, namely when modelling the evolution of spatial configurations by attraction regulated dynamics with an appropriate direct equilibrium model, in which counteracting effects of agglomeration economies and diseconomies are accounted for (see, for example, Papageorgiou, 1979; Dejon and Graef, 1983), the final stationary state may very well depend on the initial configuration. Among two competing potential centres of a developing urban agglomeration, the one that has an initial lead in terms of agglomeration economies will, under rather general conditions, develop into the centre of agglomeration, provided the land consuming populations are not so large that the second centre has to develop, too.

7. REFERENCES

Allen, P., M. Sanglier and G. Engelen, 1984, Chance and Necessity in Urban Systems, in _Stochastic Phenomena and Chaotic Behaviour in Complex Systems_, edited by P. Schuster, Berlin: Springer-Verlag, pp. 231-249.

Alonso, W., 1964, _Location and Land Use_, Cambridge, Mass.: Harvard University Press.

Benassy, J.-P., 1982, _The Economics of Market Disequilibrium_, New York: Academic Press.

Dejon, B., 1983, Attraction-regulated Dynamic Equilibrium Models of Migration of a Multiplicative Type, in _Evolving Geographical Structures_, edited by D. Griffith and A. Lea, The Hague: Martinus Nijhoff, pp. 12-23.

_____ and F. Graef, 1983, Eine Klasse von Modellen zur Beschreibung von Staedtischen Agglomerations- und Deglomerationsprozessen, in _Stadt, Region, Land_, edited by P. Maecke, Rheinisch-Westfaelische TH Aachen: Institut fuer Stadtbauwesen , pp. 32-38.

_____ and B. Gueldner, 1985, Dynamics of Wide-sense Migrational Systems: A Choice-theoretic Approach, Papers of the Regional Science Association, 55: 121-133.

_____, _____ and G. Wenzel, 1985a, The Structure of Direct Equilibrium, unpublished manuscript, Institute of Applied Mathematics, University of Erlangen-Nuernberg.

_____, _____ and _____, 1985b, On Stochastic Equilibrium with Price Dispersion, in Proceedings of the 10th Symposium on Operations Research, edited by M. Beckmann, K.-W. Gaede, K. Ritter and H. Schneeweiss, Munich, August 26-28.

Gueldner, B., 1984, Attraktivitaetsgesteuerte Dynamische Alternativenwahlmodelle: Analysen und Simulationen in drei Anwendungsbereichen, unpublished doctoral dissertation, Institute of Applied Mathematics, University of Erlangen-Nuernberg.

Hildenbrand, W., 1971, Random Preferences and Equilibrium Analysis, Journal of Economic Theory, 3: 414-429.

Leonardi, G., 1983, An Optimal Control Representation of a Stochastic Multistage-multiactor Choice Process, in Evolving Geographical Structures, edited by D. Griffith and A. Lea, The Hague: Martinus Nijhoff, pp. 62-72.

McFadden, D., 1978, Modelling the Choice of Residential Location, in Spatial Interaction Theory and Planning Models, edited by L. Lundqvist, F. Snickars and J. Weibull, Amsterdam: North-Holland, pp. 75-96.

Papageorgiou, Y., 1979, Agglomeration, Regional Science and Urban Economics, 9: 41-59.

Rogerson, P., 1983, The Effects of Job Search and Competition on Unemployment and Vacancies in Regional Labor Markets, in Evolving Geographical Structures, edited by D. Griffith and A. Lea, The Hague: Martinus Nijhoff, pp. 349-371.

TOWARDS A BEHAVIORAL MODEL OF A SPATIAL LABOR MARKET

Carl G. Amrhein
University of Kentucky
United States

Ross D. MacKinnon
State University of New York at Buffalo
United States

1. INTRODUCTION

Migration is increasingly recognized as one of the fundamental processes in geographical analysis--a process that both shapes and reflects the areal differentiation of social systems. Of particular importance is the migration that occurs in response to changes in employment (i. e., labor migration). A variety of models have attempted to describe both the general migration process (see Clark, 1982, for a complete review), as well as the more specialized labor migration process (see Rogerson, 1982). All of these models, however, are deficient in at least two areas. The first arises from the manner in which they conceptualize space; the second is their failure to recognize the importance of person-specific characteristics such as age and job tenure. The model developed in this paper will attempt to address these two shortcomings. First a formulation of space that recognizes its continuous nature will be incorporated into a general model of labor migration. Second, a disaggregate behavioral model of the migrant that recognizes the role of individual characteristics in the decision process will be developed.

While recognizing that the physical distance between locations may be one factor in a person's migration decision, recent studies suggest that an important effect of distance is the influence it has on the transmission of information between regions. Since the absence of sufficient quality or quantity of information may give rise to uncertainty about opportunities in other places, the effect of relative distance on information flows may be an important variable in the interregional movement of workers (see Clark, 1983 and 1986).

Although physical distance has been shown to have a clear impact on information flows (Brown, 1981), other effects related to location also play a role. Plane (1984) suggests incorporating a measure of 'functional space' in place of physical distance. However, such a measure only describes the observed flows rather than explains the causes of those flows. Rogerson and MacKinnon (1981), Maier (1985), and Amrhein (1985) propose that past migration flows play an important role in the movement of information between regions. While they may impact future flows, these explanations do not explain the original flows. In these models, increasing numbers of messages are equated with decreasing risk and a resulting increase in the likelihood of migration. No effort is made to distinguish among message types, or the possible correlation amongst information sources (see Clemen and Winkler, 1985).

Holt (1978) and Rogerson (1984 and 1985) suggest that elements of labor market competition and traditional views of wage differentials should be integrated into labor migration models. Cordey-Hayes and Gleave (1977) include market turnover as an important element in their model. Krumm (1983) argues for the inclusion of environmental factors in addition to job factors. All of these suggestions require that the necessary information be originally transferred from the region with the opportunities to the potential migrant in order to initiate the interregional interaction. Toward this end, Pickles and Rogerson (1984) and Amrhein (1986) propose an information emission function that includes employers and other agencies as sources of information for prospective migrants. Once available, potential migrants must be able to receive and process the information in all of the models described. This ability is implicitly assumed. Finally, Curry (1985) has introduced contact fields in a non-migration setting as an alternative information framework (see also Boorman, 1975).

Admitting the importance of individual contact fields in the migration decision requires a recognition that other individual characteristics may constrain the outcome of any job search and relocation process (McHugh, 1984). This fact has long been recognized from an empirical standpoint but has not been successfully integrated yet into a theoretical model. All of the models discussed above focus attention on the structure of the labor market and the behavior of idealized workers moving between regions, rather than on the process that might give rise to a migration decision in a 'real setting' (see Clark, 1980). Part of the problem is that such a model would have to be estimated at the micro level of aggregation, and accordingly would encounter severe, if not impossible, calibration difficulties.

Only recently have efforts been made to formulate models combining aggregate demographic characteristics of the population with classical economic models in order to study national migration patterns (Plane and Rogerson, 1985; Isserman, 1985). While important, demographic characteristics, such as vital rates, explain only part of the migrant's decision. Da Vanzo (1980) provides a review of the findings of several studies in this area of inquiry. Mueller (1982) finds job tenure, turnover, and several characteristics of a worker's current residence to be important variables in the migration equation. And, Schaeffer (1985) discusses the importance of the skill level, experience, education, and past migration history of the worker in the decision process. All of these studies are reinforced by findings in management that have traced the movement of workers up the career ladder, and from one position within a firm to another (Bartholomew and Forbes, 1979; Veiga, 1983). Even more difficult to model than individual demographic characteristics, however, are recent findings that migration decisions are affected by the migrant's view of the intended duration of the possible move (McHugh, 1984), and the perceived short- and long-run movement in the regional economies of the migration destinations (Ballard and Clark, 1981).

The model outlined in this paper is a first step toward constructing a theoretical model that incorporates individual characteristics of the migrants. These migrants live in an economy in which space is assumed to be continuous rather than segmented into discrete regions interacting over a distance. The workers receive information through two contact fields. First is their personal fields, which fluctuate in response to their success in obtaining ideal job assignments. The second contact field is controlled by the employer, and fluctuates in response to the ability to obtain profit relative to that obtained by competitors.

2. THE MODEL

The model developed here conceptualizes the labor market as two large subsystems operating in parallel. One of these systems is composed of the employees or workers segmented into groups according to their acquired level of skill. Workers engage in a search process in which they examine firms located in their activity spaces or contact fields. They compare their current wages with wages earned by workers employed by these firms. A level of stress is calculated and, if sufficiently large, the workers apply to jobs offering higher wages. Unless already hired, an offer of employment arising from an application must be accepted. The workers' contact fields generate information from two sources, namely information gathered by the worker through the search process, and information received from employers trying to hire a particular type of worker.

The parallel employer subsystem conceives of employers searching their contact fields to gain information about the levels of profit earned by other firms. If the searching employer's profit is less than that of other firms, the employer reduces wages to achieve the higher return. If the adjusted wage is zero or less, the employee is dismissed. The firm then advertises to attract a worker of the same level of skill that produced the higher return for the 'ideal' firm. The firm thus modifies its preference structure and production process in response to information gained from neighboring firms. The employer subsystem has many similarities to a classic spatial diffusion process (Hagerstrand, 1967). Firms search an 'information field' seeking information about more profitable production processes. Once encountered, firms first seek to adopt the ideal profit level. If this is not possible within their current production process, the firms dismiss their workers and adopt the new technique. As will be seen below, the rate in which an 'ideal' process diffuses among firms is partly a function of the size of the firm's information field.

The two systems communicate in a variety of ways. Employers explicitly pass information to attractive employees through invitations to apply. Employees implicitly pass information to employers by moving to higher paying jobs. Similarly, employers evaluate a worker's performance through the wage adjustment process. These systems formally interface through the hiring process. At the end of each time period, workers age, retire, enter the market, and acquire tenure with longevity on the job. Unsuccessful or dissatisfied workers and employers adjust the spatial extent of their contact fields. Figure 1 presents a flowchart of the steps involved in this model.

2.1. The Worker Sub-Model

Workers migrate in response to stress. Stress arises as a result of a perceived differential between the level of welfare a worker is currently experiencing, and the level of welfare experienced by some other worker encountered in the search process. Past efforts to model stress have focused on differences in wages (Wolpert, 1966). Recent efforts (Mueller, 1982, Amrhein and MacKinnon, 1985) have recognized the limitations of this assumption and incorporate a non-wage component. Although pleasing from a theoretical standpoint, the lack of empirical data on a worker's valuation of and response to changes in the non-wage component of the welfare function limits its usefulness. Although in the current model workers respond primarily to wage differentials, a micro-level model is perhaps better able to examine the influence of non-wage variables than other, more aggregate approaches.

In the current application, stress is a simple function of the difference between a worker's current wage, and the wage earned in another job about which the worker has knowledge (see Brummell, 1981). In the case of a vacancy, the advertised wage is the wage for comparison. If $\beta_i^{kl}(t)$ is the stress experienced by worker i in job k examining job l in period t, then

$$\beta_i^{kl}(t) = W_i^k(t) - W_j^l(t) \quad , \qquad (2.1.1)$$

where $W_i^k(t)$ is the wage earned by worker i in job k during t, k is the job the worker currently occupies, and l is the position under consideration by i. Not all workers, however, react to a given wage differential in the same manner. Although any number of additional factors could be considered here, attention is focused on the worker's tenure in a particular position. If $\gamma_i(t)$ is the effect of job tenure on worker i's experienced level of stress in response to a given wage differential, then equation (2.1.1) now becomes

$$\beta_i^{kl}(t) = [W_i^k(t) - W_j^l(t)][1.0 + \gamma_i(t)] \quad . \qquad (2.1.2)$$

Although several complex functions could be defined for $\gamma_i(t)$, in the absence of empirical evidence the one defined here will be a piece-wise linear approximation of the logistic curve, where

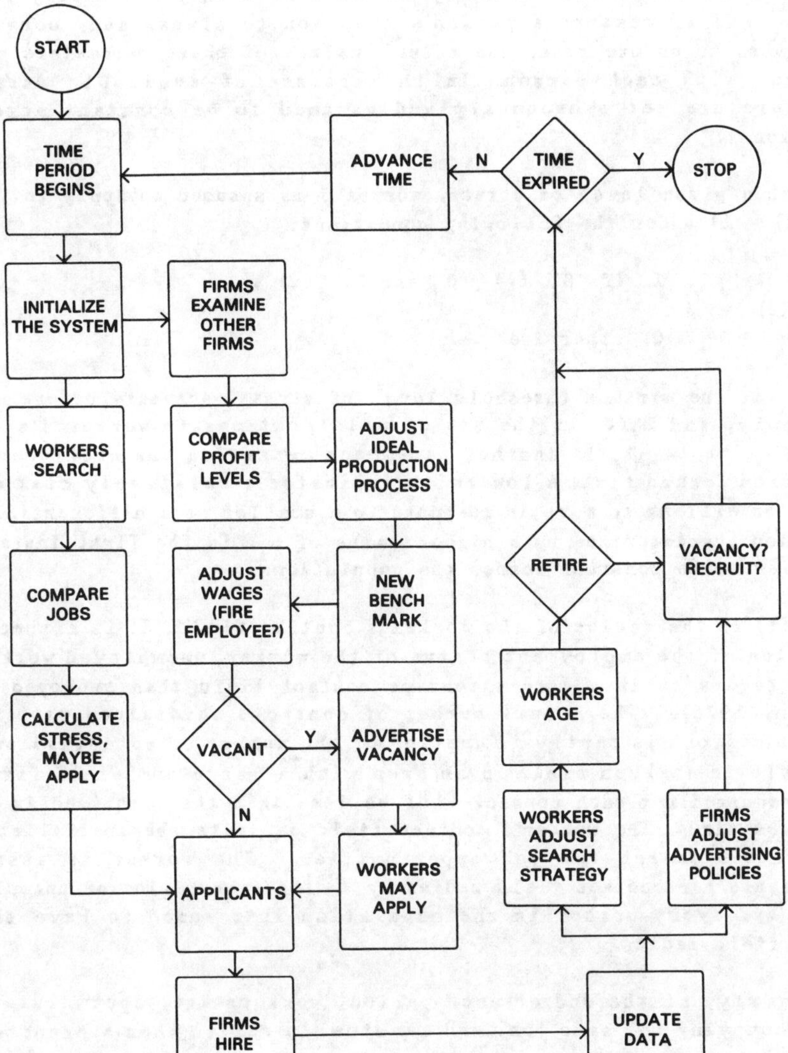

Figure 1: Flowchart of the model.

$$\gamma_i(t) = \begin{cases} \mu_1, & \text{if } T_i^k(t) \leq a_1, \\ \mu_2, & \text{if } a_1 < T_i^k(t) \leq a_2, \\ \mu_3, & \text{if } a_2 < T_i^k(t), \end{cases} \tag{2.1.3}$$

where $T_i^k(t)$ is the tenure earned by worker i holding job k during period t. Since $_i$ and a_i measure a person's reaction to stress and, conversely, willingness to endure risk, the actual values of these parameters might be different with each person. In the absence of available data these parameters are set exogenously and assumed to be constant across the population

With a given level of stress, worker i is assumed to apply to job k [i. e., $M_i^k(t) = 1$] under the following conditions:

$$M_i^k(t) = \begin{cases} 1, & \text{if } \beta_i^{kl}(t) \geq \tilde{\beta}_i \text{ and } l \varepsilon L_i(t), \\ 0, & \text{otherwise}, \end{cases} \tag{2.1.4}$$

where $\tilde{\beta}_i$ is the minimum threshold level of stress necessary before worker i will apply, and $L_i(t)$ is the set of all locations in worker i's contact field at time t. $\tilde{\beta}_i$ is another parameter measuring the attitudes of the population toward risk. A low value indicates a relatively risk neutral population willing to move in response to a smaller wage differential than a population characterized by a higher value of $\tilde{\beta}_i$. In the first instance, $\tilde{\beta}_i$ is assumed to be constant across the population.

$L_i(t)$ is the radius of the worker's contact field. It is assumed to be a function of the employment status of the worker; unemployed workers are assumed to invest in a more extensive contact field than employed workers (Boorman, 1975). The actual number of contacts maintained with a given amount of effort is partly a function of the number of potential contacts available in a given area. In an area with a large number of firms, the effort required for each contact will be less than that required in an area with fewer firms. The worker's contact field delimits the spatial extent of the worker's search for job opportunities. The worker is assumed to increase his/her contact field radius by α for each period of unemployment. Initially, every worker in the population is assumed to have the same contact field radius.

Finally, at the end of each period, workers age, accumulate tenure while occupying the same job, and eventually retire when a predetermined retirement age is attained.

2.2. Employer Sub-model

In many respects the employers engage in a process that is similar to that of the employees. Just as wage differentials motivate the behavior of workers, it is assumed that firms respond to differentials in the level of profit they earn compared to the profit earned by others in the firm's contact field. Let $\pi_i^k(t)$ be the profit earned by firm k employing worker i during period t, or

$$\pi_i^k(t) = P_i^k(t) - W_i^k(t) \quad , \tag{2.2.1}$$

where $W_i^k(t)$ is as defined before, and $P_i^k(t)$ is the productivity of worker i in job k during period t. Given a level of profit, the firm then searches other firms to determine if there is a difference in profit levels. Firm stress arises if the employer is earning less than other firms in its contact field. If stress is positive, two basic actions are open to the employer--either increase productivity or reduce wages. In the short run, productivity is difficult to change due to capital investments and a fixed set of employee skills. Therefore the employer adjusts wages, either reducing the current employee's wage sufficiently to achieve the desired level of productivity, or releasing the employee and hiring another at a reduced wage. We assume that employers first reduce wages. If a negative wage is required to attain the desired profit level, the employee is dismissed. If the employer dismisses the employee, the firm then begins to advertise the vacancy. The advertised offer reflects the most profitable production process known to the firm. Implicit in the firm's decision is a recognition that if the firm lays off its old employee, it may change production methods before filling the vacancy. Changing the production process requires adopting the previously identified most profitable process.

An advertised position reflects either the profit earned by the process used by the ideal firm, or the expected profit from the existing process with the reduced wage. The number of firms in an employer's contact field reflects the intensity with which that employer acquires information. As discussed with respect to the employee's contact field, a non-uniform distribution of firms complicates this interpretation since, with the same amount of effort, a greater density of firms will lead to a larger number of contacts. In spite of the question of density, however, an employer that extends a search over a greater area is expending greater effort (in at least a geographical sense) than would be required over a smaller area.

Information plays a crucial role in the operation of the firm. First, it allows the firm to learn the appropriate level of profit. As discussed above, in the short run this is likely to translate into wage adjustments. Secondly, as represented in the skill level of the employee, information provides the firm with knowledge about the most successful production process. Finally, in the case of a vacancy, the gathered information provides the locations of employees the firm might want to recruit. Failure to keep abreast of current market conditions may allow the firm to slip into a disadvantageous position relative to the other firms in the area. The ultimate result might be that the isolated firm goes out of business. In short, the continuous gathering of information reduces the uncertainty the firm faces in the market.

In the presence of a vacancy the firm advertises, trying to elicit applications. A policy that indiscriminately solicits applications from workers would be inefficient. Rather, the firm will target information to workers ideally suited to perform the job. We assume that a firm engages in an intensive, individual recruiting process in which specific workers are contacted.

At the end of each period, firms increase the tenure of their continuing employees, retire the employees that have reached retirement age, and adjust their search and advertising radius for the following period. Such a strategy implicitly assumes that the firm is willing to have its position filled by any worker, so long as the appropriate level of profit is maintained. Although a less than ideally qualified worker in a job may be preferred to a vacancy, in the long run, an ideally qualified worker is desirable. If an employer is unable to hire an ideally suited employee in the existing contact field, then in the following time period the search/advertising radius might be increased until the recruiting process is successful. In fact, a vacancy could be viewed as an extreme mismatch of an employee's skill level with the job's requirements. Since the desired skill level may change over time with the adoption of a new production process, an employee that at one time is ideally suited to perform a job, may be unsuited at some time in the future for this same job.

2.3. The Clearing Process

After each employee and firm has searched its respective markets, adjusted wages and then calculated stress levels, a hiring process occurs. Firms make an offer to the first applicant with the desired level of skill. Failing to hire an ideally suited worker, firms then make offers to increasingly over-qualified, followed by increasingly under-qualified, workers until either a person is hired, or there are no more applicants. Workers accept the first offer they receive.

3. THE SIMULATION EXPERIMENTS

The population for the simulation experiments presented here consists of a labor market of 100 workers and 100 firms. Laborers are partitioned into four equal groups. Membership in a group is determined by an assumed level of skill. Jobs are similarly categorized by the level of skill required to perform the job optimally. The skill level of the worker is fixed while the skill required to perform a given job optimally can vary over time as the firm adopts different production techniques.

Firms are assumed to prefer a worker with a skill level ideally suited to perform the job that is available. In the absence of such a person, it is assumed that an employer prefers an over-qualified worker to one that is underqualified, and any worker to no worker. The highest skilled worker, therefore, has an inherent advantage over the lowest skilled worker for all jobs except those requiring the lowest level of skill. Every worker is assumed to be able to perform any job well enough to return some profit to the firm, if the salary paid is sufficiently small.

The market segmentation adopted here, although simplistic, does reflect some of the market characteristics observed by others. The adopted structure is similar to the 'second type' of occupation described by Curry (1981, p. 135). An occupation's ability to compete is partly a function of the information flows and contact networks of the segment's members. For a class of workers to be successful, an occupation will have to create a region in which the preferred production process of the local firms requires the skills the workers have to offer. This regional specialization requires the segment in question to be the most productive one. Alternatively, employers might try to create a situation in which, in the absence of other opportunities known to the workers, wages are depressed, thus increasing profit levels and possibly giving rise to spatially segmented markets described in, for example, Clark (1981).

Working against this segmentation are the recruiting efforts of other firms. A distant firm observing the high profit levels attained by another firm is tempted to adopt that highly productive process and attract workers of the desired skill class. The wage adjustment process acts to integrate the market (Amacher and Sweeney, 1981). The ability of these regional markets to integrate is constrained by the spatial extent of the contact fields of the firms and workers.

The initial assignment of workers to firms involves a two stage process. First, 100 pairs of random coordinates are generated using the IMSL subroutine GGUBS. These pairs form the locations of the firms. The firms cannot move once established. A random permutation of the firms then is generated (IMSL subroutine IPER) and the firms and workers are matched sequentially.

The initial skill levels are created by assigning one of four values to four groups of 25 workers that have been numbered sequentially. For example, workers 1 through 25 are the lowest skill group while workers 75 through 100 are the highest skilled. Since the workers are randomly assigned to the firms initially, any section of the region is prevented from acquiring an imbedded preference for one production process. Figure 2 shows the initial distribution of skill groups.

The creation of wage and productivity distributions requires first establishing benchmark values for each worker class. Productivity and wages are assumed to be increasing functions of skill. Initially the profit derived from any worker/productivity match is constant. A distribution of observed values around these benchmark values is obtained by adding (subtracting) a standardized distribution (values range between $\pm$ 1.00) of 25 random normal deviates (obtained from IMSL subroutine GGNML) to these benchmark values. The initial benchmark wages are 10, 20, 30 and 40, and the initial benchmark productivities are 15, 25, 35 and 45. The initial search

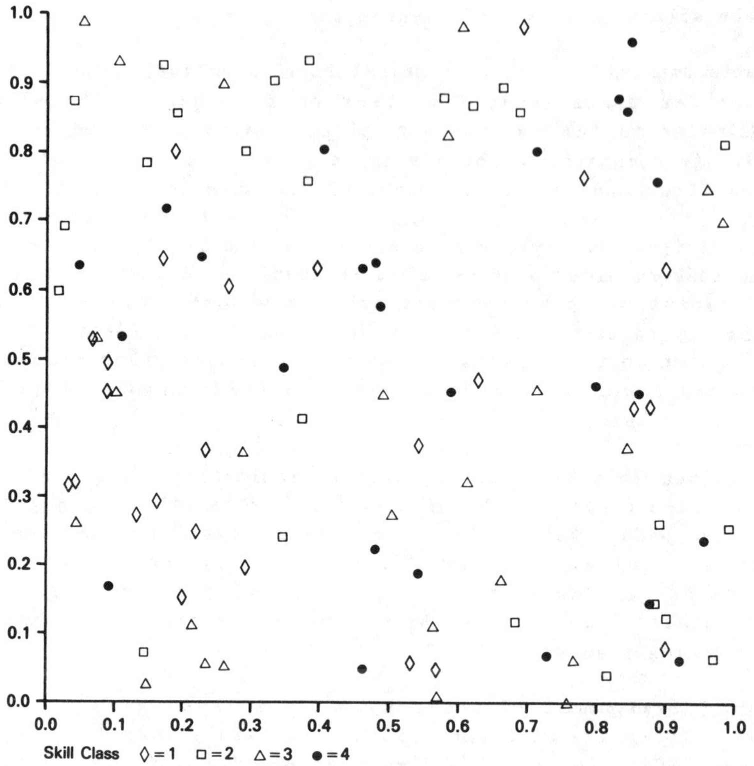

Figure 2: Desired worker skills at the beginning of time period 1.

radii for the firms and workers are arbitrarily set to 0.10, with a range of coordinates in both the x and y directions of 0 to 1. The rate of increase of search radii is set to 0.40. Thus, after the first increment the search radius will increase to 0.140. The points of discontinuity in the tenure stress function are 3 and 6 periods with retirement after the tenth period. The μ_i are 1.5 for 3 or fewer periods of tenure, 1.0 for 4 through 6 periods of tenure, and 0.5 for greater than 6 periods of tenure. The minimum level of stress required for application to occur is initially 0.10. The initial distribution of tenure is generated by again assigning a random distribution (IMSL subroutine IPER) of numbers between one and ten, sequentially, to the population. As a result, since the mandatory retirement age occurs after the tenth period, one tenth of the population will retire at the end of each period. Similarly, one tenth of the population will be new workers at the beginning of each new period after the first one. It is assumed that each worker who retires is replaced by a worker with the same skill. Recognizing that it is not possible to apply to every job in the market, workers are limited to ten applications per period. Each simulation was allowed to run for at least 40 time periods.

4. SIMULATION RESULTS

Two sets of results are generated for each experiment. The first set reports on the behavior of the aggregate economy, or the macro-economic performance of the entire population of 100 firms and employees. The second set of results describes the spatial characteristics of firms and employees.

At the macro scale, concern is focused on the aggregate measures of economic well-being. These measures include results for both the entire population and each skill group. The measures of interest here are the unemployment or vacancy rate (numerically equal since the number of firms equals the number of workers), the welfare and productivity levels, the number of workers hired each period, and the distribution of preferred skill types.

In the discussion of these results, two sets of parameters are of particular importance and will be emphasized here. The first set controls the actor's (either a firm or a worker) contact field. These parameters consist of the sizes of the minimal contact fields ($\tilde{D}_i$ or $\tilde{D}^k$), the maximum size a contact field may attain (originally assumed to be unlimited), and the rate at which the contact fields increase [$(1+\gamma_i)$ or $(1+\eta^k)$]. The effect of only the first two of these parameters will be addressed. The second important set of parameters control the application decision process through the description of the actor's willingness to assume risk ($\tilde{\beta}_i$).

Table 1 summarizes the aggregate results for a relatively risk neutral population for which the contact field can grow to reflect nearly perfect information. Risk neutrality is assumed in that before any consideration of tenure, a person will apply to a new job in response to a very small wage differential. Perfect information is achieved when the radius of the contact field exceeds the greatest dimension of the region; in other words, when every opportunity is within the actor's contact field.

The global unemployment rate climbs from zero to 28 percent in the first ten periods before reaching a fairly stable, but very high, value of around 40 percent. The distribution of unemployment among skill groups, however, is not even. Recall that the workers have no control over their wage in a particular job. In theory, if an employer reduces the wage too much, the employee will migrate. If every employer reduces wages, however, the employee has no choice but to accept a low paying job. Furthermore, even though initially every skill group generates approximately the same level of profit, if the wages for all workers are reduced to some equally

TABLE 1

AGGREGATE RESULTS EXTENSIVE CONTACT FIELD
(RISK NEUTRAL, UNLIMITED SEARCH AREA)

	End of Time Period	Skill Type 1	2	3	4	
Unemployment Rate	1	0.00	0.00	0.00	0.00	0.00
	10	.40	.48	.24	.00	.28
	20	.44	.52	.44	.28	.42
	30	.48	.40	.36	.32	.39
	40	.44	.44	.56	.24	.42
Average Wage	1	7.74	18.70	28.75	38.44	23.42
	10	2.59	2.56	4.32	3.26	2.49
	20	.51	.41	.47	.68	.27
	30	.40	.46	2.10	.52	.69
	40	.43	.63	1.91	.59	.30
Average Productivity	1	14.40	24.72	34.64	44.48	29.56
	10	8.48	12.68	26.28	44.56	18.31
	20	7.92	11.88	19.20	32.28	15.49
	30	7.64	14.88	22.20	30.40	18.34
	40	8.04	13.68	15.12	33.96	15.72
Desired Skill Level	1	15	30	34	21	100
	10	0	3	3	94	100
	20	0	1	0	99	100
	30	0	1	0	99	100
	40	0	1	0	99	100

low level, then highly productive workers become the ideal workers of choice, biasing the labor market in their favor. Aside from the entry of 10 percent of the labor force into the market each period, persistently high levels of unemployment arise from a mismatch of applications to available vacancies. Two factors contribute to this undesirable situation. Employers, aware of the profit levels earned by competing firms, dismiss workers hoping to hire better qualified employees. This course of action forces the formerly employed workers onto the unemployment roles. Similarly, workers, seeing an attractive position (possibly vacant), apply in large numbers to the same few jobs. Since the preferred job is filled by only one worker, the other applicants either remain unemployed or stay at their old jobs. At the same time, assuming that employees can apply to only a limited number of jobs, workers do not apply to other vacancies, causing those jobs to remain vacant.

Average levels of productivity, while stabilizing at the aggregate level after the initial periods, fluctuate at the skill group level. As a result of the high levels of unemployment among the unskilled workers, there is a greater reduction in the average productivity for this group than among the highest skilled workers. Overall the market is rather volatile. This is partially due to the rather high turnover rate in the labor market as employees retire, enter the labor market, and are dismissed by their employers.

With the decreasing wage levels, the highest skill group becomes the employers' preferred group. This trend is seen in the dominance of the highest skill group in the preferred production process. Initially each of the four skill levels is preferred by 25 firms. By the end of the tenth period, over half of the firms prefer the highest skill group. Although this bias preserves the employment level in the highest skill group, it does not preserve the wage level. Since the entire regional wage scale is depressed to the same low value, the highest group suffers the greatest amount of wage erosion. This downward movement in wages is not necessarily shared equally by all workers in a group. If a firm is unaware of the movement in wages in the rest of the region, there will be no pressure to reduce wages so long as the firm is earning a profit. Isolated firms thus continue to pay high salaries.

Knowledge about the behavior of other firms influences the behavior of the inquiring firm. Therefore, reducing the contact field of a firm by reducing its ability to acquire information, may change the firm's behavior. Table 2 displays the results when $\tilde{D}_i$ and $\tilde{D}^k$ (the assumed base level search radii) are reduced from 0.10 to 0.05. Although the size of the maximum possible search area is still unlimited, the initial contact field has been sharply reduced. The effect of this information reduction is dramatic. Unemployment rates, although still high, are markedly reduced. Average levels of worker welfare are much higher. The wage reduction process, however, still creates a preference for workers with the highest skills. Unemployment in this preferred group is close to zero. Information reduction has had little effect on employment levels among the lowest skilled workers. The tendency toward the adoption of a uniform production process has slowed down.

One cause of these improved aggregate results arises from the reduced diffusion of the production process requiring the highest skilled workers. Firms preferring other production processes decrease the demand for this one class of employees. More firms rely on the dominant process in their local areas, creating a segmentation of the market. It is worth noting that, in

TABLE 2

AGGREGATE RESULTS, RESTRICTED CONTACT FIELD
(RISK NEUTRAL, UNLIMITED SEARCH AREA)

	End of Time Period	Skill Type 1	2	3	4	Total
Unemployment Rate	1	0.00	0.00	0.00	0.00	0.00
	10	.48	.24	.00	.00	.18
	20	.36	.36	.16	.04	.23
	30	.32	.40	.16	.00	.22
	40	.40	.28	.16	.04	.22
	50	.24	.48	.16	.08	.24
Average Wage	1	9.19	19.26	29.39	39.09	24.23
	10	7.76	13.31	22.47	19.70	13.30
	20	5.11	11.04	12.40	11.80	7.92
	30	7.24	8.00	12.48	10.68	7.64
	40	3.08	6.44	8.52	9.92	5.07
	50	7.00	9.84	11.24	8.72	7.21
Average Productivity	1	14.40	24.72	34.64	44.48	29.56
	10	7.72	18.72	34.52	44.52	22.47
	20	9.44	15.60	29.08	42.84	22.48
	30	10.00	14.64	29.08	44.56	22.75
	40	8.68	17.68	29.08	42.80	20.53
	50	11.16	12.72	29.08	41.04	21.43
Desired Skill	1	22	31	28	19	100
	10	12	13	22	53	100
	20	6	8	19	67	100
	30	6	7	17	70	100
	40	6	6	16	72	100
	50	6	5	16	73	100

the later periods, economic conditions appear to be improving for the lower class workers, particularly in wages, at the expense of the highest skill class. It is possible that a secondary wave of diffusion, based on unskilled labor, may begin to diffuse, but simply has not yet appeared in the simulation results generated thus far.

In addition to the size of the minimal contact field, it might appear that the maximum size of the contact field also may play an important role in the evolution of the regional economy. Clearly an assumption that allows an actor to acquire knowledge about the entire economy is unrealistic. Even if it were possible to utilize such an amount of information, such a course of action would not be efficient in terms of time or cost. Since the present model assesses no cost for acquiring information, these issues cannot be addressed. However, if the maximum search area is artifically fixed in size, there would be a further reduction in information available to the actors. This reduction in information would not impact the actors in a uniform way. Recall that the search radius is increased if a worker is unemployed, or a firm has been unsuccessful in recruiting an ideally suited worker. Once these conditions are met, the actor is assumed to reduce his/her contact field to the minimal size. This process implicitly assumes that in times of high stress (unemployment or reduced profits), efforts to acquire information intensify. By recognizing a limit on the intensity of these efforts, those workers most in need of additional information are restricted and thus forced to compete more directly with those workers not in distress. Table 3 presents results for a population in which the minimal contact field is the same as that upon which Table 2 is based, but now the maximum search radius has been restricted to 0.50. An actor can now search at most 78% of rather than the entire region, as before. For most of the actors the maximum search area will actually be less, due to the boundaries of the region. The results reported in Table 3 show that, while unemployment has not changed greatly, there is a deterioration in the economic welfare of the employees. Since productivity levels are maintained, there appears to be a resulting increase in the welfare of the employers. The domination of the highly skilled production process has increased. Although the diffusion of information about profits is not impaired, the ability of employees to maintain wage levels through migration is reduced.

An examination of the aggregate results ignores the detail a micro-simulation approach provides. Figure 3a and 3b show the level of skill attained by hired workers at the end of periods 10 and 30, for the population described in Table 3. Figure 2 shows the same information at the beginning of the first period. Several interesting patterns appear in these maps. In each of the maps the location of the vacancies shift. In Figure 3a, it appears that the more isolated firms suffer vacancies. Recall that by this stage, over half of the firms preferred to hire highly skilled persons. By the end of the 20th period, pockets of unemployment begin to appear. In the following ten periods, unemployment shifts to the south (see Figure 3b) and, by the end of 40 periods it is rather evenly distributed throughout the region.

Ignoring the movement of the location of vacancies, the worker mix in various subregions appears to remain rather constant. This does not necessarily mean that the same workers occupy given positions. Considering the turnover rate described in Table 3, this sort of occupation is not

TABLE 3

AGGREGATE RESULTS, LIMITED SEARCH AREA
(RISK NEUTRAL, RESTRICTED CONTACT FIELD)

	End of Time Period	Skill Type 1	2	3	4	Total
Unemployment Rate	1	0.00	0.00	0.00	0.00	0.00
	10	.32	.24	.04	.00	.15
	20	.20	.40	.28	.00	.22
	30	.40	.28	.16	.00	.21
	40	.40	.28	.32	.24	.31
	50	.32	.40	.08	.20	.25
Average Wage	1	9.19	19.26	29.39	39.09	24.23
	10	11.03	13.31	22.08	19.08	13.99
	20	9.48	10.28	12.40	14.12	8.49
	30	6.28	7.64	12.20	7.92	6.39
	40	3.56	5.76	7.08	4.76	4.35
	50	2.12	6.84	8.96	4.92	4.39
Average Productivity	1	14.40	24.72	34.64	44.48	29.56
	10	9.92	18.72	33.16	44.52	23.23
	20	11.76	14.64	24.92	44.56	20.93
	30	8.60	17.72	29.12	44.56	22.61
	40	8.60	17.52	23.52	34.04	18.72
	50	9.88	14.60	31.84	35.76	22.37
Desired Skill Level	1	22	31	28	19	100
	10	14	13	21	52	100
	20	5	8	16	71	100
	30	6	6	12	76	100
	40	5	6	11	78	100
	50	5	4	11	80	100

likely to endure for more than a few time periods. This stability is maintained in spite of an increasingly dominant adoption of the highly skilled production process. Figures 4a and 4b show the diffusion of this production process. Beginning with the initial distribution of the preferred processes in Figure 2, the pattern of the preferred technique is

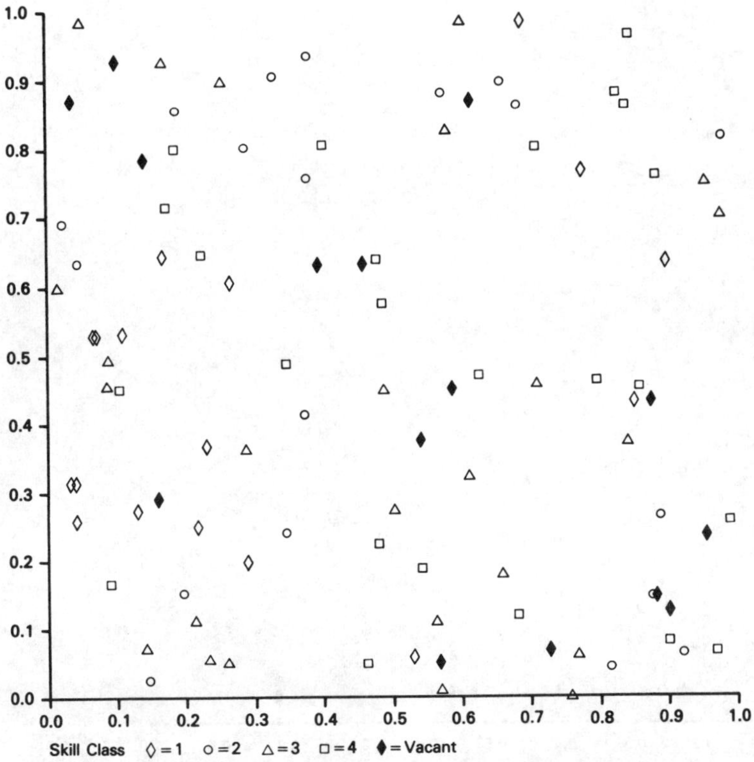

Figure 3a: Skill levels of the employees at the end of time period 10 (risk neutral population).

shown in Figure 4a after 10 periods, and 4b after 30 periods. Particularly striking is the rapid spread of the dominant process into portions of the region in which this process was not present before. As the adoption of the highly skilled production process spreads, the mismatch between the class of workers hired and the class of workers desired increases. This increasing employer 'stress' gives rise to further wage reductions and worker turnover.

By the end of 30 periods, the adoption of the skilled production process is nearly complete. The only remaining firms using an alternative process maintain minimal contact fields that do not extend to the adoptive firms. Since these isolated firms never encounter a more profitable alternative production process, they are content with their initial

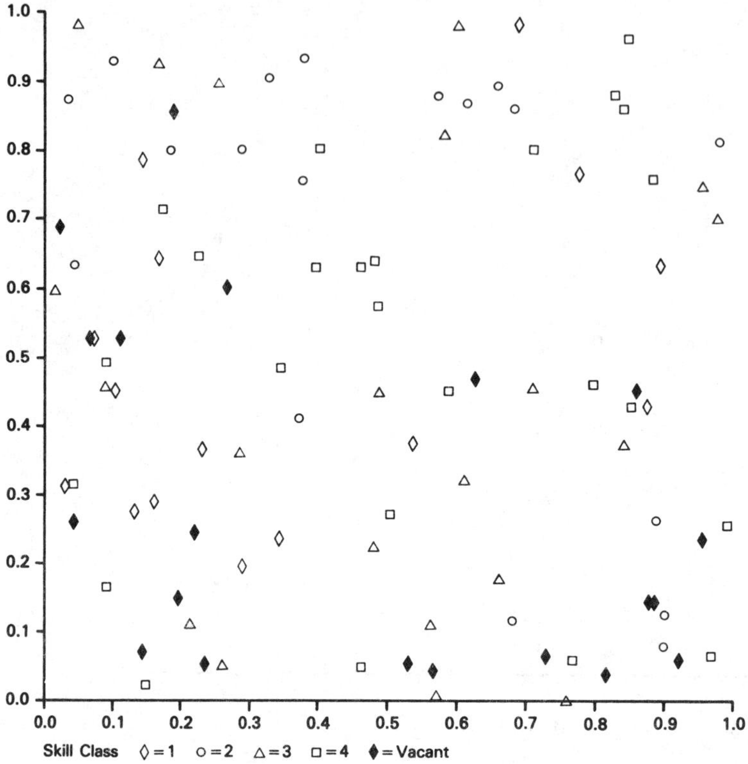

Figure 3b: Skill levels of the employees at the end of time period 30 (risk neutral population).

production technology and wage schedule. Since workers are not dismissed, vacancies are avoided eliminating the need for the firms to extend their contact fields. Three small concentrations of isolated firms appear, two in the northeast corner and a third in the southwest corner.

It is interesting to note in both Figures 3 and 4 that, as a result of the operation of the contact fields, firms with vacancies can exist in regions with generally filled positions. For this situation to occur, either the firm is not offering an attractive package, or not getting information about vacancies to those workers in need of a job. Alternatively, employees may not be applying for the job due to either a lack of information, or the presence of other, more attractive, offers. In

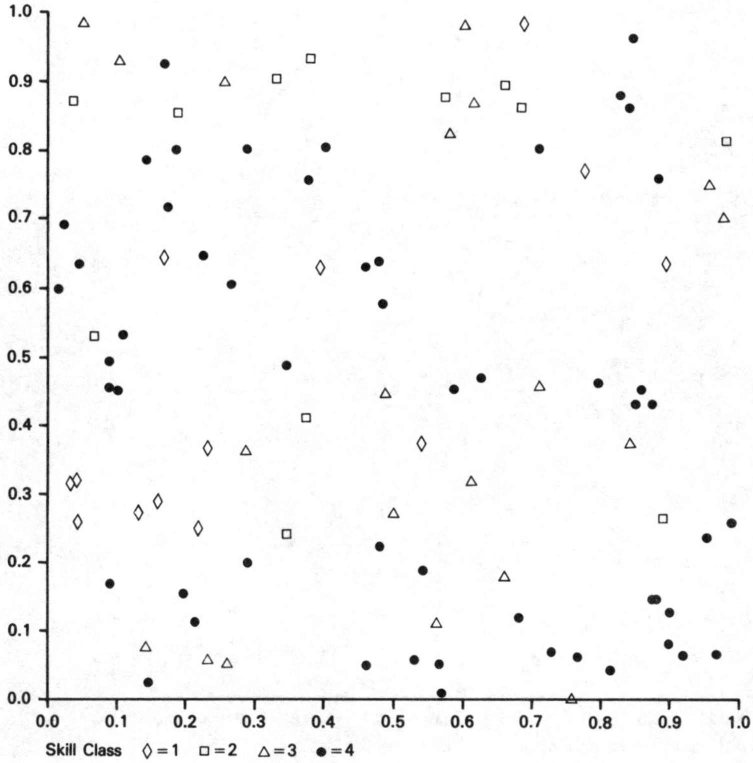

Figure 4a: Desired worker skills at the end of time period 10.

either case, as search strategies are adjusted over time, the vacancy is filled.

A second factor controlling the job application process is the attitude toward risk exhibited by employees. Results discussed so far assume a risk neutral population. Workers, before adjusting their attitudes for tenure, are willing to apply for a job with almost any difference in wages (actually, a difference as small as 0.10). This assumption is in conflict with available empirical evidence (Mueller, 1982). Since workers are required to accept a job offer for which they have applied, a more realistic assumption would require a large wage difference before making an application. The actual value of $\tilde{\beta}_i$, as discussed before, indicates a worker's willingness to assume risk. A relatively high value characterizes a risk averse worker. Table 4 shows the tabulated results for the same

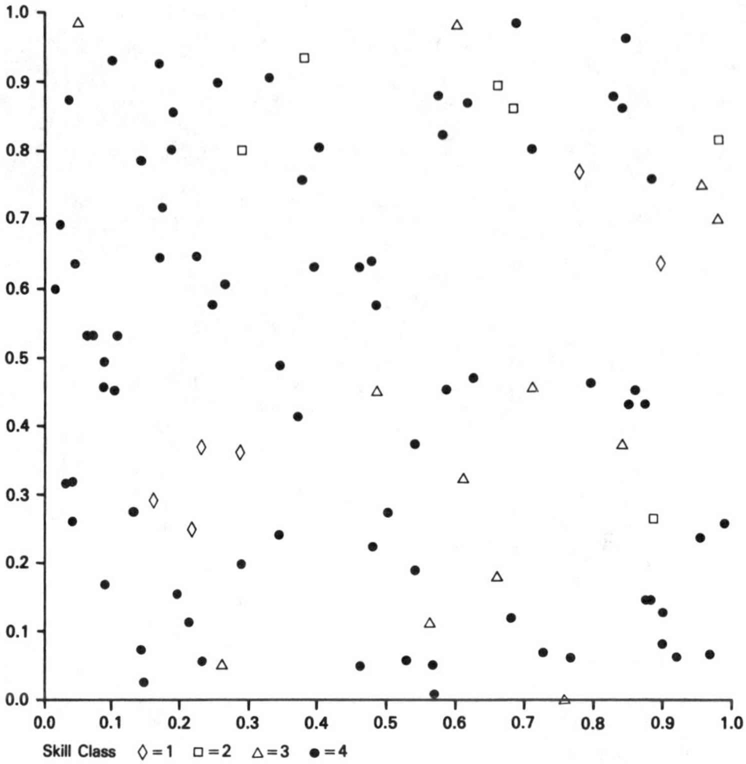

Figure 4b: Desired worker skills at the end of time period 30.

population described in Table 3, but with a value of $\tilde{\beta}_i = 5.00$ rather than 0.10. Again, the result is dramatic. Aggregate unemployment is reduced with the greatest benefit among the higher skilled classes. In the later periods average wage levels have increased for every class of workers. The turnover of workers has declined along with the dominance of the highly skilled production process. These results arise from two changes in the market. Although the employers' behavior has not changed, their ability to recruit workers with a given wage offer has been reduced. This is particularly true in the later periods as the wages offered to different classes of workers converge. At the same time, the employees are no longer applying to as many jobs. Fewer applications generate fewer moves. Since they are staying in one position longer, the workers' contact fields remain at the minimal level. Mobility is further reduced by the increasingly risk

TABLE 4

AGGREGATE RESULTS, RISK AVERSE POPULATION
(RESTRICTED CONTACT FIELD, LIMITED SEARCH AREA)

	End of Time Period	Skill Type 1	2	3	4	Total
Unemployment Rate	1	0.20	0.20	0.12	0.00	0.13
	20	.24	.16	.20	0.00	.15
	30	.24	.32	.08	0.00	.16
	40	.28	.28	.12	0.00	.17
	50	.32	.32	.12	0.00	.19
Average Wage	10	13.31	14.22	20.76	19.15	14.05
	20	9.87	14.16	13.99	15.58	11.43
	30	9.75	10.51	16.91	15.43	10.69
	40	8.35	9.87	14.39	15.10	9.84
	50	9.67	11.75	15.59	14.10	10.92
Average Productivity	10	11.68	19.64	30.44	44.48	23.97
	20	11.16	20.48	27.68	44.56	23.92
	30	10.96	16.72	31.84	44.56	24.03
	40	10.36	17.60	30.48	44.56	22.07
	50	10.00	16.68	30.44	44.56	23.79
Desired Skill Level	10	13	17	22	48	100
	20	12	13	19	56	100
	30	12	11	18	59	100
	40	12	11	17	60	100
	50	12	10	17	61	100

averse behavior arising from increased tenure. Although the highest skilled workers benefit from total employment, their wages have been reduced relative to their position at the end of the first period. Workers in the lowest skilled group, while suffering unemployment, are able to secure positions with higher than expected wages from firms filling jobs with less than ideal workers at high advertised wages.

Table 5 exhibits the results for the same population, but with a risk threshold of 20.00. This very risk averse population generates what might

TABLE 5

AGGREGATE RESULTS, EXTREMELY RISK AVERSE POPULATION
(RESTRICTED CONTACT FIELD, LIMITED SEARCH AREA)

	End of Time Period	Skill Type 1	2	3	4	Total
Unemployment Rate	10	0.28	0.08	0.04	0.00	0.10
	20	.12	.08	.16	.00	.09
	30	.16	.16	.04	.00	.09
	40	.24	.16	.16	.00	.14
	50	.16	.16	.04	.00	.09
Average Wage	10	9.75	17.12	22.39	19.35	15.20
	20	10.66	15.94	16.10	16.81	13.59
	30	9.78	15.09	18.82	15.53	13.35
	40	8.82	13.93	15.26	16.45	11.06
	50	11.70	15.94	18.34	15.49	14.14
Average Productivity	10	10.36	22.60	33.12	44.48	25.69
	20	12.88	22.60	29.12	44.56	25.28
	30	12.12	20.64	33.16	44.56	26.42
	40	11.00	20.60	29.12	44.56	23.03
	50	12.28	20.64	33.16	44.56	27.01
Desired Skill Level	10	14	18	21	47	100
	20	14	18	21	47	100
	30	14	18	21	47	100
	40	14	18	21	47	100
	50	14	18	21	47	100

be considered the 'healthiest' economy observed so far. Unemployment appears to have stabilized around the nine percent mark. Average wages again are rather high with the employed workers in the lowest skilled group earning higher than expected wages. Combined effects of increased tenure and lower intensity contact fields reduce turnover to around 24% of the market. Recalling that 10% of the workforce is comprised of new entrants during each period, the turnover among continuing employees is around 14% of the labor force. Productivity, although rather high and rising, is still not at the levels experienced under full employment.

The apparent cause of this healthy state is a complex interaction among the firms, the workers and the flow of information. The firms, operating with a restricted minimal contact field, do not learn about alternative production processes. Being content with the best process in their neighborhood, firms do not reduce wages dramatically. Thus employers are thus able to find the ideal worker nearby. Once the ideal worker is found, employers reduce their information gathering process. Workers, being risk averse people, are content to remain in the current job, especially as tenure is accumulated. The reduced mobility of the workers reduces turnover, eliminating the periodic need of employers to fill unexpected vacancies. Since wages remain high, layoffs are reduced. The economy settles into a pattern of behavior in which the hiring process is focused on new entrants. These new workers are often hired into the same positions recently vacated by retirements or the migration of mismatched workers, usually from the lower skill groups. The reduced contact fields of the employers permit the co-existence over time of several different production processes. Similar findings have been discussed in Keeley (1985). The existence of alternative production processes improves the ability of an employer to find the ideal person nearby, avoiding the creation of an inefficient competition for a limited skill group and the downward spiral in wages as firms compete for profit levels. Although not captured here, the elimination of competition for a limited supply of workers gives rise to additional savings in formerly lost productivity and reduced advertising costs for employers. Similarly, workers are spared unnecessary search and relocation costs, as well as earnings formerly lost or reduced.

5. CONCLUSIONS

The effect of individual workers' employment histories has been demonstrated. With increased tenure, workers become less mobile. The decreased labor turnover reduces the need for firms and workers to search their respective markets for opportunities. The lack of information concerning the activities of other actors in the market leads to the adoption of a wider range of production processes. Thus demand for workers is spread among all worker types rather than being concentrated in one group. However, when workers are less risk averse, the increased worker turnover requires firms to engage in a more intensive search process for the ideal worker. During this search, the employer is exposed to alternative production processes. Adopting the most productive process increases the demand for this limited supply of workers. In the closed market modelled, this competition results in a series of temporary hiring decisions followed by layoffs. The cycle continues as unemployed workers now increase their search activities, and the process repeats.

The missing pieces in this model include a more realistic wage process, effects on migration of other workers (e. g., age, sex), and firm (e. g., size, entrepreneurial skill) characteristics, and explicit recognition of the importance of the supply and demand of vacancies. Two broad areas of concern involve the assumed market structure, and the detail in the profile of the individual worker incorporated into the model. The question of empirical validity, although important, has been deferred for the time being.

Perhaps the dominant unanswered, and to this point unasked, question is the degree to which spatial job matching models, such as the one described here, can be embedded in a non-trivial way into a theory that truly generates job and population change, rather than merely distributing a given population of workers over a given set of job opportunities.

6. REFERENCES

Amacher, R., and R. Sweeney, 1981, On the Integration of Labor Markets: A definition and Test of the Radical-Segmentation Hypothesis, Journal of Labor Research, 11: 25-37.

Amrhein, C., 1985, Interregional Labor Migration with Information Flows, Environment and Planning A, 18: 1111-26.

_____, 1986, The Effect of Job Advertising on Interregional Labor Migration, Geographical Analysis, 18, forthcoming.

_____, and R. MacKinnon, 1985, The Job Matching Process Within an Interregional Setting: A Simulation Model, Regional Studies, 19: 193-202.

Ballard, K., and G. Clark, 1981, The Short-run Dynamics of Interstate Migration: A Space-Time Economic Adjustment Model of In-migration to Fast Growing States, Regional Studies, 15: 213-228.

Bartholomew, D., and A. Forbes, 1979, Statistical Techniques for Manpower Planning, New York: Wiley.

Boorman, S., 1975, A Combinatorial Optimization Model for Transmission of Job Information Through Contact Networks, The Bell Journal of Economics, 6: 216-249.

Brown, L., 1981, Innovation Diffusion, New York: Methuen.

Brummell, A., 1981, A Method of Measuring Residential Stress, Geographical Analysis, 13: 248-261.

Clark, G., 1980, Critical Problems of Geographical Unemployment Models, Progress in Human Geography, 4: 157-180.

_____, 1981, The Employment Relation and the Spatial Division of Labor: A

Hypothesis, Annals of the Association of American Geographers, 71: 411-424.

_____, 1983, Government Policy and the Form of Local Labor Markets, Urban Geography, 4: 1-15.

_____, 1986, Spatial Search Theory and Indeterminant Information, in Urban and Regional Labour Markets: Problems, Perspectives, and Policies, edited by M. Fisher and P. Nijkamp, forthcoming.

Clark, W., 1982, Recent Research on Migration and Mobility: A Review and Interpretation, Progress in Planning, 18: 1-56.

Clemen, R., and R. Winkler, 1985, Limits for the Precision and Value of Information from Dependent Sources, Operations Research, 33: 427-442.

Cordey-Hayes, M., and D. Gleave, 1977, Migration Dynamics and Labour Market Turnover, Progress in Planning, 8: 1-95.

Curry, L., 1981, Division of Labor from Geographical Competition, Annals of the Association of American Geographers, 71: 133-165.

_____, 1985, Inefficiencies in the Geographical Operation of Labour Markets, Regional Studies, 19: 203-216.

Da Vanzo, J., 1980, Microeconomic Approaches to Studying Migration Decisions, N-1201-N1CHD, Santa Monica, Calif.: The Rand Corporation.

Hagerstrand, T., 1967, On Monte Carlo Simulation of Diffusion, in Quantitative Geography Part I: Economic and Cultural Topics, edited by W. Garrison and D. Marble, Evanston, Ill.: Northwestern University, Department of Geography, pp. 1-32.

Holt, C., 1978, Wages and Job Availability in Segmented Markets, presented to International Economic Association Conference on 'Unemployment in Western Countries Today', Strasbourg, France, August 28 - September 2, 1978.

IMSL Library, 1980, Reference Manual, edition 8, Houston, Texas: IMSL, Inc.

Isserman, A., 1985, Economic-Demographic Modeling with Endogenously Determined Birth and Migration Rates: Theory and Prospects, Environment and Planning A, 17: 25-45.

Keeley, M., 1985, The Economics of Firm Size: Implications from Labor-Market Studies, San Francisco FRB Economic Review, 1984, #1 (Winter): 5-21.

Krumm, R., 1983, Regional Wage Differentials, Fluctuations in Labor Demand,

and Migration, International Regional Science Review, 8: 23-46.

Maier, G., 1985, Cumulative Causation and Selectivity in Labour Markets Oriented Migration Caused by Imperfect Information, Regional Studies, 19: 231-242.

McHugh, K., 1984, Explaining Migration Intentions and Destination Selection, Professional Geographer, 36: 315-325.

Mueller, C., 1982, The Economics of Labor Migration: A Behavioral Analysis, New York: Academic Press.

Pickles, A., and P. Rogerson, 1984, Wage Distributions and Spatial Preferences in Competitive Job Search and Migrations, Regional Studies, 18: 131-142.

Plane, D., 1984, Migration Space: Doubly Constrained Gravity Model Mapping of Relative Interstate Separation, Annals of the Association of American Geographers, 74: 244-256.

_____, and P. Rogerson, 1985, Economic-Demographic Models for Forecasting Interregional Migration, Environment and Planning A, 17: 185-198.

Rogerson, P., 1982, Spatial Models of Search, Geographical Analysis, 14: 217-228.

_____, 1984, New Directions in Modelling Interregional Migration, Economic Geography, 60: 111-121.

_____, 1985, Labor Market Competition Processes, Research Papers in Urban Systems and Policy Planning: Working Paper No. 5, Evanston, Ill.: The Technological Institute, Northwestern University.

_____, and R. MacKinnon, 1981, A Geographical Model of Job Search, Migration, and Unemployment, Papers of the Regional Science Association, 48: 89-102.

Schaeffer, P., 1985, Human Capital Accumulation and Job Mobility, Journal of Regional Science, 25: 103-114.

Veiga, J., 1983, Mobility Influences During Managerial Career Stages, Academy of Management Journal, 26: 64-85.

Wolpert, J., 1966, Migration as an Adjustment to Environmental Stress, Journal of Social Issues, 22: 92-02.

MODELING DISCONTINUOUS CHANGE IN THE SPATIAL PATTERN OF RETAIL OUTLETS: A METHODOLOGY

A. Stewart Fotheringham
University of Florida
United States

Daniel C. Knudsen
Indiana University
United States

1. INTRODUCTION

The modeling of discontinuous change in social and economic systems is a relatively recent phenomenon prompted by the pioneering work on the mathematics of catastrophe theory by Thom (1975). Previously, almost all research into system dynamics was based on the concept of continuous change, in which a continuous or smooth change in some independent variable is expected to produce a continuous change in a dependent variable. Catastrophe theory allows a discontinuous change in the dependent variable to be modeled. In urban and regional analysis there have been two major uses of this approach in the investigation of discontinuous urban dynamics. One is at the macro-scale and is concerned with the examination of sudden urban growth within a system of urban areas (Amson, 1974; Allen and Sanglier, 1979; Dendrinos, 1980). The other use is at the micro-scale, and is concerned with changes in retail outlets within a single urban area (Poston and Wilson, 1977; Harris and Wilson, 1978). This latter theme is the sole concern of this paper.

During the past 30 years, the following changes in retailing patterns have occurred:

(i) many corner grocery stores have been displaced by supermarkets;

(ii) there has been a tendency for non-grocery retail facilities to cluster into shopping malls; and,

(iii) downtown areas of major cities have lost much of their retailing dominance to the suburbs.

Whereas the first type of change involves the size of retail facilities, the other two involve the spatial distribution, or relative location, of retail facilities. All three changes have been discontinuous. Within any one city, for example, the nucleation of retailing into malls does not take place in gradual stages; rather, the advent of shopping malls represents a dramatic change in retailing patterns from a relatively dispersed to a tightly clustered pattern.

Traditional models of spatial dynamics usually can represent only continuous change--a smooth, continuous change over time in an independent variable produces a smooth, continuous change in a dependent variable. Mathematically, letting y be a dependent variable, x be an independent variable, and t be time,

$$y = f(x) \quad , \tag{1}$$

and

$$x = g(t) \quad , \tag{2}$$

where $f(\cdot)$ and $g(\cdot)$ both represent continuous functions. Such models are unrealistic in instances such as the above where $f(\cdot)$ is discontinuous.

Relatively recently, however, an investigation of modeling frameworks that can explain discontinuous change in retailing dynamics has begun, primarily at the University of Leeds in England, paralleling research undertaken at Brussels (Allen, Deneubourg, Sanglier, Boon and de Palma, 1978; Allen, 1980). These studies, however, have concentrated on the discontinuities in the size of retail outlets--the transition from the corner-store outlet to the supermarket. This paper presents a methodology for explaining discontinuities in the spatial distribution of retail facilities and for discontinuities in <u>both</u> the size and the relative location of retail facilities. The task is more complex than the mere addition of the above two themes, since trade-offs between size and relative location are possible. A large, peripheral facility, for example, may generate the same revenue and costs as a smaller, more centrally located facility.

Retail dynamics are viewed in two frameworks. One is a utility-based framework in which a combination of size and relative location is chosen that maximizes utility. The other is a classic supply-and-demand framework, where a combination of size and relative location is established when supply equals demand. In the former, consumers' utility functions associated with the attributes of retail facilities are defined. Change in these functions then is shown to produce discontinuous change in both size and relative location, under certain conditions. In the latter, demand is expressed in terms of consumers' potential expenditures at a retail facility of size x at location y. When supply and demand are not equal, changes in supply can occur until a stable equilibrium is reached. Conditions are identified under which these changes in supply (i. e., in the size and location of facilities) can be discontinuous.

While the utility framework described above has been used by Poston and Wilson (1977) to examine retail dynamics, and the equilibrium framework has been used by, <u>inter alia</u>, Harris and Wilson (1978), Beaumont, Clarke and Wilson (1981a, 1981b), Birkin, Clarke and Wilson (1983) and Wilson (1981b), all of these studies have been concerned primarily with the dynamics of facility size. Until very recently (Fotheringham, 1985), there has been no investigation of situations in which a facility may choose the option of relocating, rather than changing its size, as a response to some external stimulus (a change in social and/or economic conditions). There remains no investigation of the conditions leading to simultaneous, discontinuous changes in both the size and relative location of a retail facility. While the analysis of simultaneous changes and trade-offs is a complex task, it represents a more realistic situation than simply considering discontinuous changes in size. Studies concentrating on the latter issue do not properly address the role of relative location in determining consumers' utilities and consumers' demand, and such studies ignore the role of relative location as a determinant of supply costs through the presence of rent gradients.

2. EMPIRICAL EVIDENCE ON THE SPATIAL DYNAMICS OF NON-GROCERY RETAIL OUTLETS

To demonstrate that discontinuous changes in the spatial pattern of retail outlets occur, the distribution of clothing stores in Gainesville, Florida, was analyzed for each year of the thirty-year period 1955-1984. The names and the locations of clothing outlets were obtained from the annual city directories, and the location of each outlet was digitized from topographical maps of Gainesville.[1] For each year, the spatial pattern of outlets was described using order-neighbor analysis (Aplin, 1983). Specifically, a nearest-neighbor statistic (R), with Aplin's suggested correction for boundary effects (the underestimation of the expected nearest neighbor distance and its standard error), was used to describe the spatial pattern. While other statistical techniques, such as quadrat analysis and spatial autocorrelation analysis, could have been used, there is subjectivity associated with all three techniques. It was felt that for the present purpose, which simply is to give a brief description of the types of change in retailing patterns that have occurred, nearest neighbor analysis was sufficient. By graphing the nearest neighbor statistic over time, discontinuous change in the spatial pattern of outlets will be implied by rapid changes in the value of R. Rapid decreases in R will indicate discontinuous change from a relatively dispersed to a relatively clustered pattern of outlets, while rapid increases in R will indicate discontinuous change in the opposite direction.

The graph of R values from 1955 to 1984 is given in Figure 1. While the spatial pattern of clothing stores throughout this period always has been clustered ($R < 1$), it is clear that a strong nucleation trend has emerged, with a weakly clustered pattern in 1955 ($R = .788$) being transformed to a very strongly clustered pattern in 1984 ($R = .277$). Using the Z-test reported in Silk (1979), not until 1964 were any R values significantly different from that which would be expected if the distribution of clothing stores resulted from a random process. At least four discontinuities appear to exist in the trend over time: (1) between 1963 and 1964, when two shopping plazas opened and there was a fairly rapid increase in the spatial concentration of clothing stores; (2) between 1965 and 1967, when another two shopping plazas opened, with the same result;

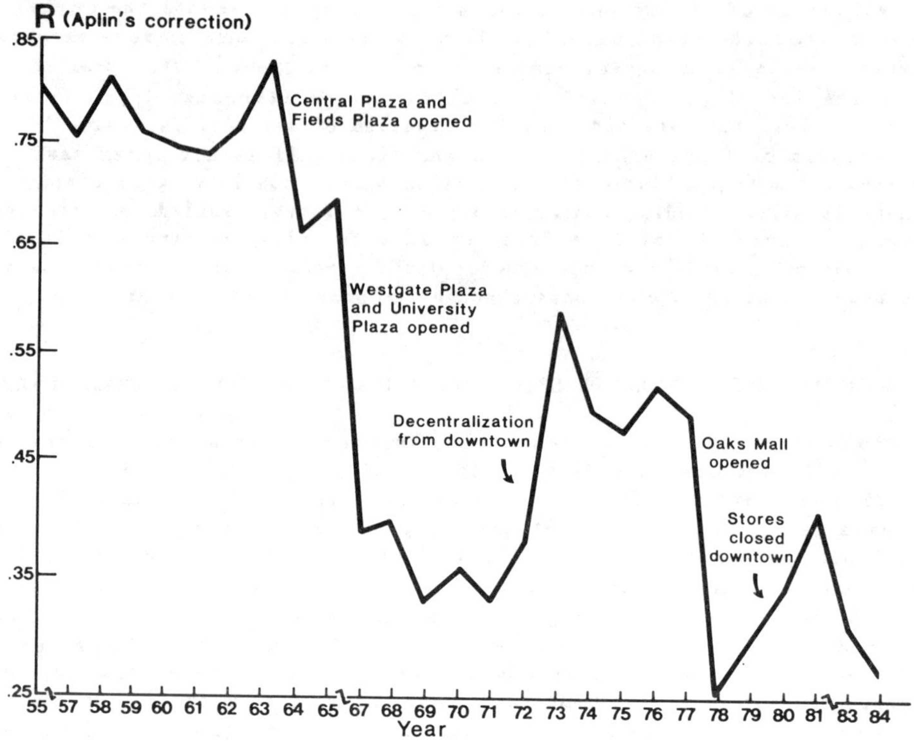

Figure 1: R (with Aplin's correction) for Clothing Stores in Gainesville, 1955-1984.

(3) between 1972 and 1973 when there was a rapid reversal of the trend toward nucleation, resulting in decentralization of downtown shopping opportunities; and, (4) between 1977 and 1978, when the major shopping mall (the Oaks Mall) in Gainesville opened. The rise in the value of R immediately following the opening of the Oaks Mall is due to downtown stores closing and the consequent further loss of dominance of this cluster of retailing outlets.

The number of clothing stores in Gainesville in each of the years 1955 to 1984 is graphed in Figure 2. The dominance of the Oaks Mall in Gainesville's retailing system is demonstrated by the rapid increase in the number of clothing stores between 1977 and 1978, when the mall opened. The addition of the Oaks Mall thus represents a discontinuous change in both the size and the spacing of retail opportunities in Gainesville, and could be modeled from either point of view. The other three discontinuities identified above, however, took place in periods when the number of clothing

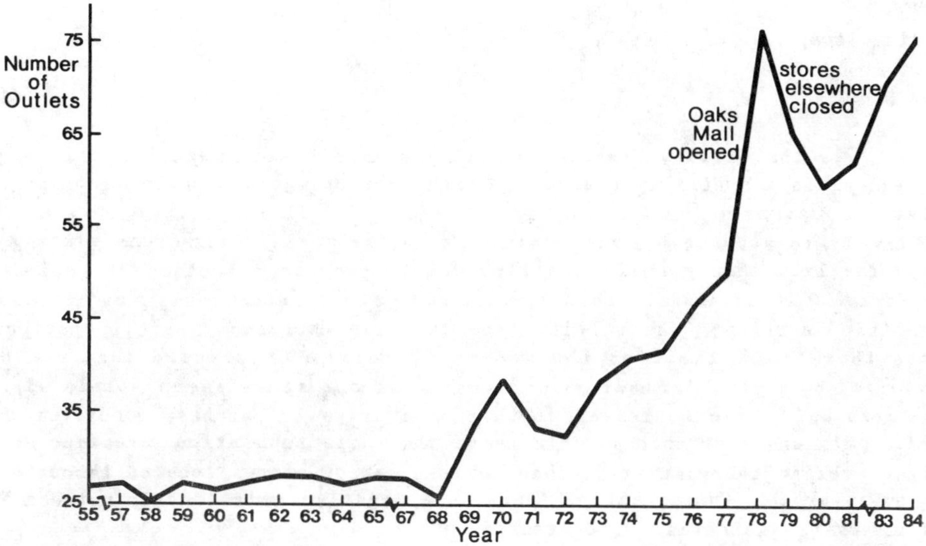

Figure 2: Number of Clothing Outlets in Gainesville, 1955-1984

outlets was relatively stable, and thus they represent pure cases of discontinuities in the spacing of outlets. Two of these changes involve nucleation, whereas the other involves dispersion and the loss of dominance of downtown retailing. The remainder of this paper focuses on ways of modeling these discontinuous changes in retailing patterns.

3. EXISTING FRAMEWORKS FOR MODELING DISCONTINUOUS CHANGE IN RETAILING SYSTEMS

Much of the research undertaken on discontinuous retail dynamics stems from two seminal papers by Poston and Wilson (1977), and Harris and Wilson (1978). Except for a recent article by one of the authors of this paper (Fotheringham, 1985), all of this research has been concerned with explaining discontinuous change in the size of retail facilities (see inter alia, Beaumont, Clarke and Wilson, 1981a, 1981b; Wilson, 1981a, 1983; Clarke and Wilson, 1983; Wilson and Oulton, 1983; Birkin , Clarke and Wilson, 1984). The basic frameworks of Poston and Wilson, and of Harris and Wilson are described briefly now, since they form the bases of the analysis used in this paper.

3.1. The Poston and Wilson Framework

Consider the utility derived from a shopping trip, U, to have two additive components, u_1 and u_2, so that

$$U = u_1 + u_2 \quad , \tag{3}$$

where u_1 is the utility associated with the size of the retail facility j, W_j, and u_2 is the utility associated with overcoming the spatial separation between consumer i and retail facility j, c_{ij}. It is assumed that u_1 is positively related to W_j; the consumer gains greater utility from visting a large facility than a small facility due to greater selection of goods in the former. It is assumed that u_2 is negatively related to c_{ij}; as the cost of a trip increases, the utility from the trip decreases, *ceteris paribus*. Since there is a limit on the amount of retail floorspace that can be supported by a given number of consumers, as one store increases in size, consumers will have to travel further on average to purchase goods (in the limit, only one super-store would serve the whole population, creating much larger average shopping trips than would a set of stores located throughout the urban area). Consequently, there is a negative relationship between W_j and u_2; as W_j increases, u_2 decreases.

Given the above scenario, Poston and Wilson (1977) demonstrate that if the relationship between W_j and u_1 is logistic, and the relationship between W_j and u_2 is linear (or approximately so), catastrophic change in W_j can occur if the relationship between W_j and u_2 is changing slowly over time. The exact mechanism for this discontinuous change is described in greater detail in Poston and Wilson (1977), and Wilson (1981b).

3.2. The Harris and Wilson Framework

Let the profit, π, associated with a retail outlet of type m at location j be defined by

$$\pi_{jm} = D_{jm} - C_{jm} \quad , \tag{4}$$

where D_{jm} is the revenue generated by a type m outlet at location j, and C_{jm} denotes the operating costs incurred by a type m outlet located at j. Further, assume that D_{jm} can be estimated by summing the potential revenue derived from a retail shopping model over all origins, so that

$$D_{jm} = \sum_i S_{ijm} \quad , \tag{5}$$

where

$$S_{ijm} = E_{im} P_i W_{jm}^{\alpha_{1m}} \exp(\alpha_{2m} c_{ij}) / [\sum_j W_{jm}^{\alpha_{1m}} \exp(\alpha_{2m} c_{ij})] \quad , \tag{6}$$

and S_{ijm} represents the amount of money spent at a type m retail outlet in location j by the residents of zone i, E_{im} represents the average expenditure on retail goods of type m by a person living in zone i, P_i represents the population of zone i, W_{jm} represents the size of the retail outlet m at location j, c_{ij} represents the cost of traveling from i to j,

α_{1m} is a parameter reflecting consumers' perceptions of the attractiveness of the size of type m retail outlets, and α_{2m} is a parameter reflecting consumers' perceptions of the disutility of overcoming the spatial separation of origins and destinations in order to purchase retail goods of type m. It is expected that $\alpha_{1m} > 0$ and $\alpha_{2m} < 0$. Henceforth, for convenience and simplification, only one type of retailing will be examined (clothing, for example) so that the index m can be dropped.

Assume that C_j is a function of W_j, so that

$$C_j = \rho W_j^{\lambda}, \tag{7}$$

where $\rho > 0$ denotes the efficiency of operating retail outlets (as ρ decreases, the operation of retail outlets becomes increasingly efficient), and $0 \leq \lambda \leq 1$ denotes the presence of internal scale economies (as λ decreases, scale economies increase). Suppose that changes in the size of an outlet will be governed by the new profit associated with that outlet, namely

$$\partial W_j/\partial t = f(\pi_j) \quad , \tag{8}$$

and that the size of the outlet increases when profits are positive, and it decreases when profits are negative. An equilibrium outlet size, $\bar{W}_j$, is achieved when profits are zero. Harris and Wilson demonstrate that only one stable equilibrium exists when $0 < \alpha_1 \leq 1$, but two stable equilibria exist when $\alpha_1 > 1$. What makes the equilibrium solutions of the differential equation in (8) interesting is that π_j is a function of D_j which, from equations (5) and (6), is a nonlinear function of W_j. The presence of this nonlinear structure generates bifurcation points in the system when $\alpha_1 > 1$, and leads to the potential for discontinuous change in $\bar{W}_j$ if it is assumed that the operation of retail outlets is becoming more efficient over time (i. e., ρ decreases over time) (for more details of this bifurcation point, see Harris and Wilson, 1978; Wilson, 1981b; and Fotheringham and Knudsen, 1984, 1985). Thus the value of α_1 in equation (6) is critical in producing the necessary conditions to explain a discontinuous change in the equilibrium size of retail outlets.

3.3. Sensitivity Analysis of the Equilibrium Values of W_j in the Harris and Wilson Framework

Fotheringham and Knudsen (1984) examined the sensitivity of the equilibrium size of retail outlets to variations in consumer behavior as represented by the parameters α_1 (perception of scale economies) and α_2 (perception of separation deterrence) in the gravity model. An urban system was constructed with ten residential zones and four existing retail outlets. The size of the existing outlets ranged from two to nine units. A location j was selected for a fifth outlet, and the cost of traveling between each residential zone and each outlet was measured by the distance from each zone centroid. An amount of purchasing power, denoted by E_iP_i, was allocated to each zone. An economic system was assumed in which ρ decreases over time and $\alpha_1 \geq 1$ so that initially, when ρ is large, $\bar{W}_j = 0$ and no retail outlet exists at j. As ρ decreases, however, a critical value is reached at which a non-zero equilibrium value of W_j exists.

To examine the relationship between $\bar{W}_j$ and consumer behavior, equilibrium values of W_j were derived for all combinations of α_1 between 1.0 and 2.5, in steps of 0.05, and for α_2 between 0 to -2.5, in steps of 0.1.[2] The resulting matrix of $\bar{W}_j$ values indicated, as expected, that the larger the size attractiveness parameter, α_1, the larger is $\bar{W}_j$ (the greater the relative attractiveness of large retail outlets over small retail outlets, the larger will be the equilibrium size of the new outlet); but somewhat unexpectedly, for any given α_1, $\bar{W}_j$ is relatively insensitive to variations in the cost-decay parameter, α_2. However, the value of α_2 does influence the relationship between $\bar{W}_j$ and α_1. When $\alpha_2 = 0$, the relationship between $\bar{W}_j$ and α_1 is exponential; when $\alpha_2 < 0$, the relationship is logistic, and becomes increasingly logistic as α_2 becomes increasingly negative.

3.4. Problems with the Poston and Wilson, and the Harris and Wilson Frameworks

While the above frameworks have been responsible for stimulating a large volume of subsequent research on retail dynamics, both suffer from some serious deficiencies. One of these drawbacks is that they largely ignore spatial processes. For instance, there is no explicit recognition of situations in which a retail facility can choose the option of relocating rather than changing its size, as a response to some external stimuli. Also, in the Harris and Wilson framework, location rent is omitted from the cost equation (7), while in the Poston and Wilson framework the utility a consumer derives from visiting a retail outlet in close proximity to other outlets is ignored in equation (3). While modifications have been made to increase the spatial content of the Harris and Wilson framework, such as replacing ρ by ρ_j in equation (7), there still remains relatively little explicit acknowledgement of spatial processes.

Another criticism of the Harris and Wilson framework as formulated in equations (4)-(8) is that any change in equilibrium is focused on the zero point. That is, discontinuous change in the system can take place only from the zero equilibrium point (non-existence of a facility) to a non-zero equilibrium point (existence of a facility), or vice versa. This mechanism does not allow a change between two non-zero equilibria, which is a more realistic situation.

The following discussion shows how these sorts of problems can be alleviated by retaining the basics of the above frameworks, while replacing the gravity demand by a competing destinations demand model.

4. REPLACING THE GRAVITY FORMULATION IN RETAILING

Consider an individual i faced with choosing a retail outlet j from a set of retail outlets. The destination choice process can be specified using the utility function

$$U_j(i) = \sum_{k=1}^{K} \alpha_k f_k(x_{ijk}) + \sum_{l=1}^{L} \omega_l f_l(y_{il}) + \varepsilon_{ij} \quad , \tag{9}$$

where $U_j(i)$ represents the utility (or benefit) individual i receives from making choice j, x_{ijk} represents the k-th attribute of retail outlet j as perceived by i, and y_{il} represents the l-th attribute (a socioeconomic characteristic) of individual i. There are K outlet attributes and L individual attributes. Each attribute is related to $U_j(i)$ by the function $f(\cdot)$ and by the associated parameters α_k or ω_l. Since utility and the determination of utility cannot be measured exactly, a random error component, ε_{ij}, is introduced into the utility function. Equation (9) simply relates the benefit an individual receives from making a particular retailing choice to the characteristics of that choice and to the socioeconomic characteristics of the individual making the choice. Equation (9) has been used to derive two retail choice models, namely the gravity model (Anas, 1983; Williams, 1977) and competing destinations model (Fotheringham, 1986). While the derivation of the gravity model formulation from random utility theory is not new, it is repeated here in brief to emphasize the assumptions, particularly about hierarchical destination choice, that are embedded in the model and which the model make unrealistic for retailing systems.

Assume, for simplicity, that the utility an individual receives from various retail attributes is not a function of the individual's attributes, that is, in equation (9) $\omega_l = 0$ for all l. Assume also that only the following two attributes of retail outlets are important in determining the utility associated with the choice of a particular retail outlet: (1) the size of a retail outlet, and (2) the cost of overcoming the distance separating i and j. The former is a useful surrogate for the variety and average price of goods being offered at a retail outlet, whereas the latter is a useful surrogate for the time and psychological cost involved when individual i patronizes retail outlet j. Let size be measured by W_j and cost by c_{ij}. It is reasonable to assume that $U_j(i)$ is a positive linear function of the logarithm of W_j, implying that $U_j(i)$ increases at a decreasing rate as W_j increases (a constant increase in the size of outlet results in a greater increase in utility when the initial size is small than when the initial size is large). For intra-urban systems it is usual to assume that $U_j(i)$ is a negative linear function of c_{ij} so that the utility in equation (9) can be represented as

$$U_j(i) = \alpha_1 \log(W_j) + \alpha_2 c_{ij} + \varepsilon_{ij} \quad . \tag{10}$$

It is well known that under the assumptions of random utility maximization, the probability of individual i choosing outlet j is given by

$$P_j(i) = \exp[U_j(i)]/\{\sum_j \exp[U_j(i)]\} \quad , \tag{11}$$

(Anas, 1983; Lioukas, 1984; McFadden, 1974; Williams, 1977; Wrigley, 1985). In terms of the utility function expressed in equation (10),

$$P_j(i) = W_j^{\alpha_1} \exp[\alpha_2 c_{ij}]/[\sum_j W_j^{\alpha_1} \exp(\alpha_2 c_{ij})] \quad , \tag{12}$$

The formulation in equation (12) commonly is referred to as either a production-constrained gravity model (Wilson, 1971), or a multinomial logit model of destination choice (Domencich and McFadden, 1975). A major problem with this model, however, is that it has what is known as the Independence from Irrelevant Alternatives (IIA) property, which implies that all retail outlets in the choice set are evaluated and compared equally and that no hierarchical choice process takes place (Domencich and McFadden, 1975; Sobel, 1980; Tye, et al., 1982; Wrigley, 1985). That is, it is assumed that consumers do not perceive of retail outlets in clusters (such as shopping malls), but rather solely as individual units. If, however, consumers do not consider all possible retail outlets, and instead concentrate their search on a restricted set of outlets, then the choice of retail outlets is said to be hierarchical and the formulation in equation (12) is a poor specification of reality. Accordingly inaccurate forecasts of consumption patterns and biased model parameter estimates are likely to result. To demonstrate the inaccuracy of the gravity formulation in equation (12) caused by the IIA property, consider the ratio of the probabilities of selecting two retail outlets, 1 and 2, which is $[W_1^{\alpha_1} \exp(\alpha_2 c_{i1})/W_2^{\alpha_1} \exp(\alpha_2 c_{i2})]$. This ratio is independent of the location of any other retail outlet so that if a new outlet is located immediately adjacent to Outlet 2, the ratio would remain unchanged. This seems contradictory to our expectations of reality. In essence, Fotheringham (1986) has shown that variations in the choice probabilities of individuals due to differences in the relative clustering of retail outlets cannot be modeled by the gravity model described by equation (12).

The gravity formulation is predicated on the belief that an individual first compares and evaluates <u>all</u> retail outlets prior to the selection of a particular outlet. Whether one assumes that the consumer is making independent decisions each time a good is purchased, or that an initial decision that is being made and thereafter the consumer simply stays loyal to the chosen outlet does not matter. What is important is that at some point the consumer makes a decision regarding which retail outlet to patronize. A more realistic assumption is that individuals select retail outlets in a hierarchical manner; that is, customers first select a cluster of retail outlets (say, a shopping mall), and then select an individual outlet from within the chosen cluster. Fotheringham (1986) recently demonstrated how a competing destinations model can model hierarchical spatial choice. The model consists of the gravity formulation in equation

(12), coupled with an additional variable A_j that denotes the accessibility of relative location of outlet j to all other outlets. This model version is written as

$$P_j(i) = W_j^{\alpha_1} \exp(\alpha_2 c_{ij}) A_j^{\alpha_3} / [\sum_j W_j^{\alpha_1} \exp(\alpha_2 c_{ij}) A_j^{\alpha_3} \quad , \qquad (13)$$

where A_j usually is defined as a potential measure,

$$A_j = \sum_{p \neq j} W_p c_{jp}^{\beta} d_{jp} \quad . \qquad (14)$$

In this general representation of a potential measure, p represents a retail outlet other than j, $d_{jp} = 1$ if p is within a prespecified range of j and $d_{jp} = 0$ otherwise, and β is a spatial decay parameter ($\beta < 0$). There is no a priori expectation regarding the sign of α_3. A positive value of α_3 indicates the presence of consumer agglomeration forces between retail outlets; the more accessible outlet j is to other outlets, the greater is the probability that individual i will select outlet j, ceteris paribus. A negative value of α_3 indicates the presence of consumer competition forces between retail outlets; the more accessible outlet j is to other outlets, the smaller is the probability that individual i will select j, ceteris paribus. Consumer agglomeration forces can result from either multipurpose shopping, comparative shopping, or from an anticipated search process; consumer competition forces can result from destination choice being hierarchical, as is described by Fotheringham (1986). For purposes of this paper, A_j is implicitly defined as the accessibility of a retail outlet to outlets of the same type. By the 'same type' we have in mind a simple division between grocery and non-grocery outlets, although much finer divisions could be made within this latter category. A finer division would involve several A_j variables in equation (13), and also later in equations (15) and (17). Then there would be a set of A_{jm} variables, where m denotes a type of retail outlet, so that A_{jm} would reflect the accessibility of store j to outlets of type m. If the equilibrium conditions for a type n store are being investigated, the parameter associated with the A_{jn} variable in equation (13) would reflect the strength of comparative shopping by consumers (Eaton and Lipsey, 1979), while the parameters associated with A_{jm} variables ($m \neq n$) would reflect the strength of multipurpose shopping by consumers (Mulligan, 1983; O'Kelly, 1981; Thill, 1985).[3]

The derivation of the competing destinations model from random utility theory can be achieved simply by including the accessibility variable in the utility function given in equation (10), so that

$$U_j(i) = \alpha_1 \log(W_j) + \alpha_2 c_{ij} + \alpha_3 \log(A_j) + \varepsilon_{ij} \quad . \qquad (15)$$

The use of the logarithmic function of accessibility assumes that the utility gained from choosing outlet j increases or decreases, at a decreasing rate, as A_j increases. A justification for the addition of this variable to the utility function could be made in terms of the availability of alternative opportunities to those in outlet j. For example, should the opportunities in outlet j fail to satisfy the needs of the individual, utility is gained from having alternative outlets in close proximity (in which case, α_3 would be positive). Alternatively, the presence of other outlets in close proximity to j may result in congestion costs around j, or in a reduction of the diversity and quality of merchandise in j (in which case, α_3 would be negative). The form of the competing destination model also has been justified by its ability to model hierarchical destination choice (Fotheringham, 1986). Clearly this model does not contain the undesirable IIA property that is present in the gravity model; the ratio of the probabilities of patronizing two retail outlets, say 1 and 2, is now a function of A_1 and A_2, which in turn are functions of the location of all other potential destinations. Consequently, the location of a new outlet is likely to have different effects of A_1 and A_2, and hence on $p_1(i)$ and $p_2(i)$.

The competing destinations model is a more accurate and a more general specification of consumer behavior than the gravity model. The following section illustrates how the replacement of the gravity hypothesis by the competing destinations hypothesis in the Poston and Wilson, and in the Harris and Wilson frameworks allows the modeling of discontinuous change in the spatial pattern of retail outlets, such as was described in Section 2.

5. MODELING DISCONTINUOUS CHANGE IN SPATIAL PATTERNS

The Poston and Wilson, and the Harris and Wilson frameworks described in Section 3 can be termed two-dimensional frameworks. Revenues and costs associated with individual retail outlets are assumed to be a function of only one variable, facility size. More realistically, revenues and costs are a function of both facility size and facility location. Adding the locational dimension not only makes the analysis more realistic, but also allows the modeling of both the size and the spatial distribution of retail outlets.

5.1. Extension of the Poston and Wilson Framework

It seems reasonable to assume that consumers derive utility from visiting retail facilities in close proximity to other facilities. Such behavior minimizes transportation costs either in terms of multipurpose shopping or if demand is not satisfied at the first choice of retail facility. A shop may have sold out of some product, for example, or a suitable choice may not be available. Thus equation 3 should be rewritten as

$$U = u_1 + u_2 + u_3 \quad , \tag{16}$$

where u_1 and u_2 are defined as before, and u_3 represents the utility derived from patronizing facilities in close proximity to other facilities. Hence the utility u_3 is related to the accessibility of an outlet to other outlets, and the relationships between U and u_1 and u_2, and U and u_3 form a three-dimensional surface corresponding to combinations of W_j and A_j. This surface will increase as both W_j and A_j increase, as is shown in Figure 3.

As a retail facility becomes increasingly clustered with and accessible to other facilities, the average travel costs over the geographic landscape to reach some facility will increase. In the limit, all facilities would be clustered together in one location, maximizing travel costs within the urban area. Consequently, as A_j increases, u_2, the utility associated with having to overcome the spatial separation between consumers and retail facilities, will decrease. There will be a three-dimensional surface representing the relationship between U and u_2 at each combination of W_j and A_j. This surface will decrease when both W_j and A_j increase, as is depicted in Figure 3.

If it is assumed that the ease of travel is increasing over time, and the relationship between u_1 and W_j, and between u_3 and A_j is logistic, then combinations of W_j and A_j yielding maximum utility to consumers can involve discontinuous change. The exact non-zero combination of W_j and A_j resulting from this procedure depends upon social and economic conditions specified in the system. Social conditions are specified via the utility functions relating u_1 and u_3 to W_j and A_j, respectively. Economic conditions are specified via the relationship between u_3 and W_j, and between u_3 and A_j. The combinations of W_j and A_j yielding maximum utility to consumers can be examined under a variety of social and economic conditions with the aid of numerical simulation. The conditions under which the combinations of W_j and

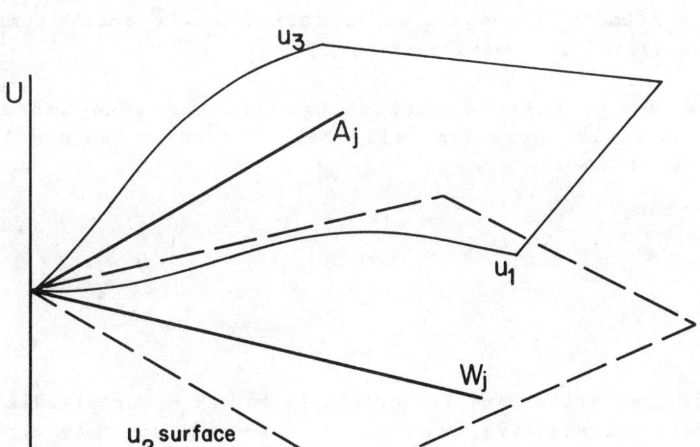

Figure 3: Three-dimensional Utility Surfaces.

A_j can exhibit catastrophic change between non-zero values also can be examined. In this way it would be possible to answer questions such as those that ask about the mix of social and economic conditions that prompt retailing to move from small, downtown facilities to large, suburban facilities, or about the different mixes of social and economic conditions that yield similar patterns of retailing or that produce similar changes in retailing patterns.

5.2. Extension of the Harris and Wilson Framework

The Harris and Wilson framework, described in Section 3, employs a gravity demand formulation to model discontinuities in the size of retail outlets. In Section 4, however, it was shown that the gravity model appears to be unrealistic when applied to retailing and consumer choice systems, and accordingly should be replaced with the competing destinations model given in equation (13). The cost equation used in the Harris and Wilson framework [equation (7)] also is unrealistic in that the costs of operating a retail outlet are assumed to be merely a function of size, with locational factors being ignored. The cost equation should incorporate the relationship between costs of operating a retail outlet and its location with respect to its competitors; that is,

$$C_j = k_1 W_j^{\lambda_1} + k_2 A_j^{\lambda_2} \quad , \tag{17}$$

where $k_1 > 0$ denotes the efficiency of operating retail outlets (as k_1 decreases, the operation of retail outlets becomes increasingly efficient), λ_1 denotes the presence of suppliers' internal scale economies ($0 \leq \lambda_1 \leq 1$, and as λ_1 decreases, scale economies increase), k_2 represents retail rent gradients around existing outlets (as k_2 decreases, locating in close proximity to existing retail outlets becomes cheaper), and λ_2 denotes the presence of suppliers' agglomeration economies (external scale economies). As λ_2 decreases, agglomeration economies increase.

Therefore, it can be postulated that changes in the size and/or relative location of an outlet over time will be governed by the profit associated with that outlet. Moreover,

$$\partial W_j/\partial t = f(\pi_j) \quad , \tag{18}$$

and

$$\partial A_j/\partial t = f(\pi_j) \quad , \tag{19}$$

such that both the size of an outlet and its proximity to other outlets will increase when its profits are positive, whereas the size and proximity will decrease when its profits are negative. Equilibrium values of W_j and A_j, $\bar{W}_j$ and $\bar{A}_j$ respectively, are achieved at zero (or normal) profits. Such equilibria, however, can be stable or unstable, as was described in Section 3.

Both the demand and the cost surfaces now are functions of W_j and A_j. The exact shapes of these surfaces depend on the parameters α_1 and α_3 in the demand equation, and on the parameters k_1, k_2, λ_1 and λ_2 in the supply equation. The parameters α_1 and α_3 denote social conditions; α_1 represents consumers' scale economies and α_3 represents consumers' agglomeration economies. The parameters k_1, k_2, λ_1 and λ_2 denote economic conditions as described above. Interaction between these parameters will lead to a variety of discontinuous changes in the equilibrium combinations of W_j and A_j. To demonstrate such changes, a demand surface can be constructed from a spatial system in which W_j and A_j are independently increased. A cost surface also can be derived as a function of k_1 and k_2 such that the cost surface lies above the demand surface, except at the origin, as shown in Figure 4.[4] Increasing (decreasing) efficiency in the operation of retail facilities over time, and decreasing (increasing) rent gradients can be simulated by lowering (raising) the cost surface by decreasing (increasing) the values of k_1 and k_2. If the cost surface is being lowered onto the

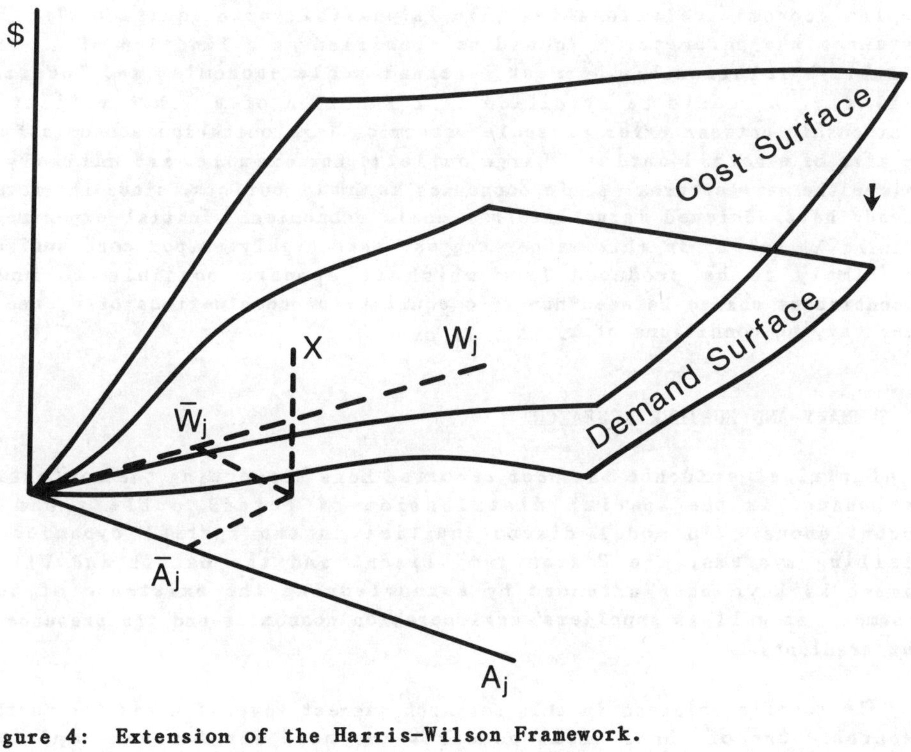

Figure 4: Extension of the Harris-Wilson Framework.

demand surface, the point of contact between the two surfaces, X, represents a discontinuous change in $\bar{W}_j$ and $\bar{A}_j$ from the origin. Conversely, if the cost surface is being raised, the point at which the two surfaces separate represents a discontinuous change in $\bar{W}_j$ and $\bar{A}_j$ toward the origin. By changing only one of the cost parameters, either k_1 or k_2, it would be possible to examine discontinuous change in either $\bar{W}_j$ or $\bar{A}_j$. Thus, by extending the Harris and Wilson framework through the replacement of the gravity model by the competing destinations model, it is possible to explain the discontinuous changes in the spatial distribution of retail outlets that was described in Section 2. Discontinuous change in $\bar{A}_j$ away from the origin represents the rapid nucleation of outlets into malls; discontinuous change in $\bar{A}_j$ toward the origin represents the decentralization of retail outlets and the loss of downtown retailing dominance. Equilibrium values of W_j and A_j could be examined under a variety of social and economic conditions through a systematic study of the parameters of the supply-and-demand equations. It then would be possible to determine the conditions giving rise to a retailing pattern of small, clustered outlets, or large, dispersed outlets. It also would be possible to examine what changes (economic or social) in the system would promote required changes in the retailing system with the least effort, or what would be required to counteract an expected change in some condition and to maintain the status quo.

The extended cost equation in (17) also allows the modeling of more complex economic relationships than is possible with equation (7). For instance, the parameter λ_2 could be specified as a function of A_j that reflects a relationship between internal scale economies and location. Similarly, λ_2 could be specified as a function of W_j that reflects a relationship between external scale economies (agglomeration economies) and the size of a retail outlet. Large outlets, for example, are unlikely to achieve the same external scale economies as small outlets, since the former already have achieved large internal scale economies. Initial experiments defining λ_1 and λ_2 in this manner suggest that highly warped cost surfaces are likely to be produced from which it appears possible to model discontinuous change between non-zero equilibrium combinations of W_j and A_j under varying conditions of k_1, k_2, λ_1 or λ_2.

6. SUMMARY AND FURTHER RESEARCH

Empirical evidence has been reported here supporting the contention that change in the spatial distribution of retail outlets can be discontinuous. To model discontinuities in the spatial dynamics of retailing systems, the Poston and Wilson, and the Harris and Wilson frameworks have been extended by acknowledging the existence of both consumers' as well as suppliers' agglomeration economies and the presence of rent gradients.

The results obtained in this research suggest several areas for further research. One of these areas concerns the need for further empirical research into discontinuities in retail dynamics. It would be useful to expand the initial empirical evidence presented here and to investigate the existence of possible discontinuities in spatial pattern for types of retail outlets other than clothing. Have grocery stores, for example, exhibited any spatial discontinuities over the same time period?

Until now, it has been possible only to model discontinuous change either away from or toward a zero equilibrium point, whereas in reality discontinuities can occur between non-zero equilibria. Such discontinuities possibly can be modeled by expanding the cost equation (17). For example, if λ_2 is a function of W_j, a more complex cost surface would result from which more complex discontinuities are more likely to result. Experiments in utilizing more realistic cost surfaces presently are underway.

Future research should address the issue of what combinations of social variables (α_1, α_3) and economic variables (k_1, k_2, λ_1, λ_2) lead to particular equilibrium combinations of size and relative location. Also, how sensitive are such equilibrium combinations of size and relative location to changes in social and economic conditions? Are equilibrium points more sensitive to economic (supply) or social (demand) conditions? In a similar manner, information can be provided on the amount and direction of effort needed to maintain the status quo in retailing patterns. For example, under changing social conditions, what can be done in terms of changing conditions to retain retailing in downtown locations? Can changes in uncontrollable parameters be balanced by changes in controllable parameters? As Wilson (1983, p. 21) states:

> From the modeller's point of view, an important research topic is the articulation of variables and parameters which are wholly or partly controllable ... in a variety of different circumstances.

Finally, the type of research reported here leads into the area of 'strange attractors' (Feigenbaum, 1980; Hofstadter, 1981). Certain configurations of retailing, for example, may be controlled by one or two critical parameters so that no matter what the initial state of the system is, when certain parameter values occur, one type of retailing pattern will emerge. Although this is a very complex topic mathematically, the analysis of retail dynamics described in this paper provides a starting point.

7. REFERENCES

Allen, P., 1980, Planning and Decision Making in the Human System: Modelling Self Organization, paper presented to the NATO Advanced Research Institute, Oxford, England, September, 1980.

______, and M. Sanglier, 1979, A Dynamic Model of Growth in a Central Place System, Geographical Analysis, 11: 256-272.

_____, J. Deneubourg, M. Sanglier, F. Boon, and A. de Palma, 1978, The Dynamics of Urban Evolution, vol. 2: Intraurban Evolution, final report, Washington, D. C.: U. S. Department of Transportation.

Amson, J., 1974, Equilibrium and Catastrophic Models of Urban Growth, in Space-Time Concepts in Urban and Regional Models, edited by E. Cripps, London: Pion, pp. 108-128.

Anas, A., 1983, Discrete Choice Theory, Information Theory and the Multinomial Logit and Gravity Models, Transportation Research B, 17: 13-23.

Aplin, G., 1983, Order-Neighbor Analysis, Concepts and Techniques in Modern Geography, 36, East Anglia, England: GEO Abstracts Ltd. .

Beaumont, J., M. Clarke and A. Wilson, 1981a, Changing Energy Parameters and the Evolution of Urban Spatial Structure, Regional Science and Urban Economics, 11: 287-315.

_____, _____ and _____, 1981b, The Dynamics of Urban Spatial Structure: Some Exploratory Results Using Difference Equations and Bifurcation Theory, Environment and Planning A, 13: 1473-1483.

Birkin, M., M. Clarke and A. Wilson, 1984, Interacting Fields: Comprehensive Models for the Dynamical Analysis of Urban Spatial Structure, Working Paper 385, School of Geography, University of Leeds.

Clarke, M., and A. Wilson, 1983, The Dynamics of Urban Spatial Structure: Progress and Problems, Journal of Regional Science, 23: 1-18.

Dendrinos, D., 1980, Catastrophe Theory in Urban and Transport Analysis, Washington, D. C.: U. S. Department of Transportation, Research and Special Programs Administration.

Domencich, T., and D. McFadden, 1975, Urban Travel Demand: A Behavioral Analysis, Amsterdam: North-Holland.

Eaton, B., and R. Lipsey, 1979, Comparison Shopping and the Clustering of Homogeneous Firms, Journal of Regional Science, 19: 421-435.

Feigenbaum, M., 1980, Universal Behavior in Nonlinear Systems, Los Alamos Science, Summer: 4-27.

Fotheringham, A., 1983, A New Set of Spatial Interaction Models: The Theory of Competing Destination, Environment and Planning A, 15: 15-36.

_____, 1984, Spatial Flows and Spatial Patterns, Environment and Planning A, 16: 529-543.

______, 1985, Spatial Competition and Agglomeration in Urban Modelling, Environment and Planning A, 17: 213-230.

______, 1986, Modeling Hierarchical Destination Choice, Environment and Planning A, forthcoming

______, and D. Knudsen, 1984, Critical Parameter Values in Retail Shopping Models, Modeling and Simulation, 15 (Part 1): 75-80.

______, and ______, 1985, Discontinuous Change in the Relative Location of Retail Outlets: Further Results, Modeling and Simulation, 16 (Part 1): 111-115.

Harris, B., and A. Wilson, 1978, Equilibrium Values and Dynamics of Attractiveness Terms in Production-constrained Spatial-interaction Models, Environment and Planning A, 10: 371-388.

Hofstadter, D., 1981, Strange Attractors: Mathematical Patterns Delicately Poised Between Order and Chaos, Scientific American, 245: 22-43.

Lioukas, S., 1984, Thresholds and Transitivity in Stochastic Consumer Choice: A Multinomial Logit Analysis, Management Science, 30: 110-122.

McFadden, D., 1974, Conditional Logit Analysis of Qualitative Choice Behavior, in Frontiers in Econometrics, edited by P. Zarembka, New York: Academic Press, pp. 105-142.

Mulligan, G., 1983, Consumer Demand and Multipurpose Shopping Behavior, Geographical Analysis, 15: 76-81.

O'Kelly, M., 1981, A Model for the Demand for Retail Facilities Incorporating Multistop, Multipurpose Trips, Geographical Analysis, 13: 134-148.

Poston, T., and A. Wilson, 1977, Facility Size Versus Distance Travelled: Urban Services and the Fold Catastrophe, Environment and Planning A, 9: 681-686.

Silk, J., 1979, Statistical Concepts in Geography, London: George Allen and Unwin.

Sobel, K., 1980, Travel Demand Forecasting by Using the Nested Multinomial Logit Model, Transportation Research Record, 775: 48-55.

Thill, J., 1985, Demand in Space and Multipurpose Shopping: A Theoretical Approach, Geographical Analysis, 17: 114-129.

Thom, R., 1975, Structural Stability and Morphogensis, Reading, Mass.:

Benjamin.

Tye, W., L. Sherman, M. Kinnucan, D. Nelson and T. Tardiff, 1982, Application of Disaggregate Travel Demand Models, Report 253, Washington, D. C.: Transportation Research Board, National Cooperative Highway Program.

Williams, H., 1977, On the Formulation of Travel Demand Models and Economic Evaluation Measures of User Benefit, Environment and Planning A, 9: 285-344.

Wilson, A., 1971, A Family of Spatial Interaction Models, and Associated Developments, Environment and Planning, 3: 1-32.

_____, 1981a, Oscillation in Dynamic Models of Shopping Centres and Other Urban Structures, Sistemi Urbani, 3: 391-401.

_____, 1981b, Catastrophe Theory and Bifurcation, London: Croom Helm.

_____, 1983, Making Urban Models More Realistic: Some Strategies for Future Research, WP 358, School of Geography, University of Leeds.

_____, and M. Oulton, 1983, The Corner-shop to Supermarket Transition in Retailing: The Beginnings of Empirical Evidence, Environment and Planning A, 15: 265-274.

Wrigley, N., 1985, Categorical Data Analysis for Geographers and Environmental Scientists, London: Longman.

8. FOOTNOTES

1 City directories for Gainesville were unavailable for 1956, 1966, and 1982, and these breaks are marked in Figures 1 and 2.

2 To analyze a range of cost-decay parameters generally representative of reality, the exponential cost function in equation (3) was replaced by a power function $c_{ij}^{\alpha_2}$. The usual values of the parameter in this latter function are known to range between 0 and -2.5, since they are independent of the units in which costs are measured.

3 The authors are especially indebted to an anonymous reviewer for his comments here.

4 For the cost surface to lie above the demand surface initially, α_1 and α_3 must be greater than unity, resulting in a logistic demand surface. Thus catastrophic change in $\bar{W}_j$ or $\bar{A}_j$ can occur only when there are either strong consumer scale economies or strong consumer agglomeration economies (Harris and Wilson, 1978; Fotheringham, 1985a).

SECTION 4

TRANSFORMATIONS INVOLVING AUTOREGRESSIVE DEPENDENCIES

The spatial science paradigm in geography placed an emphasis on the development of theory and models relating to geographical entities, taking into account the geographical or spatial relationships between those entities. It follows, therefore, that in examining and testing those models it has become an important research objective to develop statistical methods appropriate for the analysis of spatial data. The geographer/regional scientist is not only aware of the spatial nature of his data, but is in a far better position now (than, say, twenty years ago) to analyse such data.

Spatial statistics is a broad, somewhat fragmented area of research where there is still much to be done. One area focusses on the problem of statistical inference in those situations where classical assumptions of independence in the population no longer hold. Another area focusses on the problem of establishing relationships between variables (correlation, regression, and the such) in situations of univariate spatial dependence. These areas in turn merge into the broader (and more controversial) fields of spatial modeling, pattern recognition and the development of spatial languages.

The last two papers in the proceedings address specific problems in the analysis of spatial data--evidence, if such is needed, that there are still many unanswered questions at a technical level. Arbia deals with a purely spatial problem, namely the estimation of the spatial autocorrelation function. In time series analysis the autocorrelation function is a basic tool for model specification. However the lack of a unique ordering in space means that the use of the spatial autocorrelation function is not without difficulty. He notes that the often observed tendency for spatial autocorrelation to decay with increasing distance may be due to statistical and geometrical factors rather than data attributes. This undermines the usefulness of this estimator for model specification, and perhaps explains why models of second or higher order are rarely fitted. The severity of the problem is indicated and a systematic examination is undertaken using simulation experimentation methods.

All too often spatial analysis encounters the problems of incomplete data records and in a study of the U. S. consumer price index (CPI), Griffith deals both with the problem of modeling space-time variation in the index (for some 30 U. S. cities) and the problem of consistent estimation of missing values. The approach he adopts is to specify models that incorporate space-time dependencies between CPI values. Missing data estimation then may take place within the framework of the chosen model.

Model selection, parameter estimation and missing data estimation are treated simultaneously, which constitutes the approach that is adopted in this paper. In addition, Griffith emphasizes the importance of practical spatial data analysis in a research agenda that purports to address real world problems and issues. The difficulties he highlights are a timely reminder, as the present volume closes, that although elaborate theory (statistical or otherwise) is the medium by which we will shed light on the complexities of geographical organization, it is not on its own the message.

PROBLEMS IN THE ESTIMATION OF THE SPATIAL AUTOCORRELATION FUNCTION ARISING FROM THE FORM OF THE WEIGHTS MATRIX

Giuseppe Arbia
University of Rome
Italy

1. INTRODUCTION

Space-time autoregressive moving average models (STARMA) have been introduced into statistical literature to model the temporal evolution of data recorded at a network of spatial sites (Cliff, et. al., 1975; Martin and Oeppen, 1975; Bennett, 1975; Pfeifer and Deutsch, 1980; Cliff and Ord, 1981). The large amount of work in the field in the last decade demonstrates the interest that statisticians, geographers, and social and natural scientists have found in such models (for a recent review see Bennett, 1984; Bennett and Haining, 1985).

In a STARMA model the observation of a random variable at site i and time t is expressed as a weighted linear combination of past observations and random errors lagged both in time and space (Pfeifer and Deutsch, 1981). The usual procedure in model building derives directly from the time series analogue (Box and Jenkins, 1970), and seeks to identify the most parsimonious model in terms of parameters to be estimated, their estimation, and diagnostic checking. For time series data that are Gaussian and weakly stationary the main idea is that instead of estimating the whole set of covariance between the random variables at different points in time, we can identify the lag order after which the covariance becomes negligible. Invoking this restriction usually allows the number of parameters that is to be estimated to be considerably reduced.

However, for the STARMA family this procedure creates a number of problems. For this reason we will confine our attention to purely spatial models (Whittle, 1954), which will allow us to simplify the discussion and to attack the core of the problem.

The main difficulty is that in space there does not exist a complete and unique ordering of observations such as we have in time. Instead, ordering must be introduced using some a priori information. For a Gaussian stationary process on the plane this is usually achieved by defining a set of neighbours for each site i, say N(i), and we then consider the interaction to exist only within this set. We also define a weight matrix $\underline{W}$ with zero elements in its diagonal, non-negative off-diagonal elements scaled to sum to unity in each row, with this off-diagonal elements being non-zero only if $j \in N(i)$. Given these definitions, the term

$$L(X_i) = \sum_{j \varepsilon N(i)} w_{ij} X_j \qquad (1.1)$$

plays the role of the spatially lagged variable. An extension of equation (1.1) to lags higher than a first order one is also possible by defining a hierarchical spatial ordering of the neighbours of each site in a sequence of sets N(i), and, accordingly, a set of g-th order weights matrices $\underline{W}(g)$.

Starting from this spatial lag definition, different autoregressive models may be constructed. The usual tool to identify a model prior to its estimation is (by analogy with time series) the spatial correlogram (Cliff and Ord, 1981). However, in most of the empirical studies where this approach has been utilized significant correlation after the first lag is only very rarely observed (Bennett, 1975; Hooper and Hewings, 1981; Martin and Oeppen, 1975; Pfeifer and Deutsch, 1980). In this paper we suggest that empirical results may be due to statistical reasons as well as to a genuine distance decay of spatial interaction. The identification of spatial autoregressive models in non-lattice schemes with the usual estimates of the spatial correlogram is, in fact, strongly affected by the definition of the matrices $\underline{W}(g)$. Hence model identification may become very unreliable under certain conditions.

2. THE REDUCTION IN VARIANCE IN THE SPATIALLY LAGGED VARIABLE

Suppose we are given a system of n sites and a vector of observations $\langle x_1, x_2, \ldots x_n \rangle$ that we can consider as a single realization of a discrete stochastic process $\{X_1, X_2, X_3, \ldots, X_n\}$. This process is assumed to be stationary and Gaussian, with

$$E(X_i) = 0 \quad , \text{ for all } i \quad , \qquad (2.1)$$

$$E(X_i^2) = \sigma^2 \quad , \text{ for all } i \quad , \qquad (2.2)$$

$$E(X_i X_j) = \sigma^2 c|i-j| \quad , \text{ for all } i \text{ and } j, \qquad (2.3)$$

where c is a function depending only on the distance between any pair of sites i and j.

The sample spatial correlogram is based upon the sample covariance between the observed series and the spatially lagged series [defined as in equation (1.1)]. However, two different scalings are possible, and accordingly we have two estimates for the spatial autocorrelation function. If we use as a scaling the variance of the series itself, we get

$$I(g) = (\Sigma z_i {}^g z_i)(\Sigma z_i^2)^{-1} \quad , \qquad (2.4)$$

where $z_i = (x_i - \bar{x})$, $\bar{x}$ is the sample mean, and ${}^g z_i = L^g(z_i)$ is the g-th order lagged series. Alternatively we can scale the covariance with the geometric average of the variance of the original and of the lagged series, leading to

$$R(g) = (\Sigma z_i {}^g z_i)(\Sigma z_i^2 \ \Sigma {}^g z_i^2)^{-1/2} \quad , \tag{2.5}$$

Equation (2.4) represents the well known Moran Coefficient (Cliff and Ord, 1981), while (2.5) is usually referred to as a modified spatial autocorrelation coefficient (Cliff and Ord, 1981; Bennett, 1979), and is based on a formula originally due to Whittle (1954).

A property that links these two preceding estimates is the inequality

$$|I(g)| \leq (\Sigma {}^g z_i^2 / \ z_i^2)^{1/2} \quad , \tag{2.6}$$

and therefore the Whittle estimate can be viewed as the Moran Coefficient divided by its theoretical maximum, namely

$$R(g) = I(g)/\mathrm{Max}|I(g)| = I(g) \ (\Sigma z_i^2 / \Sigma {}^g z_i^2)^{1/2} \tag{2.7}$$

This last equation shows why in a time series the two estimators are equal. In fact, for a time series $\Sigma z_i^2 \cong \Sigma {}^g z_i^2$ unless the number of observations is very small. This 'identity' is generally false for a spatial series.

Next consider the variates Z_i corresponding to their observed values z_i. Their expectations are (Cliff and Ord, 1981)

$$E(Z_i) = 0 \quad , \text{ and} \tag{2.8}$$

$$E(Z_i^2) = (n-1/n)\sigma^2 \quad , \tag{2.9}$$

The corresponding first-order lagged variate $'Z_i$ have expectations

$$E('Z_i) = E(\Sigma w_{ij} Z_j) = \Sigma w_{ij} E(Z_j) = 0 \quad , \tag{2.10}$$

$$E('Z_i^2) = E(\Sigma w_{ij} Z_j)^2 = E(\Sigma w_{ij}^2 Z_j^2 + 2 \sum_{j,k \varepsilon N(i)} \sum w_{ij} w_{ik} Z_j Z_k) \quad . \tag{2.11}$$

Beginning with the simple assumption that each site has the same number of neighbours, say η, equation (2.11) reduces to

$$E[\Sigma (1/\eta^2) Z_i^2] + 2E[\sum_{j,k \varepsilon N(i)} \sum -(1/\eta^2) Z_j Z_k] \quad . \tag{2.12}$$

The first summation contains η non-zero terms, while the second is defined relative to all pairs of sites j and k that share a common neighbour i. There are C(n,2) such pairs of sites, so that

$$E('Z_i^2) = E(Z_i^2)/\eta + [(\eta-1)/\eta] \ E(Z_j Z_k) \quad , \tag{2.13}$$

where $E(Z_j Z_k)$ is limited to those j and k sites included in N(i).

Let us now define the expected reduction in variance with the ratio

$$H = E({}'Z_i^2)/E(Z_i^2) \quad , \tag{2.14}$$

In our elementary case we have

$$H = 1/\eta + (\eta-1)E(Z_jZ_k)/\{\eta E(Z_i^2)\} \quad . \tag{2.15}$$

In the simplest case that the process X_i is white noise, we have that $E(Z_jZ_k) = -\sigma/n$ (Cliff and Ord, 1981), so that

$$H = [1-(\eta-1)/(n-1)]/\eta \quad . \tag{2.16}$$

This equation shows that H is negatively related to the average number of neighbours η. The reasons for this negative relationship is that if the number of terms involved in a moving average increases, the steeper will be the decline in variance. Clearly then, if the number of neighbours increases monotonically as the spatial lag increases,

$$E({}^gZ_i^2) \leq E({}^{g-1}Z_i^2) \leq \ldots\ldots\ldots\ldots \leq E(Z_i^2) \quad , \tag{2.17}$$

and, as a consequence,

$$H(g) \leq H(g-1) \leq \ldots\ldots\ldots\ldots\ldots \leq 1 \quad , \tag{2.18}$$

where H(g) measures the reduction in variance at lag g.

Table 1 displays the average number of neighbours for various lattices of different dimension and edge structure. All lattices are mapped onto a torus. It can be seen that the average number of neighbours increases up to a maximum of one-half of the diameter of the associated graph, and then decreases. Figure 1 shows the same pattern for the average number of neighbours in a rectangular grid lattice of 20 x 25 cells, for the Mercer-Hall data (Mercer and Hall, 1911), which have been studied several times in the spatial literature.

A similar pattern is likely to occur for non-lattice schemes and distance-based neighbours (Ripley, 1977). Figure 2 shows the contiguity-based number of neighbours for the Italian provinces. Figure 3 displays the same Italian provinces represented by their centroids, and the new measure of distance-based neighbours at each lag. The plot of the number of neighbours calculated by the two methods closely resembles that of Figure 1.

Equation (2.16) shows that the average number of neighbours determines H when the process is white noise, but it is only one element in determining H when the process is an autocorrelated one, as in equation (2.15). The degree to which the presence of an autocorrelated process can influence the rate of decrease in variance will be examined further in Section 3 through Monte Carlo methods. However, it already is possible to notice that, if the

TABLE 1

AVERAGE NUMBER OF NEIGHBOURS IN DIFFERENT LATTICES THAT HAVE BEEN MAPPED ON TO A TORUS

Lag Order	Rook's Case		Queen's Case	
	6-by-6	8-by-8	6-by-6	8-by-8
1	4	4	8	8
2	8	8	16	16
3	10	12	11	24
4	8	14	--	15
5	4	12	--	--
6	1	8	--	--
7	--	4	--	--
8	--	1	--	--

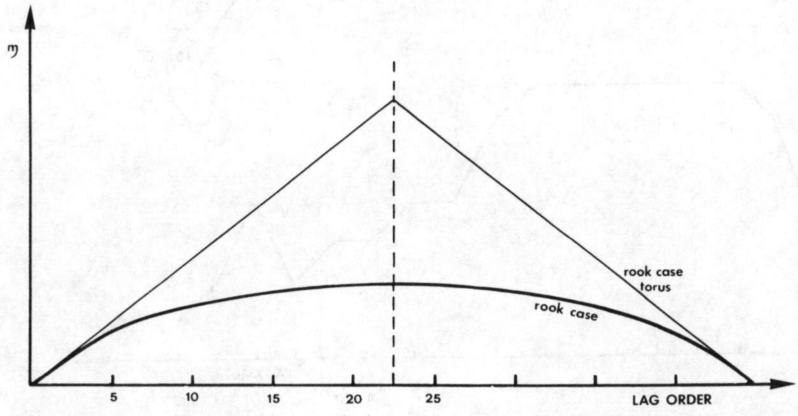

Figure 1: Average number of neighbours in a 20-by-25 rectangular grid lattice

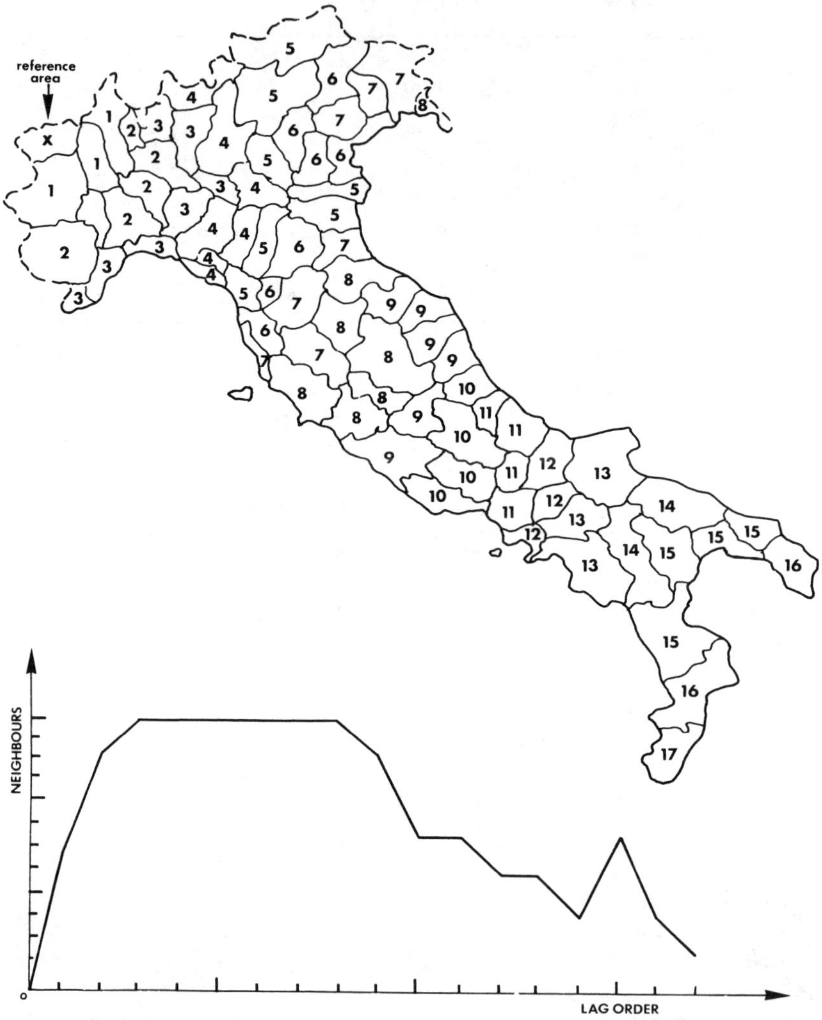

Figure 2: Various order neighbours for the Italian provinces study area.

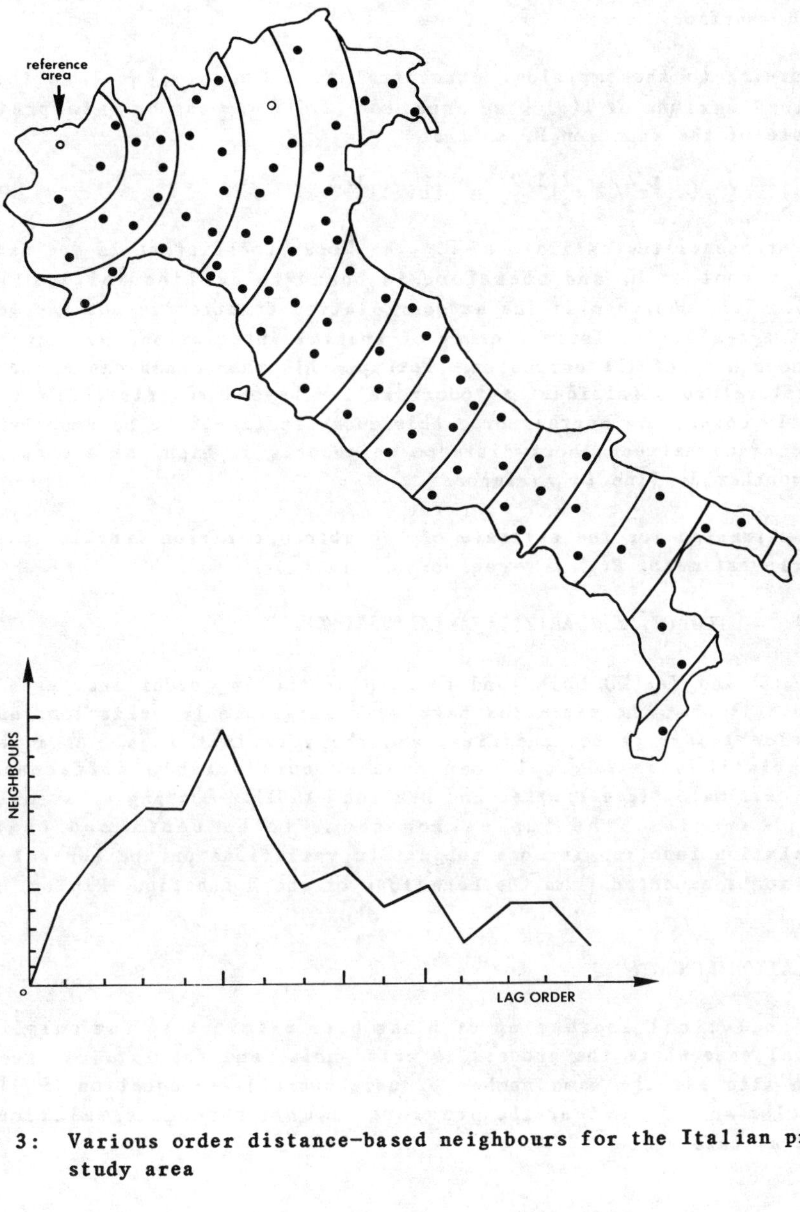

Figure 3: Various order distance-based neighbours for the Italian provinces study area

correlation between sites with a common neighbour is positive, then H will be inflated and hence the reduction in variance is moderate compared with the white noise case. In contrast, when $E(Z_j Z_k)$ is negative, H will be depressed and hence the reduction in variance is more marked. This change in relationship happens because when pairs of sites with a common neighbour are positively correlated, the average will contain values that tend to be similar and the reduction in variance will not be as sharp. Conversely if the same sites are negatively correlated the average will dramatically smooth the surface.

Returning to the empirical autocorrelation function, we have that the theoretical maximum of I(g) [see equation (2.6)] now can be interpreted as an estimate of the function H, so that

$$|I(g)| \leq (\Sigma\, {}^{g}z_i^2 / \Sigma\, z_i^2)^{1/2} = [\hat{H}(g)]^{1/2} \quad , \tag{2.19}$$

where $\hat{H}$ represents the estimate of H. The Moran Coefficient is dominated by the square root of H, and therefore is bound to decline initially (see Figure 4). This decrease in the autocorrelation function is not the genuine result of a realistic distance decay of spatial interaction, but merely due to the inadequacy of the estimator. Perhaps this phenomenon can explain why in the literature significant autocorrelation beyond the first lag is only very rarely observed. Furthermore, this decay is likely to be smoother when the correlation between short distance neighbours is high, as a consequence of the smoother decline in variance.

The situation for the estimate of the autocorrelation function based on the Whittle estimator R(g) is even worse. In fact,

$$R(g) = [COV(Z,'Z)/VAR(Z)][VAR(Z)/VAR('Z)]^{1/2} \quad , \tag{2.20}$$

and COV(Z,'Z) and VAR('Z) both tend to zero as the lag order increases. The consequence is that the estimates have very large sample variations and are very unreliable except for the first and the very last lags. As a result, small variations in a sample can produce considerable differences in empirical estimates (see Pfeifer and Deutsch, 1981). Looking at a series of empirical studies, the impression seems to be confirmed that the autocorrelation function is more subject to variations on the central lags, which is to be expected from the behaviour of the H function (Ripley, 1981).

3. SIMULATION RESULTS

The analytical expression of H has been examined so far only in the theoretical case where the process is white noise and the lattices are such that each site has the same number of neighbours [(see equation (3.7)]. In this section we will extend the previous results through simulation to a more general case where

(i) the process is spatially autocorrelated,
(ii) the lattices are irregular, and
(iii) the number of observations is variable.

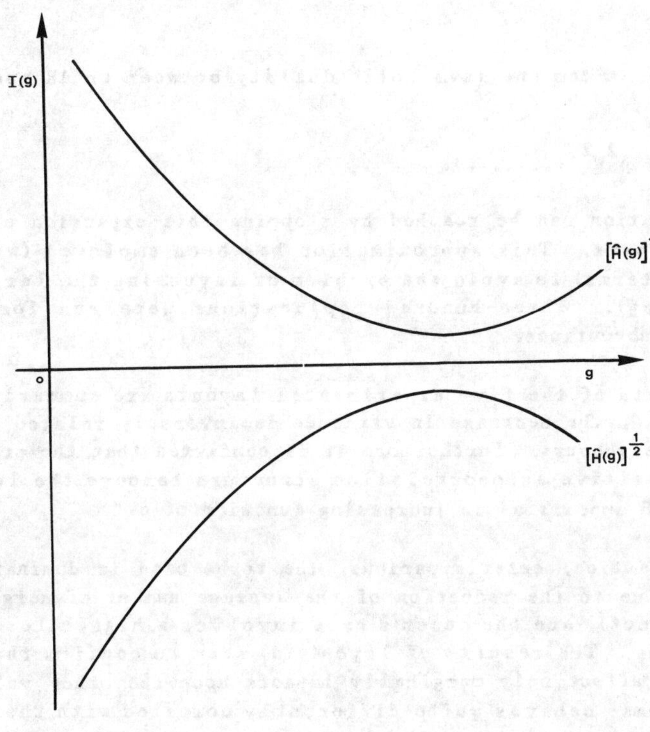

Figure 4: Bounds of the Moran-based autocorrelation function.

To isolate these three effects, five different experimental layouts were considered, namely

(a) a 6-by-6 regular lattice on a torus with the rook's case joins,
(b) a 6-by-6 regular lattice with the rook's case joins,
(c) a 6-by-6 regular lattice with the queen's case joins,
(d) a 8-by-8 regular lattice with the rook's case joins, and
(e) a 36-cell irregular lattice representing southern Italian provinces.

In all five cases standardized binary weights have been employed so that invertibility is guaranteed when $|\rho| < 1$.

A number of approaches exist for the generation of observations drawn from autocorrelated distributions (Haining, Griffith and Bennett, 1983; Larimore, 1977; Chambers, 1977). For the particular purpose of simulating a simultaneous model, the observations x_i can be expressed as

$$\underline{x} = (\underline{I}-\rho\underline{W})^{-1}\underline{e} \quad , \tag{3.1}$$

with iid $e_i \sim N(0;\sigma_e^2)$. From the asymptotic duality between an AR and a MA process, we have that

$$\underline{x} = (\underline{I} + \rho\underline{W} + \rho^2\underline{W}^2 + \ldots\ldots.)\underline{e} \quad , \tag{3.2}$$

and a good approximation can be reached by stopping this expansion after a finite number of terms. This approximation has been employed (with an expansion up to 25 terms) to avoid the problem of inverting the large and sparse matrix $(\underline{I}-\rho\underline{W})$. Three-hundred replications were run for each experiment, using subroutines.

The main results of the five experimental layouts are summarized in Table 2. As expected, the decrease in variance is inversely related to the average number of neighbours. Furthermore it is confirmed that the presence of a first-order positive autocorrelation structure tempers the loss in variance, and thus H appears as an increasing function of ρ.

In a regular lattice, ceteris paribus, the torus case is dominated by the non-torus one (due to the reduction of the average number of neighbours for the boundry effect), and the queen's case involves a higher loss than does the rook's case. The results of layout (d) seem to confirm that the size of the lattice affect only marginally impacts upon the other value of H. The irregular case behaves quite differently compared with the other cases, in that the introduction of a dependence structure among the observations has a greater moderating effect. In the degenerate case, when $\rho = 0.99$, the values of H, however, appear unaffected by η except in the irregular case.

Table 3 compares the simulated values of H with those expected from the

TABLE 2

PERCENTAGE ESTIMATES OF H
FROM 300 REPLICATIONS OF THE SIMULATION EXPERIMENT

Lattice	η	$\rho = 0$	$\rho = 0.5$	$\rho = 0.99$
6-by-6 Rook's case	3.3	29.6	29.8	32.5
8-by-8 Rook's case	3.5	28.6	30.2	34.6
36-cell irregular lattice	3.9	26.7	45.3	53.6
6-by-6 Rook's case on a torus	4.0	22.8	23.0	35.6
6-by-6 Queen's case	6.1	15.9	17.1	32.2

TABLE 3

THEORETICAL AND ESTIMATED VALUES OF H FOR $\rho = 0$

Lattice	η	σ^2_η	Estimated H	Theoretical H	% Error
6-by-6 Rook' case	3.3	0.44	29.6	28.3	4.3
8-by-8 Rook's case	3.5	0.37	28.6	27.4	4.2
36-cell irregular lattice	3.9	1.22	26.7	23.5	12.0
6-by-6 Rook's case on a torus	4.0	0.00	22.8	22.8	0.0
6-by-6 Queen's case	6.1	3.20	15.9	14.0	11.9

Note: The estimated values were calculated from the numerical results of the simulation experiments. The theoretical values were calculated with equation (2.16).

theoretical analysis of Section 2 when the process is white noise. As expected, when all the sites have the same number of neighbours (i. e., the variance of the number of neighbours, say σ^2_η, is zero), equation (2.16) provides the exact value of H. When this condition is not met, the formula provides an approximation that appears related to the amplitude of σ_{η}^{2}.

4. CONCLUSIONS

In this paper we have discussed some properties of what can be termed the spatially lagged variable by drawing analogies with time series modelling efforts. Problematic aspects of its definition for empirical spatial autocorrelation function work also have been addressed. The spatially lagged variable in a spatial autoregressive model (or in the space component of a mixed space-time model) is a moving average of the random variables that constitute the process from which the observations are drawn [see equation (1.1)]. As a consequence, the variance of the lagged variates is reduced in comparison with the (stationary) variance of the process. This phenomenon already has been observed by Cliff and Ord (1981), Ripley (1981), and Besag (1974). However, it has never received a formal treatment.

The main purpose of this paper was to present the major results of a simulation study that has been used to analyze, in a preliminary fashion, elements that affect this 'reduction in variance.' Through the examination of the resulting simulation data sets it is possible to argue that the decline in variance in the lagged variates is related to the configuration of the lattice over which the observations are laid (whether it is regular or irregular, finite or infinite, mapped onto a torus or not), the rules employed to define the neighbours, and hence the weights, and the true autocorrelation structure of the underlying process. The number of observations, however, does not seem to have any influence on the variance reduction in question. In particular, the analysis undertaken here leads to the following conclusion. First, the decline in variance is likely to be sharper when the average number of neighbours is high. This fact raises the dilemma, already pointed out by Cliff and Ord (1981), between a choice of the weights that covers a wide range of possible spatial patterns of autocorrelation (in which case η has to be high), and a choice that seeks to moderate the decrease in variance (where η is small). Second, related to the previous point, the variance decrease is likely to be sharper as the lag order increases since the number of neighbours is larger. And third, the decline in variance is moderated by the presence of positive autocorrelation between sites that share a common neighbour, while it increases when this correlation is negative.

The consequences of variance reduction on the identification of an autoregressive model to describe underlying processes was briefly examined, too. First, the I-Moran estimate is bounded by this reduction in variance. Hence it tends dramatically to zero as soon as a lag is introduced. This implies that the identification of a model of higher than first-order is extremely unlikely, strictly because of statistical and geometrical factors. Second, the estimate based on Whittle's R supplies very erratic and hence unreliable estimates.

The variance reduction phenomenon seems to be worthy of subsequent research efforts. An exploration of how spatial autocorrelation procedures, other than model identification, are affected by the reduction in variance should be of interest. In particular, it would be worth examining consequences on Moran's I statistic within the context of a spatial autocorrelation test. Second, a generalization to higher order studies of results reported in this paper would prove interesting. Third, distribution theory is needed for H in order for it to be useful in statistical testing formats. Finally, solutions are needed that would allow the problem of smoothing of spatial series estimators to be overcome.

5. REFERENCES

Bennett, R., 1975, Dynamic Systems Modelling of the North-West Region, Environment and Planning A, 7: 525-566, 617-636, 887-898.

_____, 1979, Spatial Time Series, London: Pion.

_____, 1984, Advances in Spatial-time Series, in Spatial Statistics and Models, edited by C. Willmott and G. Gaile, Dordrecht: D. Reidel, pp. 235-251.

_____, and R. Haining, 1985, Spatial Structure of Geographical Data, Journal of the Royal Statistical Society, Series B, 46: 1-48.

Besag, J., 1974, Statistical Analysis of Non-lattice Data, The Statistician, 24: 179-195.

Box, G., and G. Jenkins, 1970, Time Series Analysis: Forecasting and Control, New York: Wiley.

Chambers, J., 1977, Computational Methods for Data Analysis, New York: Wiley.

Cliff, A., and K. Ord, 1981, Spatial Processes, London: Pion.

______, P. Haggett, J. Ord, K. Bassett and R. Davies, 1975, Elements of Spatial Structure, London: Pion.

Haining, R., D. Griffith and R. Bennett, 1983, Simulating Two-dimensional Autocorrelated Surfaces, Geographical Analysis, 15: 247-255

Hooper, P., and A. Hewings, 1981, Some Properties of Spatial Interaction Models, Geographical Analysis, 13: 203-233.

Larimore, W. 1977, Statistical Inference on Stationary Random Fields,

Proceedings of the Institute of Electrical and Electronical Engineers, 65: 961-970.

Martin, R., and J. Oeppen, 1975, The Identification of Regional Forecasting Models Using Space-time Autocorrelation Functions, Transactions of the Institute of British Geographers, 70: 330-349.

Mercer, W., and A. Hall, 1911, The Experimental Error of Field Trials, Journal of Agricultural Science, 4: 107-132.

Pfeifer, P., and J. Deutsch, 1980, A STARIMA Model Building Procedure With Application to Description and Regional Forecasting, Transactions of the Institute of British Geographers, 70: 330-349.

_____, and _____, 1981, Variance of the Sample Space-time Autocorrelation Function, Journal of the Royal Statistical Society, Series B, 43: 28-33.

Priestley, R., 1981, Spectral Analysis of Time Series, New York: Academic Press.

Ripley, B., 1977, Modelling Spatial Pattern (with discussion), Journal of the Royal Statistical Society, Series B, 39: 172-212.

_____, 1981, Spatial Statistics, New York: Wiley.

Whittle, P., 1954, On Stationary Processes in the Plane, Biometrika 41: 434-449.

MODEL IDENTIFICATION FOR ESTIMATING MISSING VALUES IN SPACE-TIME DATA SERIES: MONTHLY INFLATION IN THE U. S. URBAN SYSTEM, 1977-1985

Daniel A. Griffith
State University of New York at Buffalo
United States

1. INTRODUCTION

Dealing with the problem of missing values in data sets has received increasing attention by applied statisticians over the last decade (see Bennett, et. al., 1984). While there are many conceptual and practical statistical reasons for developing ways of coping with this particular problem, the motivation for work reported in this paper stems from a need to interpolate missing data values in order to calibrate optimization models, and other sorts of non-statistical models, as was done in Griffith (1983). In contrast, for the most part statisticians have been concerned with obtaining unbiased parameter estimates in the presence of missing data. Hence the thrust of this paper is more closely akin to the aims and goals of forecasting. Not surprisingly, then, the forecasting literature will be drawn upon heavily, as the primary basis for developments reported in this paper.

2. THE MISSING VALUE PROBLEM

Much of the pertinent literature, especially for spatial data series, has been reviewed by Bennett, et. al. (1984). Since the appearance of this timely paper, though, the number of additional articles appearing in the statistics literature has at least doubled. Three articles of interest here are by Damsleth (1980), who has discussed the interpolation of missing values in time series, Haining, et. al. (1984), who have discussed the interpolation of missing values in spatial series, and Griffith, et. al. (1985), who have discussed the interpolation of missing values in space-time series. One salient characteristic of solutions proposed in these three papers is that missing values can be estimated by their conditional expectations. This specific finding is consistent with numerical results implicit in calculating maximum likelihood estimates of the accompanying statistical model parameters.

A solution to the missing value problem involves three steps, namely model identification, parameter estimation and diagnostics, and missing value interpolation. The three aforementioned papers have focused on parameter estimation and missing value interpolation. They have tended to deal with contrived data sets and quite simple models.

The application appearing in Griffith (1983) assumed that these sorts of simple models are valid descriptors of reality, without exploring issues in model identification and diagnostics. The objective of this paper is to help rectify this misconception, and hence to address aspects of model identification for space-time data series containing missing values. Cliff, et. al. (1975), provide some insight into issues of parameter estimation for this class of models.

3. THE CONSUMER PRICE INDEX

One space-time data set that is of wide interest, and also contains missing values, is part of the U. S. Consumer Price Index (CPI). This index is a statistical measure of the change in prices of selected goods and services frequently and regularly purchased by two specified population groups. One group has been labelled 'All Urban Consumers,' and covers about 80% of the national civilian non-institutional population of the U. S. This will be the index group considered in this paper, mostly because the second group covers only about 40% of the same population.

3.1. The CPI and Inflation in the U. S. Urban System

The CPI monitors price changes for food, clothing, automobiles, rents, household furnishings, fuel, recreational goods, fees paid to doctors, beauty shops, and the like, sales, real estate and excise taxes, and many other items people buy for living. However, it does not include income or personal property taxes. Meanwhile, this index is computed by differentially weighting prices of the goods and services in question. The 'All Urban Consumers' index was first computed in October of 1977. It is calculated monthly for 5 major urban areas, and bimonthly for 23 other urban areas. The base year for all but 2 urban areas is 1967, while one of the urban areas--Northeastern Pennsylvania--is not a consolidated urban place. Accordingly, these latter three urban areas have been excluded from the data base explored here. Of the 23 urban areas monitored bimonthly, approximately one-half are indexed each month, with these two sets alternating between consecutive months. This set of bimonthly CPIs is why missing values occur in the space-time data set.

As and aside, one should note that individual urban place index values measure how much prices have changed over time, but do not measure intercity differences in prices.

That part of the U. S. urban system represented by this data set includes that single city from the first-level of the hierarchy (i. e., New York), 7 cities from the second level of the hierarchy (i. e., Los Angeles, Chicago, Philadelphia, Detroit, Boston, Washington, and San Francisco), and 17 cities from the third level (see Yeates and Garner, 1980, p. 68). This total set of cities seems to be reasonably representative of the overall U.

S. urban system, although some of these cities that more recently achieved a status of large, such as San Antonio, are conspicuously absent from the set. Consequently, this data set offers more potential for use than does its predecessors, which consisted of 16 cities, mostly from the northeastern and midwestern regions of the country, and has been compiled only quarterly. This earlier data set was divided into three sets of cities, whose index calculations sequentially alternated over consecutive months, yielding 4 CPI values per city each year.

3.2. Modelling Attempts for the CPI

While the CPI has received much attention as a data base, relatively few studies have exploited its latent space-time information. One very recent attempt was made by Clark (1984a, 1984b), who in part modelled the earlier CPI data as a set of univariate time series, then tested for the presence of significant residual patterns. In addition, he divided his data into three time period subsets, namely the 1950s, 1960s, 1970s, and then fitted an analysis of variance model in an attempt to discriminate between the contribution of time and space to variations in overall inflation.

Three drawbacks are noteworthy in Clark's analysis. First, no attempt was made to determine whether or not the variance was homogeneous over these various time series. Inferential statements are hampered when the variance of a series is not constant over time. Second, no seasonal terms were evaluated. Possibly seasonal terms might not appear, since the time intervals are noticeably longer, but, this seems unlikely. Hence, there is a possibility of model mis-specification. Third, his analysis of variance totally ignores the autocorrelation latent in these data. Again, proper inferential statements are obstructed (see Griffith, 1978).

4. TEMPORAL COMPONENTS OF THE CPI MISSING VALUES MODEL

The first step in model identification is to describe each urban area's time series with an appropriate ARIMA model. Following Vandaele (1983), each series was divided into three non-overlapping segments, approximately equal in length, and then the maximum F-ratio was calculated from the resulting three possible variance pairs. Those variables for which F-ratios were found to be significant then were subjected to a variance stabilization power (β) transformation, following the Box-Cox procedure (1964). Because this research is merely exploratory in nature, the search procedure employed was incremented by 0.5, without subsequent refinement. Those values of β for which the mimimum maximum pairwise F-ratio was achieved appear in Table 1.

Once variance homogeneity through time was established, a time series plot was constructed for each city. These plots indicated that each time series is non-stationary in its respective mean, such that a linear trend was apparent. This non-stationarity was removed by a single differencing,

TABLE 1

PARAMETER ESTIMATES FOR THE INDIVIDUAL URBAN AREA ARIMA MODELS

City	β	Order of Model	AR	D	MA	Cycle	SAR	SD	SMA	Constant	Wilk-Shapiro Statistic
Chicago	3	1	0.2238*	1	—	6	—	—	0.2844*	1.5613*	0.972*
		2	0.0633						−0.2585*		
		3	−0.3074*								
Detroit	3.5	1	—	1	—	12	—	—	−0.2196*	1.4067*	0.976*
Los Angeles	2.5	1	0.3407*	1	—	6	—	—	0.1804*	1.0115*	0.997
New York	1 (2)	1	—	1	−0.2592*	3	—	—	0.1047	1.3540*	0.991
		2							0.1909*		
		3							−0.2794*		
Philadelphia	2	1	−0.6766*	1	−0.8996*	12	0.3834*	—	—	0.8910*	0.990
		2	0.3503*								
		3	0.2149*								
Baltimore	3	1	0.5348*	1	—	3	—	—	−0.5570*	—	0.993
		2	−0.2938*						−0.5844*		
		3							−0.7579*		
Boston	1 (2)	1	—	1	−0.4319*	12	—	—	0.1883*	0.1574*	0.995
Cincinnati	3.5	1	—	1	—	4	0.3928*	—	—	1.7388*	0.973*
Denver	3	1	—	1	—	—	—	—	—	3.6556*	0.986
Milwaukee	3.5	1	—	1	—	3	—	—	0.3018*	3.3472*	0.994
		2							0.0397		
		3							0.5676*		
Portland	4.5	1	—	1	0.4062*	—	—	—	—	2.7148*	0.973*
St. Louis	3.5	1	—	1	—	6	−0.5630*	1	−0.1520	—	0.973*
		2							0.6661*		
San Diego	1 (2.5)	1	0.2255	1	—	3	0.3570*	—	—	2.9347*	0.975*
		2	−0.3509*								
Seattle	4	1	—	1	0.0473	3	—	—	0.6270*	3.0993*	0.993
		2			−0.7523*						
Washington	1 (2.5)	1	−0.0100	1	—	12	—	—	0.8153*	2.8720*	0.996
		2	−0.2786*								
		3	0.3286*								

TABLE 1 (continued)

City	β	Order of Model	AR	D	MA	Cycle	SAR	SD	SMA	Constant	Wilk-Shapiro Statistic
Atlanta	1 (2)	1	—	1	−0.3329*	3	—	—	0.3246*	2.8541*	0.952*
		2							−0.6046*		
		3							−0.3612*		
Buffalo	1 (2.5)	1	—	1	—	12	—	—	0.7246*	2.7258*	0.983
Cleveland	3	1	—	1	—	3	—	—	0.2695*	3.5451*	0.934*
Dallas	1 (2.5)	1	—	1	—	6	—	—	−0.6024*	3.3271*	0.965*
Honolulu	2.5	1	—	1	0.1768	6	—	—	−0.3931*	2.3926*	0.986
		2			−0.5182*						
Houston	3.5*	1	−0.5743*	1	−0.7531*	6	—	—	−0.5551*	6.7270*	0.986
		2	−0.8560*		−0.8912*						
Kansas City	3	1	—	1	—	8	—	—	0.3485*	3.0275*	0.987
Minneapolis	2*	1	0.1719	1	—	—	—	—	—	2.1483*	0.954*
		2	−0.1437								
		3	0.3033*								
Pittsburgh	3	1	−0.6559*	1	−0.8861*	4	—	—	0.5187*	7.5642*	0.980
		2	−0.8065*		−0.9630						
San Francisco	1 (1.5)	1	—	1	−0.2999*	8	—	—	0.3877*	3.3130*	0.983

* Denotes a significant parameter estimate.

Note: Those variables subjected to a power transformation there were multiplied by a judiciously selected constant that preserved the original variation displayed by each variable.

corroborating part of Clark's findings.

Next ARIMA models were calibrated using Box-Jenkins procedures. Parameter estimates for the resulting 'best' models are presented in Table 1. Virtually all of the models include seasonal components, which mostly are multiples of 3 months. Again some corroborating evidence is provided for Clark's findings, for his models had 3-month time intervals. With regard to diagonistics, many of the residuals behaved as normal random variates, and those that did not pass the Wilk-Shapiro test have statistics that are very close to the critical value. One model failed to achieve absolute convergence, three models had disturbing patterns in their residual autocorrelation function, five autocorrelation functions had significant spike within the first 12 lags, and no parameter sets had marked correlations. As an aside, none of the residual autocorrelation function spikes resulted in the inclusion of an additional significant parameter in the corresponding model, though. Consequently, those models that appear in Table 1 are believed to provide reasonably good descriptions of their respective time series.

4.1. Transfer Function Models

In order to assess the role of urban structure in the evolution of inflation in the U. S. urban system, cross correlations between selected differenced time series were calculated. Those time series to be correlated were selected in accordance with the hierarchical structure mentioned in Section 3.1, and presented here in Table 2. Almost every urban pair that is hierarchically adjacent displays at least one significant non-zero lag cross correlation, supporting the contention that an hierarchical effect is present. A more accurate picture would be obtained by the cross correlation function for the accompanying ARIMA model residuals. But since few of the parameters for these models are markedly different from zero, and this is an exploratory exercise, those results presented in Table 3 suffice.

As an aside, cross correlations between the CPIs for San Francisco, on the one hand, and Portland and Seattle, on the other hand, are not available, since these two sets of cities do not have CPIs calculated for the same months.

5. SPATIAL COMPONENTS OF THE CPI MISSING VALUES MODEL

Spatial components for the model are of three types. First, absolute regional trends, analogous to non-stationary trends found in time series, may be represented by trend surface terms. Second, spatial autoregressive and moving average terms account for the presence of spatial autocorrelation in a data set. Haining, et. al. (1984), discuss these two model components. Third, designated regional centers act somewhat like seasonal cycles in time series data. Because none of the monthly data are complete across the set of urban places at any one point in time, tentative geographical model

TABLE 2

STRUCTURE OF THE U. S. URBAN SYSTEM

Level 3 Cities	Level 2 Cities	Level 1 City
	Boston	New York
	Detroit	New York
	Philadelphia	New York
	Washington	New York
Atlanta		New York
Baltimore		New York
Buffalo		New York
Cincinnati		New York
Cleveland		New York
Dallas		New York
Houston		New York
Pittsburgh		New York
Denver	Chicago	New York
Kansas City	Chicago	New York
Milwaukee	Chicago	New York
Minneapolis	Chicago	New York
St. Louis	Chicago	New York
Honolulu	Los Angeles	New York
San Diego	Los Angeles	New York
Portland	San Francisco	New York
Seattle	San Francisco	New York

results will not be explored here. Rather, information from the cross correlations functions will be further exploited.

5.1. Spatial ARIMA Models

Returning to Table 3, all but one of the pairwise urban places cross correlations are significant at lag zero. This result implies that spatial autocorrelation is present in the data. Thus, the conditional or

TABLE 3

LAGS HAVING SIGNIFICANT CROSS CORRELATIONS

Hierarchy Lead City Lag City Lags

Hierarchy Levels	Lead City	Lag City	Lags
1, 2	New York	Chicago	0, 6, 9
		Detroit	0
		Los Angeles	0, 1, 8, 9, 11
		Philadelphia	0, 1, 2, 12
		Boston	0
		Washington	0, 1, 12
		San Francisco	0
1, 3	New York	Baltimore	0, 1
		Cincinnati	0, 5
		Atlanta	0, 3, 12
		Buffalo	0, 6, 12
		Cleveland	0, 4
		Dallas	0, 5, 6, 9, 12
		Houston	0, 2
		Pittsburgh	0, 4
2, 3	Chicago	Denver	0
		Milwaukee	0, 9, 11
		St. Louis	0, 5, 6, 8, 9, 11
		Kansas City	0, 9
		Minneapolis	1, 7, 9, 10
	Los Angeles	San Diego	0
		Honolulu	0, 2
	San Francisco	Portland	unknown
		Seattle	unknown

simultaneous autoregressive model, and some spatial moving average model (see Haining, 1978), need to be explored. Furthermore, since the zero-order cross correlations exhibit somewhat of an absolute decline from the New York City area westward, a first- or second-order trend surface model more than likely needs to be considered, too.

Haining, et. al. (1984), outline a procedure for obtaining maximum likelihood estimates, in the presence of missing values, for the conditional model. This procedure can be extended quite easily to the simultaneous spatial autoregressive model. The complication here is that incomplete data are being used to make a decision about the underlying spatial process, whereas this problem was not encountered in the time series approach. Nevertheless, given the prevailing data constraints, this would be the best approach to take. Unfortunately, ways of incorporating the trend surface component have not been spelled out explicitly, yet.

5.2. The U. S. Urban Hierarchy

One conclusion reached in Section 4.2 is that an urban hierarchy effect is present. Conceptually seasonality in time series alludes to certain higher-order influences, which appear cyclical only because, due to physical phenomema and human behavior, time trajectories move through repetitive sequences. However, the important idea here is that of higher-order influences. Haggett (1976) illustrates that spatial autocorrelation need not arise strictly from the planar juxtaposition of places, and in fact could emanate from the non-planar hierarchical structure of an urban system. Representing this factor by a spatial linear operator, it can be combined multiplicatively with the more traditional spatial linear operator associated with spatial autocorrelation, exactly like the seasonal factors are entered into ARIMA models. The sole difference here, though, is that the order of multiplication becomes important, since sources of this higher-order influence do not occur repeatedly over the geographical landscape. Presumably the urban hierarchy operator is the premultiplier, since influences should originate from the top of the urban hierarchy, and then be operated on locally.

Consequently, a table comparable to Table 1 would need to be constructed to summarize results for the 95 individual spatial models.

6. IMPLICATIONS

The fundamental implication that can be derived from the foregoing analysis is that the model proposed by Griffith, et. al. (1985), is far too simple to serve as a good missing values estimator. Employing such a model would fail to fully exploit the information latent in the CPI space-time data series. Results reported in Table 1 clearly demonstrate both that a homogeneous spatial process is not operating over the U. S., and that a first-order time series structure would be quite inadequate. Findings summarized in Table 3 further support this non-homogeneity conclusion. In addition, these latter results strongly suggest that the appropriate missing values model should have an urban hierarchy component in it.

In conclusion, the model partially identified here is highly complex. It will consist of three basic ingredients. First will be ARIMA terms, including first differences and seasonal terms. Because only the first-difference term is prominent, possibly this component could be simplified by setting all other parameters to zero. Second, it will consist of a spatial autoregressive/moving average term. The order of these terms is yet to be determined, but zero lag cross correlations imply that this will be a prominent ingredient. Finally. it will have an urban hierarchy term. Again, possibly parameters affiliated with this term could be set to zero, since none of the non-zero lag cross correlations were found to be exceptionally large. The simplifications suggested in this section could produce a model much like that evaluated in Griffith, et. al. (1985), suggesting that their model may well provide a reasonable first-approximation for estimating those missing values in question. However, it would be plagued by problems of specification errors. The seriousness of these specification error problems remains unknown for now.

7. REFERENCES

Bennett, R., R. Haining and D. Griffith, 1984, The Problem of Missing Data on Spatial Surfaces, Annals of the Association of American Geographers, 74: 138-156

Box, G., and D. Cox, 1964, An Analysis of Transformations, Journal of the Royal Statistical Society B, 26: 211-252.

Bureau of Labor Statistics, 1977-1985, CPI Detailed Report, Washington, D. C.: U. S. Department of Labor.

Clark, G. 1984a, Does Inflation Vary Between Cities?, Environment and Planning A, 16: 513-527.

______, 1984b, Price Shocks and the Components of Urban Inflation, Environment and Planning A, 16: 1633-1648.

Cliff, A., P. Haggett, J. Ord, K. Bassett and R. Davies, 1975, Elements of Spatial Structure: A Quantitative Approach, Cambridge: Cambridge University Press.

Damsleth, E., 1980, Interpolating Missing Values in a Time Series, Scandinavian Journal of Statistics, 7: 33-39.

Griffith, D., 1978, A Spatially Adjusted ANOVA Model, Geographical Analysis, 10: 296-301.

______, 1983, Phasing-out of the Sugar Industry in Puerto Rico, in Evolving Geographical Structures, edited by D. Griffith and A. Lea, The Hague: Martinus Nijhoff, pp. 196-228.

______, R. Haining and R. Bennett, 1985, Estimating Missing Values is Space-time Data Series, in Time Series Analysis: Theory and Practice 6, edited by O. Anderson, J. Ord and E. Robinson, Amsterdam: North-Holland, pp. 273-282.

Haggett, P., 1976, Hybridizing Alternative Models of an Epidemic Diffusion Process, Economic Geography, 52: 136-146.

Haining, R., 1978, The Moving Average Model for Spatial Interaction, Transactions, Institute of British Geographers, New Series, 3: 202-225.

______, D. Griffith and R. Bennett, 1984, A Statistical Approach to the Problem of Missing Spatial Data Using a First-order Markov Model, Professional Geographer, 36: 338-345.

Vandaele, W., 1983, Applied Time Series and Box-Jenkins Models, New York: Academic Press.

Yeates, M., and B. Garner, 1980, The North American City, 3rd ed., New York: Harper and Row.

EPILOGUE

Considerable work has been published in recent years by geographers, spatial economists, regional scientists, spatial statisticians, and scholars from other allied disciplines, on the topic of space-time transformations. A main emphasis of this work is description of the spatio-temporal trajectories followed by geographic systems. Good examples of affiliated dynamic spatial models are provided in Griffith and MacKinnon (1981), while interesting examples of some initial attempts to formulate evolutionary geographical models are furnished in Griffith and Lea (1983). Unfortunately many of these models tend to be linear in form, tend to treat time as reversible, and tend to view locations in space as independent entities. But spatial autocorrelation mechanisms, the irreversibility of space-time processes, and the multiplicative or exponential nature of descriptors of change highlight the inappropriateness of most dynamic as well as 'evolutionary' spatial models. As soon as non-linear formulations embracing interaction effects are coupled with bifurcation possibilities, the modelling game becomes dramatically changed.

To better understand the rules of this more advanced version of the modelling game, those transformations through space and time that are embedded in dynamic and evolutionary spatial models need to be further explicated. The utlimate goal is to construct bona fide evolutionary spatial models. An immediate goal is to understand how change occurs through space and time. Transformations are the heart of pursuits aimed at achieving both of these goals. Such endeavors supply a fertile research topic area.

This volume has attempted to move towards the attainment of both of these aforementioned goals by treating a number of themes, including non-linear structures, bifurcation points, and autoregressive structures. These themes were addressed within the context of conceptual models as well as empirical analyses. They sampled the spectrum of geographical issues, including urban and regional modelling, spatial interaction, equilibrium models, diffusion, and spatial autocorrelation. In most cases only a foundation has been laid, though, and sufficient theoretical work remains to be undertaken.

Several prominent themes have been overlooked in this volume, in an attempt to provide some focus to the discourse assembled here. Certainly work in the areas of growth pole theory, general equilibrium analysis, dynamic input/output modelling, cognitive and behavioral geography, and a wider variety of spatial search situations all merit attention. In fact some of these topics were touched upon in the Institute lectures, but for one reason or another their source papers did not find their way into this volume. Regretably, though, others of these topics had to be foregone in order to preserve what little cohesion is displayed here.

It is hoped that those papers found in this Institute Proceedings have provided solid stepping stones to, and perhaps some motivation or stimulus for, future developments in an evolving field of analysis.

REFERENCES

Griffith, D., and R. MacKinnon (eds.), 1981, Dynamic Spatial Models, Alphen aan den Rijn: Sijthoff and Noordhoff.

______, and A. Lea (eds.), 1983), Evolving Geographical Structures, The Hague: Martinus Nijhoff.

INDEX

INSTITUTE PARTICIPANTS

Country	Participant	Affiliation
Belgium		
	D. Peeters	Universite Catholique de Louvain
	J. Thill	Universite Catholique de Louvain
Canada		
	B. Boots	Wilfrid Laurier University
	L. Curry	University of Toronto
	P. Kanaroglou	McMaster University
	Y. Papageorgiou	McMaster University
England		
	G. Arbia	University of Cambridge
	R. Bennett	University of Cambridge
	M. Birkin	University of Leeds
	G. Clarke	University of Leeds
	R. Haining	University of Sheffield
	P. Longley	The University of Wales Institute of Science and Technology
	W. MacMillan	University of Oxford
Federal Republic of Germany		
	J. Broecker	Institut fur Regionalforschung, Kiel
	B. Dejon	Universitaet Erlangen-Nurnberg
	H. Fassmann	Austrian Academy of Sciences (Austria)
	G. Haag	Universitaet Stuttgart
	H. Koerner	Universitaet Erlangen-Nurnberg
	M. Munz	Universitaet Stuttgart
	R. Reiner	Universitaet Stuttgart
	C. Schoenebeck	Universitaet Dortmund
	W. Weidlich	Universitaet Stuttgart
France		
	F. Durand-Dastes	University of Paris
	P. Durey	University of Paris
	G. Feyt	Centre National de la Recherche Scientifique
	F. Guerin	University of Paris
	B. Marchand	University of Paris
	D. Pumain	University of Paris
	S. Richardson	ZINSERM, Paris
	L. Sanders	University of Paris

Greece

A. Makris	National Technical University

Israel

M. Sonis	Bar-Ilan University

Italy

A. Bellacicco	Universita di Roma
S. Griguolo	D. A. E. S. T., Venezia
G. Grimaldi	University of Torino
S. Lombardo	DPTU, Rome
G. Rabino	Instituto Ricerche Economico-Sociali del Piemonte, Torino
M. Zerbi	Universita Cattolica del Sacro Cuore

The Netherlands

T. ten Raa	Erasmus Universiteit
D. Reitsma	University of Nijmegen
A. Rima	Free University

Portugal

M. Correia	Direccao Geral do Planeamento Urbanistico
F. Ferrao	Direccao Geral do Planeamento Urbanistico
M. Poeira	Universidade Nova de Lisboa

Spain

J. Linacero	Universidad del Pais Vasco

Sweden

N. Sarafoglou	University of Umea

Turkey

C. Erol	Gukurova Universitesi
M. Guvenc	Middle East Technical University
I. Tekeli	Middle East Technical University
S. Togan	Middle East Technical University

United States

C. Amrhein	University of Kentucky
T. Barringer	U. S. Geological Survey
J. Barringer	U. S. Geological Survey
H. Couclelis	University of California/Santa Barbara
D. Dendrinos	University of Kansas
T. Eagle	Dartmouth College
S. Fotheringham	University of Florida
A. Getis	University of Illinois
D. Griffith	State University of New York at Buffalo
D. Hanink	University of Connecticut
J. Huff	University of Illinois

L. Leifer	State University of New York at Buffalo
B. Lentnek	State University of New York at Buffalo
R. MacKinnon	State University of New York at Buffalo
G. Mulligan	University of Arizona
M. O'Kelly	Ohio State University
C. Pavlik	University of Minnesota
D. Plane	University of Arizona
K. Roeder	Pennsylvania State University
E. Sheppard	University of Minnesota
D. Straussfogel	Pennsylvania State University
D. Straussfogel	Pennsylvania State University
D. Wartenberg	Harvard School of Public Health
A. Williams	Pennsylvania State University